...nes in the same work

...ORIGIN AND FOUNDATION, 1946–62
...I Sendall)

...EXPANSION AND CHANGE, 1958–68
...I Sendall)

...COMPANIES AND PROGRAMMES, 1968–80
...Potter) (*in preparation*)

...emy Potter

...NG RICHARD?

...ERS

...ding **Order**

...ld like to receive future titles in this series as they are
...you can make use of our standing order facility. To place a
...rder please contact your bookseller or, in case of difficulty,
...at the address below with your name and address and the
...e series. Please state with which title you wish to begin your
...rder. (If you live outside the United Kingdom we may not
...ghts for your area, in which case we will forward your order
...isher concerned.)

...Services Department, Macmillan Distribution Ltd
...s, Basingstoke, Hampshire, RG21 2XS, England.

INDEPE
TELEV
IN BRI

Volu

Politics and Cc

SERVIRE

Other

Volume
(*by Ber*

Volume
(*by Ber*

Volume
(*by Jere*

Also by

GOOD

PRETEN

Series S
If you w
publishe
standing
write to
name of
standing
have the
to the pu

Custome
Houndm

INDEPENDENT
TELEVISION
IN BRITAIN

Volume 3
Politics and Control, 1968–80

JEREMY POTTER

MACMILLAN

First published 1989

Published by
THE MACMILLAN PRESS LTD
Houndmills, Basingstoke, Hampshire RG21 2XS
and London
Companies and representatives
throughout the world

Typeset by Footnote Graphics,
Warminster, Wilts

Printed in Great Britain by
The Camelot Press Plc, Southampton

British Library Cataloguing in Publication Data
Independent Television in Britain.
Vol. 3 : Politics and Control, 1968–80.
1. Great Britain. Commercial television services,
to 1980
I. Potter, Jeremy
384.55′4′0941
ISBN 0–333–33019–6

CONTENTS

LIST OF PLATES

PREFACE AND ACKNOWLEDGEMENTS

Illness tragically curtailed Bernard Sendall's authorship of this history after completion of the first two volumes. Enjoying few of his advantages of personal involvement in the events which he recorded, I have taken up his pen with some diffidence: mindful too of a closer approach to the present time and of Macaulay's warning about the 'great and obvious objections to contemporary history'.

Volumes 1 and 2 – *Origin and Foundation* and *Expansion and Change* – covered the years between 1946 and 1968. Volumes 3 and 4 will carry the story of Independent Television forward from the start of a new contract period in the summer of 1968 to the announcement of new contracts at the end of 1980. I had thought to cover a span of no more than twelve and a half years in a single volume but the richness of the material dictated otherwise.

The period was marked by a profusion of official investigations and committees of enquiry into broadcasting, culminating in the comprehensive report of the Annan Committee. But until the passage of the Broadcasting Act 1980 it was a decade of debate with few decisions. Politically, broadcasters lay under constant suspicion as usurpers of the powers of parliament and press. Socially, much public concern was voiced about the influence of television in the home. There were campaigns against violence, sex and bad language on the screen and campaigns for greater accountability and wider access to the medium. Politicians and broadcasters argued over coverage of events in Northern Ireland, over the size and mechanism of the Exchequer levy on the companies' profits, and over the number and length of party political broadcasts. Controversy about its character, cost and control delayed the launch of the fourth channel for more than twenty years. Meanwhile colour television came to Britain, and the whole system was re-engineered from VHF to UHF transmission. Restriction on hours of broadcasting was relaxed, bringing a growth in programme production and afternoon viewing.

These years of recession and inflation, of expansion and prosperity – above all, of enquiry and uncertainty – offer no natural break in chronology, but ITV is a two-tier system. To governments and in the eyes of the law Independent Television is the Independent Broadcasting Authority to which parliament has entrusted responsibility and control; to the viewer it is one or other of the companies which make and present the programmes. Volume 3 therefore concentrates on the IBA and central issues, while Volume 4 will focus on the companies and their programmes.

This work has been commissioned jointly by the Independent Broadcasting Authority and the Independent Television Companies Association (now the Independent Television Association). To that extent it is an official history. On the other hand, I have been free to write as I pleased, and in that sense it is entirely personal. Employed as a publisher within ITV during most of the period under review, I can make no claim to having been either a central participant or a wholly detached observer. Mine is a view of ITV not from the stage nor from the auditorium, but from the wings.

Those from whom I have solicited information have imparted it readily, and the acknowledgements I have to make are many. I am indebted to Bernard Sendall for the legacy of some drafts relating to this period, but an even more prized inheritance has been Joseph Weltman, who worked with him on the first two volumes and generously agreed to provide some continuity by assisting me in turn. He was the Authority's Head of Programme Services from 1967 to 1975 and his knowledge and wisdom, as well as his contributions to the text, have been invaluable.

The other half of my part-time team has been Anne von Broen, who cheerfully mastered both my handwriting and the word-processor. My particular thanks are due too to Joan Shilston for her work on the appendices and references; to Lewis Breakspear and his staff in the IBA's central registry for conjuring up so many files from the archives; and to Barrie Macdonald, the IBA's librarian, for the index.

Others to whom I am grateful for information, permission to quote or other assistance are: Lord Annan, Lord Aylestone, Bernard Bennett, Christopher Bland, Kenneth Blyth, Christopher Booker, William Brown, Lord Campbell of Eskan, Chris Daubney, Jonathan Dimbleby, Ray Fitzwalter, Sir Denis Forman, Paul Fox, David Glencross, Archie Graham, Bernard Green, Dr Barrie Gunter, Dr Ian Haldane, Don Harker, Lord Harris of Greenwich, Pat Hawker, Brian Henry, Fred Hotchen, Jeremy Isaacs, Lady Littler, Kathie May, Hon. David Montagu, Anne Newman, Michael Peacock, Lady Plowden, Anthony Pragnell, Rt Hon. Merlyn Rees, Tom Robson, Bryan Rook, Colin Shaw, David Shaw, Anthony Smith, Iain Sproat, Mike Taylor, Brian Tesler, Harry Theobalds, Dr Bob Towler, Dr Boris Townsend, Sir Ian Trethowan, Ben Whitaker, Phillip Whitehead, Viscount Whitelaw, Lord Windlesham, Dr Mallory Wober, Sir Brian Young and Paul Young. If others have been omitted inadvertently, I offer them sincere apologies together with my thanks.

No less than television, historiography is a channel of communication. In the hope of attracting a wider public than media students and broadcasters themselves, I have assumed as little knowledge as possible among readers and endeavoured to avoid jargon and baffling initials. (A list of abbreviations appears on p. 320). But the nomenclature is sometimes unavoidably confusing. For example, the Independent Television Authority became the Independent Broadcasting Authority on 12 July 1972 when the Sound Broadcasting Act imposed upon it responsibility for an Independent Local Radio service. Before that date it was commonly referred to as the ITA; afterwards the IBA. In this volume both abbreviations are used, but more often it has seemed best to adopt a familiar ITV usage and refer simply to 'the Authority'.

The ITV year, too, requires some elucidation. The Authority's financial years run to the end of March, but most of the companies' financial years run or ran to the end of July or near the end of July, the day on which, until 1982, contract periods commenced. ITV's programme years run from the first week in September. For some company and industry calculations the calendar year is used. Comparisons between annual figures should therefore be made with caution.

The 1970s will, I believe, be seen by future historians as ITV's high noon, but in the text I have for the most part refrained from subjective judgments of this kind, whether flattering or critical. The contemporary historian's job, or so it has seemed to me, lies in the choice and analysis of material from records and recollections and its shaping into theme and narrative so that readers may form their own judgments.

Oxford, March 1988   J.P.

1

TELEVISION AND SOCIETY IN THE 1970s

It was typical of the mood of Britain during the 1970s, when pride in achievement was out of favour and success gave grounds for suspicion, that international praise for British television should be reflected at home with the belittling and ungrammatical acknowledgement that it was 'the least worst television in the world'.

In 1947 television had been described as a bomb about to burst, and its impact on society was indeed scarcely less explosive than that of nuclear power. In many parts of the world this unprecedentedly influential medium of mass communication was welcomed less as a novel means of enlightenment than as an instrument of thought-control. Like print, it was appropriated by governments – communist, military and 'one party' – for the benefit of the ruling élite. In such countries television stations were as closely guarded as presidential palaces and among the first objectives of rebel forces. Even in de Gaulle's France television became the official voice of the nation. 'How can one govern without television?' André Malraux is reported to have enquired.

In this context the organisation of broadcasting in Britain was unique. Formally the powers of the government of the day over the BBC were absolute, but in practice, in its programme-making and day-to-day business, the Corporation operated independently of government and parliament. This position was set out clearly in the Beveridge Committee's report in 1949 and noted without dissent.[1] Except during the Second World War, governments intervened in the BBC's affairs at their peril. As in the General Strike of 1926, so in the Suez crisis thirty years later, the BBC demonstrated that it could and would fight successfully to preserve its freedom of action, however precariously held during national emergencies.

Elsewhere in the world where freedom was permitted and encouraged, most notably in the United States of America and Japan, television developed into a commercial free-for-all in which the interests of advertisers dominated those of the viewer. In Britain the ambitions of those who lobbied in the early 1950s for a fully commercial service were frustrated by reservations within the Conservative government and party.

1

Government policy was to break the BBC's monopoly with a service not dependent on a licence fee or any other call on the public purse. There was to be financing by revenue from advertisements, but the Independent Television Authority was established as a statutory body with powers to control those advertisements and the programme output, shareholdings, board membership and financial activities of the commercial companies which were to make the programmes. So ITV was born of an arranged marriage between public authority and private enterprise. Properly described, it was not commercial television, but semi-commercial television.

Thus, by two unrelated arrangements of indirect control, Britain's legislators, pursuing a tradition of moderation and compromise in an apparently haphazard manner, pioneered a middle course between state broadcasting and uninhibited commercialism: the path of regulated independence. BBC and ITV were different faces of a similar reconciliation between the opposing desiderata of freedom and control. A crucial ingredient in what evolved into the best of all systems was rivalry for audiences but not for finance: regulated competition.

The mechanism was finely balanced and dependent for success on a number of compromises. Freedom from political pressure was subject to the obligation of responsibility for fairness and impartiality. Licence for programme-makers was conditional on respect for viewers' rights. Programmes designed to entertain had to be blended with programmes designed to inform and educate. Devolution devised to meet regional needs moderated the strength of centralisation.

The first volume of this history records the conception, birth, early traumas and prodigious growth of Independent Television in Britain during its first six years (1955–61). So rapidly did ITV become attuned to popular taste that the television service of the long-established BBC was soon deserted by two-thirds of the audience and the initial financial losses of the pioneering companies were amply redeemed.

As recounted in Volume 2, which carries the story on to 1968, ITV's triumph was altogether too sweeping, its hubris too manifest, to be borne by those who found the profit motive distasteful and its introduction into the world of television inappropriate. Nemesis arrived in the shape of the puritan-minded Pilkington Committee, which effectively pronounced a verdict of 'BBC good, ITV bad'. Triviality, a natural vice of television so shamelessly indulged in by the Independent Television Authority and the companies it licensed, was (so the committee declared in what it took to be an apt quotation) 'more dangerous to the soul than wickedness'.[2] But one man's triviality is another man's entertainment. As one critic commented: 'The Pilkington Report thought that reform could be achieved by creating a service that was supported by advertising but not dependent on popular appeal. It is an attractive notion to those who dislike advertising and popular appeal.'[3]

Although Pilkington's proposals for a radical restructuring of ITV were rejected, the government consented to handicap the infant prodigy by awarding a second channel to the BBC only. A second ITV channel was withheld until a chastened ITA took the necessary steps to ensure that, as the angry populists saw it, the public got less of what it wanted (or, in its ignorance, thought it wanted). ITV had boasted of being 'people's television', a *Daily Mirror* of the air, but it was now required to take cognisance of the fact that everyone belonged to minorities. To provide a comprehensive television service on a single channel with limited hours of broadcasting was not an attainable ideal, but the attempt had to be made to cater more for specialised tastes and for *Times* and *Guardian* readers.

The BBC meanwhile was moving briskly in the opposite direction. 'However good our case might be,' wrote Sir Hugh Greene, the Director-General, 'there would be no political or public support for any recommendations the Pilkington Committee might make along the lines we urged if people were still turning to ITV in an overwhelming majority. Why should they pay a licence fee if they were not using the BBC? I therefore told the television service that without any abandonment of BBC standards they must aim at increasing our share of the television audience from its lowest ration of 27:73 to 50:50 by the time the Pilkington Committee reported. That was exactly achieved at the beginning of 1963.'[4]

With the rude blast of competition forcing the BBC to make its programmes more appealing to more people, the two organisations grew less and less distinguishable in terms of programme content during the 1960s and 1970s. The BBC imported American soap operas to bolster its ratings while ITV took the lead in news and current affairs coverage. Attracted by higher salaries, less bureaucracy and new creative opportunities, a steady trickle of senior BBC staff flowed into ITV bringing BBC attitudes and aspirations with them, while some ITV notables (Ian Trethowan, a future BBC Director-General, among them) journeyed the other way. BBC governors and members of the ITA continued to be chosen from a common pool of honourable non-broadcasters. Lord Hill moved with no evident change of style or view from the chairmanship of the ITA to that of the BBC. Lady Plowden, Vice-Chairman of the BBC, was appointed Chairman of the IBA (as the ITA became) and brought values traditionally associated with the BBC.

Despite this trend a 'holier than thou' attitude persisted. The BBC arrogated to itself the accolade of 'public service broadcasting', although in truth the BBC was operating a licence-financed public service and ITV an advertising-financed public service. Only in the 1980s, when cable, satellite and other new means of broadcasting began to threaten what the Peacock Committee described as the 'comfortable duopoly', was ITV publicly embraced as a fellow public-service broadcaster.

Throughout the 1970s, as its programme output broadened, ITV progressed steadily towards respectability and parity of esteem, and the verdict of the Annan Committee in 1977 was very different from Pilkington's in 1962. In opinion-forming circles it became less fashionable to decry ITV programmes while asserting that one never watched them. Internationally ITV shared the acclaim long accorded to the BBC: clear proof that a structural balance between regulation and independence produced systems within which creativity could flourish. As demonstrated by the rich annual harvest of prizes for their programmes gathered by both organisations, British television at its best was widely acknowledged to be second to none – even if at its worst it could be as trite and unimaginative as any.

But legislators, however liberal-minded and well-intentioned, could not forbear to tinker with this rival power in the land, and broadcasters remained trapped in long-running political controversy. Television saturates its viewers with reflections of society, reflections on society and a means of escape from society, and in a democratic community such a potential force for good or ill rightly attracts constant suspicion and scrutiny. The penetrative power of the medium, intruding into the homes of almost the entire population, has made the history of Independent Television part of the political, social and cultural history of Britain.

Lord Swann, a former Chairman of the governors, has reckoned that in thirty years there were twenty public enquiries into the BBC and that no fewer than twelve of those years had been spent either living with the enquiries or awaiting government decisions on them.[5] Such was the fate of both broadcasting organisations in the 1970s, and it was not unique. Broadcasters in nearly every democratic society suffered similar turmoil. During this decade there were government enquiries into television in West Germany and Austria, Sweden and Switzerland, Canada, Australia and New Zealand. French television was reorganised and in Italy the monopoly of the RAI state system was broken.

So ITV came of age during an unsettled and unsettling period. The contracts awarded in 1967 for six years (as recorded at the end of Volume 2) lasted for nearly thirteen and a half from the summer of 1968, but this could never have been foreseen. They were extended unpredictably in stages: for two years from 1974 to 1976, for another three until mid-1979, and for a further two and a half until the end of 1981. Uncertainty about the future prevailed throughout the decade as successive enquiries investigated and reported, and the need for stability became ever more acute if the system was to function properly.

Governments of whichever complexion first sought to establish facts, but when findings were presented ministers commonly found themselves unable to come to a decision. So a White Paper was issued and public discussion stimulated, but for the most part little emerged from the debates except radically differing opinions and resultant dither. Decisions were then postponed pending an assessment of important technical advances – a permanently available excuse. Meanwhile short-term expedients like contract renewal became inescapable.

The problems and instability of the broadcasters at this time were accentuated by the problems and instability of the whole nation. To fulfil its proper role television has to be responsive to changes in society, and these were bewilderingly rapid. In 1972, when questioned by a committee of the House of Commons about ITV's alleged permissiveness, Lord Aylestone, the ITA's Chairman, replied disarmingly: 'To be absolutely frank, I think programmes would be shown or would get through now that would not have been shown in 1967 when I first joined the Authority, but don't blame me!'[6] Such was the turbulence in Britain during the 1970s that guardians of the public interest and programme-makers alike might be forgiven if their responses sometimes appeared too small or too great, to come too early or too late. For Britons were struck by a hurricane of economic recession, unprecedented inflation and industrial unrest, seemingly insoluble financial crises, growing violence and puzzling self-doubts.

Governments stumbled from crisis to crisis in an atmosphere of deepening despondency, and there can be little understanding of the story of ITV without some appreciation of the demoralising economic and social conditions which it was expected to illuminate with dispassionate information in news bulletins, studio discussions and documentaries, to make relevant in drama and relieve with light entertainment. The responsibility was all the greater because 'watching the box' had become the nation's favourite leisure-time occupation during a period when leisure was expanding, with shorter working hours and growing numbers of pensioners and unemployed.

The excitement and excesses of the heady post-rationing, pre-inflation period were followed by a 'morning after' mood of pessimism. By 1968 the swinging sixties had stopped swinging and 'never had it so good' was succeeded by 'never bright confident morning again'. Rises in the price of oil from the autumn of 1973 caused hyperinflation, and prices on average trebled during the 1970s. Fortunes were made in property and real estate while those on low and fixed incomes fell into despair. In 1974 the balance of trade deficit on visible exports reached £5.3 billion. Divisions in society were seemingly unbridgeable. The permissive society came under attack from upholders of traditional values. The poverty gap between north and south widened. While the Americans were putting a man on the moon, Britain was strikebound and bleeding from self-inflicted injuries. Management and unions were enemies, not partners: during the month of March 1977 a million working days were lost in strikes.

The political parties too were in poor shape. By 1980 Labour had torn itself apart in an unfraternal struggle between marxists and social democrats, and the break-away Social Democratic Party was launched with its Limehouse Declaration in January 1981. The Conservatives held together as a party, but divided between traditional 'One Nation' 'wets' and radical monetarists for whom inflation and mounting public expenditure and trade-union demands for more and more money for less and less work were the enemy within. The electoral system allowed the Liberal and other minority parties no significant representation in parliament, although in the general election of October 1974, for example, a quarter of the electorate voted for candidates who were neither Labour nor Conservative.

The so-called United Kingdom itself appeared in danger of falling apart, with much agitation in Scotland and Wales and referenda on devolution. In Northern Ireland demands for civil rights for the Catholic minority became a movement for Irish unity. Hostility between Protestant loyalists and Catholic republicans gave rise to a terrorist campaign rivalling the excesses of the Basque separatist movement in Spain, the Baader-Meinhof gang in Germany, the Italian Red Brigade and the Japanese Red Army. Airey Neave MP was assassinated in the palace of Westminster and Earl Mountbatten in the Republic of Ireland. The question of Ulster remained a demanding topic for broadcasters throughout this period; its coverage in the public interest a persistent dilemma. Acts of terrorism could not be ignored, the public had a right to be informed on all aspects and all views, but at what cost? The activities of the IRA by no means abated as a result of the publicity accorded them by television.

1974 was a year of disaster, with IRA bombings in England, runaway inflation rising to a rate of more than 25 per cent per annum, the pound heading for collapse, a miners' strike and two indecisive general elections. 1979 was another. During a winter of discontent and a 'dirty jobs' strike, schools and hospitals were closed, uncollected refuse was piled high in the streets, ambulances were left unmanned and bodies unburied. *Der Spiegel* took the opportunity to publish a series of articles on 'Sick England'.

Indeed the defeated enemies of the Second World War, West Germany and Japan, had become models of prosperity while Britain, which in the 1940s stood alongside the USA and USSR as one of the Big Three, was falling precipitously from world power and influence. Never in history had an unconquered nation suffered such a steep and rapid decline. No longer rich by comparison with North America, Japan and the leading western European countries, by the mid-1970s Britain had slid down the scale of international prosperity to occupy twentieth place in the world economic league as assessed by gross national product per head.

Yet Britons not only believed that they had some inherent right to an undiminished standard of living; they even deluded themselves with expectations of ever greater affluence, which some sections of the population continued to enjoy. In a democracy, where free elections must be won, politicians are understandably reluctant to spell out unpalatable truths. 'We are a world power and a world influence or we are nothing,' Harold Wilson proclaimed in a speech in the City of London as late as 1964. Three years later falling reserves, ominous trade figures and the highest level of unemployment since the 1930s, heralded the troubles to come and forced his government into a belated devaluation of the pound.

Directly and indirectly, the affairs of Independent Television were greatly affected by the complexion of the government in power. The fact that ITV had become the staple diet of Labour voters did not moderate the enmity of intellectuals on the left. It was the Labour government of the latter half of the 1960s under Harold Wilson which nearly throttled the system by turning the screw of the Exchequer levy too tightly. The Conservatives approved of ITV in principle, while not much caring for it in practice. Under the Heath government between 1970 and 1974 the levy was halved and the restriction on broadcasting hours lifted.

When the miners had brought Heath down, one of the first acts of the new Wilson administration was the revival of a full-scale inquiry into broadcasting under Lord Annan which the Conservatives had aborted in 1970 and which was to cast its shadow over broadcasters for five years. Its work began amid the encircling gloom which foreshadowed Wilson's retirement in 1976, overwhelmed by the collapse of his incomes policy and ever higher public expenditure and government borrowing.

The premiership then passed to James Callaghan, and a huge IMF loan rescued sterling from collapse. For Wilson's failed Social Contract with the unions on prices and incomes Callaghan substituted an equally ineffective Concordat. Towards the end of 1978 BBC technicians threatening industrial action were awarded a pay increase of 15 per cent – three times the government's norm – amid accusations of an abandonment of pay policy to save *The Sound of Music* on Christmas Day. Ford workers, offered a similar 15 per cent, went on strike for 30 per cent. A rank-and-file revolt against wage restraint in a period of inflation burst into a rash of strikes, and Labour supporters brought down their own government. No administration could have hoped to survive long without the good will of the trade unions. The new Thatcher government, determined to prove otherwise, was soon faced with other problems. By the summer of 1980, following the fall of the Shah of Iran, oil prices moved sharply upwards once more, the rate of inflation touched 20 per cent, and the number of unemployed rose from 1.5 to 2 million.

In these fraught circumstances it is not surprising that hard decisions about the future of broadcasting, debated continuously from Annan's resurrection in 1974 to the fall of the Callaghan government in 1979, were given no priority. Had they been, ITV might have lost its long-sought second channel to an Open Broadcasting Authority. The decisive Thatcher administration awarded it to the IBA without further ado.

Although, to the end of this period and beyond, no Labour government has ever produced a major piece of broadcasting legislation, Harold Wilson was the first prime minister to grasp and fully exploit the political opportunities offered by television. Others followed his example, and while politicians sought to use the medium to explain their policies and gain and retain power, they and senior civil servants became convinced that broadcasters, and the BBC in particular, held Westminster and Whitehall in contempt and were intent on losing no opportunity to ridicule those who ran the country. The BBC infuriated Home, Wilson and Thatcher in turn. So long sedate, 'Auntie' swung with society and, infected with the mood of the times, discarded deference. Interviews not infrequently became confrontations, and in dramas and documentaries she sank her teeth into nearly every sacred cow in sight: the British Empire, the Church of England, the medical profession.

Critics on the right fulminated against the Anti-British Broadcasting Corporation, while *Yesterday's Men*, a wounding documentary about Wilson and other Labour ex-ministers screened in 1971, was never to be forgotten or forgiven on the left. It was denounced by his press officer as 'carefully calculated, deliberate, continuous deceit'. It may, wrote the BBC's historian in 1979, 'have vindicated editorial independence, but it led to a distrust among Labour politicians which has affected all subsequent BBC attitudes to documentary productions'.[7]

Attacks from the left were more usually concentrated on ITV, although its documentaries often drew fire from the right. Within the Conservative Party ITN's news service was seen as a welcome alternative to the BBC's, but there was disappointment that the political colour of much of the ITV companies' current affairs and drama output appeared no less pink. To the marxist, by contrast, both broadcasting organisations were supine lackeys of a right-wing establishment. The 1960s and 1970s witnessed the rise of the academic specialist in media studies and the heyday of the marxist sociologist, fervently critical of television practitioners and of the ideologically objectionable ITV in particular.

The attitude of individual Members of Parliament of whatever party was believed by cynics to be determined by the number of their personal appearances on the screen, and here its regional structure gave ITV an advantage over the BBC. During 1974, it was estimated, politicians made 2,700 appearances on ITV, most of them in regional programmes discussing public affairs and problems with their constituents.[8] The main forum of political debate had moved from the chamber of the House of Commons to the television studio.

Unlike democratic institutions in other countries, the House of Commons repeatedly resisted moves to allow the British electorate to see its representatives at work. The concept of a club for 'members only' lingered on from the eighteenth century, when the penalty for reporting parliamentary proceedings was imprisonment. The case for televising parliament had first been seriously pressed in 1959 by Aneurin Bevan, Robin Day and others. In February 1975 a motion in favour of experimental radio broadcasts of proceedings was carried by a large majority in the House, while a similar motion to experiment with television coverage was narrowly defeated. A regular service of radio coverage was inaugurated in 1978, but television cameras remained excluded.

Politicians were therefore in a weak position to complain, as they did, of television usurping the function of parliament and television interviewers assuming an Opposition role in questioning ministers. 'How,' asked Robin Day, 'can parliament expect to maintain its position as the prime forum of debate if it shuts itself off from the prime medium of communication?'[9]

'Prime medium of communication' was a fair claim. Standards in Fleet Street were in decline. Publication of the national newspapers was repeatedly interrupted by labour disputes and in 1979 *The Times* suspended publication for eleven months. Some of the best journalists left what seemed an industry heading for suicide to join an enterprise which was looking to the future, not the past, and whose news coverage was growing in volume and authority.

How adequately, then, did television's journalists rise to the challenge of keeping the public well informed? Television reported – but how far did it explain? – the state of the economy, inflation and pay policy and strikes, poverty and the housing shortage, the argument over nuclear disarmament and the troubles in Northern Ireland, the Middle East and Vietnam. A bias was detected in television journalism – a bias not against any particular political party or point of view, but a bias against understanding. This view was advanced in the columns of *The Times* in 1975 by John Birt and Peter Jay, the executive producer and presenter of London Weekend Television's *Weekend World*, who even suggested that television coverage might be aggravating the difficulties of society in trying to resolve its problems: 'Three industrialists discussing illiquidity or three party spokesmen quarrelling over inflation or three Ulstermen exchanging rhetoric about power-sharing will not succeed in fifteen minutes in communicating anything other than confusion.' A man, they argued, might know everything yet understand nothing.[10]

This thesis attacked the narrow focus of television news bulletins and the emphasis of current affairs programmes on society's sores without examination of root causes. It pointed to the presentation of discrete facts without context or background and the dressing up of important issues as human interest stories. It held the distinction between news and news-analysis – between hard news and current affairs documentaries – to be a fundamental misconception, a distinction without a difference. It saw television journalism as a misbegotten child and demanded a break from its ill-matched antecedents – newspaper journalism and documentary film-making – in favour of a new concept of programme-making and a new

structure. The major specific Birt–Jay recommendation was for the replacement of the daily half-hour evening news bulletin by a programme which
lasted for a full hour and, after brief headlines, concentrated on a few main
stories analysed in context and depth.

Critics were not slow to denounce this theory as patronising, élitist quasi-
philosophy. The Chairman of the BBC gave it his blessing, but the
Director-General described the remedies proposed as brain-washing. The
Editor of ITN thought the malady exaggerated and the root cause
incorrectly diagnosed. He maintained that the process of integrating news
and analysis had been going on at ITN for years: *News at Ten* already
included background material and *First Report* had been devised to meet
precisely the points raised.[11] Others suggested that the fault lay not in
concepts and structures but in the medium itself which could not accommodate many words without boring its viewers into switching off. What
advantage would there be in doubling the length of news bulletins if
audiences were to be halved?

Much argument ensued. It ranged over techniques of presentation, the
number and scheduling of information programmes, audience receptivity
and response, and the location of editorial control. But in provoking this
important debate the thesis failed to gather support, or even respect,
among most broadcasters. The 'superior pair' were accused of overween-
ing pomposity and pedagoguery and even of a bias against communication.
While *Weekend World* continued to develop its format as an adult
education lecture, meticulously explaining the general background before
examining a particular issue, other programme-makers remained unshaken
in their belief that the best approach for television was inductive; that it
should not attempt to do what print could do more effectively, but should
instead illuminate a whole issue through particular cases which were
broadly representative. The ideas of Birt and Jay persisted nonetheless as
an influence on attitudes in news and current affairs departments, and the
teasing phrase 'bias against understanding' which infuriated so many of
their colleagues became a permanent addition to the vocabulary of
broadcasting.[12]

The path of illumination was strewn with hazards, for explanation of
changing conditions and moods seldom command general assent. One
focus of dissension was on law and order, where television was thrust willy-
nilly into the argument. The much-loved British bobby all but disappeared
during the period, falling victim to the controversial role of the police at a
time of rising crime and violence on the streets. In attempting to keep the
peace in the face of angry demonstrations of protest they were denounced
as defenders of the *status quo*: 'pigs' became the fashionable term of abuse.
Did the presence of television cameras encourage the violence and the
abuse?

The word 'mugging' entered the language. In the 1960s it had still been safe to walk the streets of London by night: during the 1970s more and more Londoners stayed at home behind locked doors. Sir Robert Mark, former Commissioner of Metropolitan Police, deplored 'blatantly perverse' acquittals by juries and told the public that it could no longer expect the police to protect them from burglaries. Had so-called 'anti-police' television programmes contributed to this by undermining public confidence in the force, many people were asking.

While Britain's image as a united and peaceful society was vanishing, television offered not only reports and explanations but also a means of escape. One commentator wrote:

I do not believe it is wholly accidental that one of the most popular TV serials in the Britain of the seventies has been that cosy, nostalgic evocation of life in an aristocratic Edwardian household, *Upstairs, Downstairs*. Viewers have obviously derived some obscure comfort from this daydream vision of a world in which the two realms, upper and lower, could still coexist in happy, interdependent harmony – such a striking contrast to the bleak, envious, divided, class-racked society of today.[13]

A Race Relations Act and a Sex Discrimination Act were found necessary. The general election of 1979 gave the nation its first woman prime minister, but the number of women elected to the House of Commons fell from twenty-seven to nineteen. There were no black or coloured members. Yet some economic trends towards equality were real. The consumer boom which began in 1955 following the austerity of wartime and post-war rationing led to a gradual bourgeoisification of society. Affluence spread down the social scale; distinctions in dress, food and household furnishing dwindled; an annual foreign holiday became the norm for those who had never been abroad before. Middle-class comforts such as cars, refrigerators, telephones and colour television sets became common in all homes. By 1983, 59 per cent of households had a car, 76 per cent a telephone and 97 per cent a television set.

The brighter side of life in Britain received less attention from media geared to reports of dissension and woe, attacks on ailing institutions and predictions of impending disaster. As one Home Secretary conceded regretfully, television pictures of strikers streaming out of factories were more newsworthy than pictures of workers working.[14] But in 1977 the country was briefly its old nostalgic self in celebrating the Queen's Silver Jubilee. Like television, the royal family was one of a diminishing number of British institutions admired abroad. From the Queen's coronation onwards televised displays of royal pageantry attracted immense audiences. The wedding of the Prince of Wales in July 1981 was watched by a world-wide television audience estimated to number 750 million. The popularity of royalty owed much to television and vice versa. Yet it was characteristic of this period that millions of viewers were denied ITV coverage of the Queen's Silver Jubilee by a squabble over pay affecting a small number of well-paid staff in one company.

The discovery of gas and oil in the North Sea offered some reassurance that God was still an Englishman. What wool had brought to the country in the Middle Ages and coal in the Industrial Revolution, oil might well supply in the years ahead: prosperity without too much effort. While Britain's manufacturing base deteriorated, expenditure on imported goods rose. The pound suffered accordingly but, despite all the grumbling, Britons were surviving not uncomfortably in a world of hunger and real poverty. Even the sick and the deprived were partly cushioned by the National Health Service, redundancy payments, unemployment benefits and the other social services of a caring society. Since Elizabeth II had come to the throne in 1952 the country had been engaged in no war more serious than the humiliating Cod War with Iceland and, although the fact was seldom acknowledged, the standard of living for most of the population had doubled in the twenty-five years.

During so little counting of blessings and so much turbulence television retained its popularity and enhanced its reputation as a cheap and ready bearer of entertainment and an undemanding source of information. Mr and Mrs Briton became Mr and Mrs Viewer. The average citizen could do little in the face of economic recession, industrial unrest and sometimes capricious public services, but he and she had a comfortable retreat at home where they could amuse themselves inexpensively through ITV or BBC channels. The more dire the state of the economy, the less other amusements could be afforded and the greater the nation's dependence on the box in its sitting room. For many, television had displaced the cinema (whose annual admissions were down from 1585 million in 1945 to 193 million in 1970) and, as the Arts Council told the Annan Committee, television drama had developed 'a plausible claim to be for many people outside London the National Theatre of Britain'.[15]

The national decline has sometimes been judged to be partly connected with the 'omnipresence of the television camera, which has played such a subtle part in diminishing, trivialising and making everything seem less "real" '.[16] But this may be a case of flattering the medium with dispraise. Possibly more pertinent is the criticism that by its very success television turned Britain into a nation of uninvolved spectators with sensitivities dulled: passive watchers instead of active participants. Yet there must have been many whose interests and activities were stimulated by television viewing.

The notion of television as a harmfully escapist medium may be complemented by another attributing to it a more sinister role as the agent of Britain's degeneration. 'Because of the power of the media, people were assaulted by one idea after another, foreign to the basic concepts of our society, but those who would have challenged them were silenced and kept off the medium.'[17]

This was the complaint of Mary Whitehouse, most vociferous of the crusaders campaigning for traditional values. Like Margaret Thatcher, the counter-revolutionary triumphant in 1979, she believed that the ideal of the permissive society – or, as others preferred to regard it, the civilised society – was permissive claptrap. Insulted, ridiculed and kept off the air by the BBC, she was treated with more politeness and forbearance, but scarcely more sympathy, by ITV. The dilemma of television in meeting its obligations to satisfy different sections of a divided society is most vividly illustrated in the chasm between the mood of the broadcasters themselves and the backlash of opinions which Mrs Whitehouse represented so articulately.

Thus, speaking in the House of Lords debate on his committee's report, Lord Annan declared:

> I have very considerable respect for Mrs Mary Whitehouse. It is common, among both the intelligentsia and the broadcasters, to sneer at her. Let no one forget that on the evidence which our committee received she speaks for millions, and the force for which she speaks is matched by her personal modesty and lack of rancour. Our committee judged that she, and others of course, had made out a case which the broadcasters had not answered.[18]

The contrary view was expressed by Lord Willis (Ted Willis, the writer) in the same debate:

> It may be true that she represents a current of feeling... but the members and people behind that organisation [the National Viewers' and Listeners' Association] are reactionary in the extreme. They are the floggers, the bring-back-the-censorship people. They are the worst kind of people, and Mrs Whitehouse was clever enough in her PR to see that this would not wash with the public and the television companies, so she appears to be all sweetness, light and reason... My opinion is that she is a dangerous woman.

With so little common ground it was far from clear what stance the broadcasters should adopt in obedience to their duties to society.

Accused of promoting permissiveness and violence, of imposing an unreal world on society, of reducing a once-great Britain to a nation of goggle-eyed puddings, ITV was comforted by one finding of Lord Annan's committee above all: that competition between ITV and BBC had raised the standards of British television. This was a long overdue recognition of achievement and performance, justifying the very existence of Independent

Television, which the whole Labour party and influential sections of the Conservative party had been intent on strangling at birth because it would debase standards and drag a (relatively) pure and noble BBC down into the gutter of commerce. It was with an excusable measure of self-satisfaction, and some relief, that the Chairman of the Authority reported in 1980 'recent expressions of confidence, both by Labour and Conservative governments, in the Authority's conduct of its affairs'.[19]

Twenty-five years of growth was reaching fruition in the 1970s. For ITV the decade was a peak period. There was an upsurge of good programmes. The range was enlarged. Production in the companies' studios increased from 6,500 to 8,500 hours a week. Annual expenditure on programme-making rose from £41 million to £230 million – a doubling in real terms. The programme service, transformed from black and white to colour, was extended from 70 to 100 hours a week. At the beginning of the decade the Authority had nine UHF transmitters in service for the introduction of its 625-line service; by the end, another 400. In undertaking an additional responsibility imposed by parliament it had also inaugurated an Independent Local Radio network with nineteen stations on the air.

But whatever the quality of the service broadcasters could not afford complacency. From the political right they were threatened by free marketeers: enemies of duopoly urged on by the complaints of advertisers. On the political left lurked those who believed that television had replaced religion as the opium of the people: until brought under tighter control and greater accountability, an obstacle to the creation of a better society. Wider access was demanded on all sides. Many who did not share his political opinions agreed with Tony Benn that broadcasting was too important to be left to broadcasters.

2

RECESSION AND THE LEVY

At the end of July 1968 the kaleidoscope of Independent Television formed a new pattern in conformity with decisions taken by the Independent Television Authority the previous year. The fourteen regions and fifteen programme companies are listed overleaf.

Within ITV's two-tier system of an Authority and its programme contractors the second tier had always been subdivided. Previously four of the contractors, now five, were known as central, major or network companies. Under the terms of their franchises they were jointly responsible for programmes networked throughout the country. These were ATV (in Birmingham), Granada (in Manchester), London Weekend Television and Thames (in London) and Yorkshire Television (in Leeds). The other ten, so-called regional, companies were without network obligations. They differed in size and were usually distinguished as large regionals and small regionals. The large were Anglia (in Norwich), Harlech (in Bristol and Cardiff), Scottish (in Glasgow), Southern (in Southampton) and Tyne Tees (in Newcastle). The small were Border (in Carlisle), Channel (in Jersey), Grampian (in Aberdeen), Ulster (in Belfast) and Westward (in Plymouth). Independent Television News, a company jointly owned by the fifteen contractors, provided national and international news to the network. An account of each of these companies and their programmes between 1968 and 1980 will be found in Volume 4 of this history.

During 1968 there were, by ITA edict, three deaths in the family: those of Rediffusion, ABC and TWW. Rediffusion was a heavy loss. At a time of financial crisis, when his partner, Lord Rothermere of Associated Newspapers, lost heart and withdrew, John Spencer Wills, Associated-Rediffusion's Chairman, had had the faith and courage to soldier on and save the infant Independent Television from an early grave. The other companies had grown up in its shadow. Rediffusion had introduced schools television to Britain and pioneered the use of videotape. Its summary relegation to junior partnership in a forced merger with ABC to form a new company had little to do with natural justice.

The Authority's contract decisions and the reasons for them are set out

Independent Television 1968

ITA Regions	Programme Companies	Population Served
1. London	Thames Television (weekdays to 7 p.m. Friday) Lord Shawcross (Chairman) Howard Thomas (Managing Director) London Weekend Television (weekends from 7 p.m. Friday) Aidan Crawley (Chairman) Michael Peacock (Managing Director)	13,940,000
2. Midlands	ATV Network Lord Renwick (Chairman) Sir Lew Grade (Deputy Chairman and Joint Managing Director) Robin Gill (Joint Managing Director)	10,620,000
3. Lancashire	Granada Television Sidney Bernstein (Chairman) Cecil Bernstein (Joint Managing Director) Denis Forman (Joint Managing Director)	8,040,000
4. Yorkshire	Yorkshire Television Sir Richard Graham (Chairman) G. E. Ward Thomas (Managing Director)	6,200,000
5. Central Scotland	Scottish Television James Coltart (Chairman) William Brown (Managing Director)	4,008,000
6. Wales and West of England	Harlech Television Lord Harlech (Chairman) A. J. Gorard (Managing Director)	3,294,000 plus 2,174,000 (Welsh Service within Wales)

7. South of England	Southern Independent Television John Davis (Chairman) C. D. Wilson (Managing Director)	4,989,000
8. North-East England	Tyne Tees Television G. H. J. Daysh (Chairman) J. A. Jelly (Managing Director)	2,720,000
9. East of England	Anglia Television The Marquess Townshend of Raynham (Chairman) Aubrey Buxton (Principal Executive Director)	6,102,000
10. Northern Ireland	Ulster Television The Earl of Antrim (Chairman) R. B. Henderson (Managing Director)	1,375,000
11. South-West England	Westward Television Peter Cadbury (Executive Chairman) W. H. Cheevers (Managing Director)	1,601,000
12. The Borders and Isle of Man	Border Television John Burgess (Chairman) James Bredin (Managing Director)	579,000
13. North-East Scotland	Grampian Television Captain Ian Tennant (Chairman) Lord Windlesham (Managing Director)	1,847,000
14. Channel Islands	Channel Television Senator W. H. Krichefski (Chairman) K. A. Killip (Managing Director)	107,000

in an earlier volume of this history.[1] But whatever the justification, the resultant upheaval left Independent Television dangerously buffeted, and Sir Sydney Caine, the Authority's Deputy Chairman in 1967, subsequently wrote that he had no doubt that the decisions were wrong.[2]

Only two of the central companies survived intact, and the wings of these were clipped. With the map redrawn, applications had had to be made for different contract areas and times. Lew Grade's essentially metropolitan ATV Network had lost its combined London/Midlands contract and was to serve the Midlands for seven days a week. This entailed the expense of new office buildings, studios and equipment in Birmingham. It also meant a reduction in coverage, as measured in population and days. The Bernsteins' Granada Television, which had served the Northern area with a population of 14.5 million for five days a week, was now to be confined west of the Pennines and serve a population of 7.5 million for seven days a week. This represented an even heavier loss of coverage than ATV had suffered and brought an extra cost in weekend working.

The competition for the largest and most profitable contract, serving the Greater London area during weekdays, had been won by Howard Thomas and his team from ABC. They took control of the newly formed Thames Television, jointly owned by ABC and Rediffusion. Taking the pick of former ABC and Rediffusion staff, Thames was well placed to assume a dominant network role, but start-up costs were heavy.

Its teething troubles were small compared with those of the other new London contractor, London Weekend Television. The saga of LWT's disastrous endeavour to lead public taste to higher ground is told in the next chapter. In LWT and Yorkshire Television the network had untried performers in leading roles. The initial impact of their inexperience was to lessen the appeal of ITV programming overall. Moreover, the introduction of a fifth major company increased the cost of the whole system in facilities and staff, while diminishing the role and profitability of other companies. Yorkshire Television's financial vulnerability was dramatically exposed in March 1969 when its main transmitting mast collapsed.

By outside observers the shake-up of ITV was applauded. No one doubted that such a rich industry could well afford and easily absorb the extra costs, and that the introduction of new companies and new talent would be beneficial in recharging the batteries of creativity. The contracts were regarded as valuable prizes. Subscribers registered handsome capital gains before the new companies had even come on air. Non-voting shares in Yorkshire Television were quoted at more than four times their nominal value.[3]

In Wales and the West the ousted TWW, headed by the Earl of Derby, unwisely spurned the Authority's offer of a 40 per cent interest in the new contract. It retired in a huff without its consolation prize, handing over to its successor two months earlier than the due date. Thus on 20 May 1968 the Chairman of the Authority attended the much-heralded offical opening night of the first newcomer to appear before the public. It proved an embarrassing occasion, marred by mishaps and the non-appearance of the expected stars, Richard Burton and Elizabeth Taylor, whose names had glamourised Harlech's application.

In July members of the Authority attended, amongst other functions, the laying of the foundation-stone for Southern Television's new studio building in Southampton and a formal luncheon in Leeds to celebrate the opening of the new Yorkshire station. The Granada and Yorkshire contracts began on 29 July; the remainder on the following day. But the euphoria of the opening nights was short-lived.

The surgical operation by Lord Hill and his colleagues had left ITV employees resentful and insecure. Successful production teams had been broken up; colleagues of long standing dispersed. The trade unions were bitterly angry. They were determined that every one of their members employed during the previous contract period should have a job in the reshaped industry, and that redundancy money should be paid not only to those who were forced to move but even to those who stayed doing the same job in the same place for a higher wage under new management.

Within a few days of the launching date the Association of Cinemato-graph and Television Technicians (ACTT) had called out all its members among the companies' staff in a national strike. For two weeks in August the programme schedule was disrupted and programme production brought to a halt. An emergency service had to be mounted, and programmes planned for the autumn rescheduled or cancelled. It was an ominous start to an unhappy period. For during 1969 and 1970 ITV was to experience a decline in popularity and its biggest financial setback since 1956.

Following the strike, audience measurement, which had been taken over by Audits of Great Britain Limited (AGB), recorded a disastrous reduc-tion in ITV's share of the total television audience. In April 1968 this had been 58 per cent. By October it was down to 47 per cent, recovering to 53 per cent in March 1969. AGB's first figures had been too low to be credible and had to be fudged. Interim figures which did not meet the required specifications were then produced until the new Setmeters and Watcher-graphs were in proper working order.

Advertisers complained that the network programme schedules were wearing a distinctly less appealing look. Out had gone such veteran audience winners as *Sunday Night at the London Palladium, Double Your Money* and *Take Your Pick*, lamented by millions of viewers, but not by the ITA. In had come the experimental and the unresearched, most prominently a weekly trio of programmes featuring David Frost and a feast of new-style plays and 'specials' at weekends. ATV's long-running serial, *Crossroads*, was dropped from the London area and had to be brought

back in response to furious public demand. Unfamiliar styles of presentation appeared on the screen in areas where new companies had been
appointed. Taking full advantage of the opportunity, the BBC put on a
burst of competitive popular programming.

Programme problems could be, and were, overcome as soon as the new
creative teams found their feet: with, for example, popular thriller series
like *Callan, Department S* and *Special Branch*. But, more fundamentally,
both ITV's principal growth factors during the 1960s had ceased to
operate. As television coverage of the country approached saturation level
the total potential audience was no longer growing, and the size of ITV's
share of that audience could not be sustained when BBC2 became more
popular.

Growth in revenue had slowed down, and in 1969 ITV's revenue actually
fell. Against the disconcerting possibility of several years of static income
had to be set the certainty of imminently rising costs. Re-equipment to a
625-line standard and for programme-making and transmission in colour
involved the companies in capital expenditure of the order of £20 million,
in substantial additional programme costs and in higher annual rentals to
the ITA to pay for the construction of the new network of Ultra High
Frequency (UHF) transmitting stations. Colour came to the screen on 15
November 1969. It was a technical triumph, but so long as the amount of
colour viewing remained below a level at which an increase in advertisement rates could be justified, no compensatory rise in income was
forthcoming. In fact, the fall which had begun in the summer and autumn
of 1969 became more pronounced during the winter.

What had been a highly profitable industry was now paying nearly half
its total gross income to the Exchequer. The ITA's Annual Report
recorded:

> In its discussions with the Postmaster-General, the Authority forecast
> that the profit before tax of the companies as a whole, which had been
> some £19 million in 1967–68, would fall to £12 million in 1968–69, and
> that even without any increase in the levy it would fall by a further
> substantial amount to £5 million or less before tax in 1969–70. In the
> Authority's view any increase in levy would adversely affect the com-
> panies' ability to spend the money necessary to maintain and improve
> programme standards, would shake the confidence of the system at a
> time when it should be finding stability after the radical changes effected
> by the Authority in 1968 and ... would prejudice the Authority's own
> ability to devlop its UHF system in accordance with national policy and
> in the national interest.[4]

The warning went unheeded by a government itself in financial straits.
Neither the weight of the Authority's opposition nor urgent representa-

tions by the companies succeeded in preventing an increase in the levy, and this was not an impost which the ITA could see any means of alleviating in its rental charges to the companies, since it was necessary for ITV's programme of UHF station building to keep pace with that of the BBC. Yet, if the Authority was to continue charging rentals sufficient for the proper discharge of its functions, some companies would be left with insufficient funds to meet their programme obligations and still maintain an acceptable return on capital employed.

For whom should the blow be softened? The extent to which, in a situation of this kind, the financial well-being of the companies lay outside their own control is revealed in a report by the Director General to members of the ITA in April 1969:

> At a meeting with the Chancellor of the Exchequer, the Financial Secretary to the Treasury and the Postmaster-General which the Chairman and I attended during the week before last, we were informed that it was the government's intention to raise £3 million and the suggestion was made that this should be raised exclusively in altering the levy on advertising revenue above £10 million. The aim of this proposal was to spare Yorkshire Television, but it had the result of reducing the return on capital employed of the other four large companies (the only ones affected by the increase) to an unacceptably low level compared with those of the regional companies. The return on capital employed for the four largest companies as a group would have been reduced to just over 13 per cent on the estimated 1968–69 figures, while the return for the nine regional companies as a group (excluding Channel who are not liable) would have been over 31 per cent.[5]

The contention that because the companies had only a limited contract life there was an element of risk which justified a return of not less than 25 per cent on capital employed, had fallen on deaf ears. But the ministers had invited the Authority to submit alternative proposals, provided they produced the £3 million (originally £6 million) which the government required, and those proposals were adopted. They were generally more equitable, but it was foreseen that, even so, Yorkshire Television's modest profit would be reduced almost to break-even. ITA staff were therefore enjoined to consider 'all aspects of Yorkshire's position to see what, if anything, should be done on our side about the company's finances'.[6]

'Additional payments', commonly called the levy, were assessed on each contractor on the basis of advertising receipts. They were payable to the Exchequer's Consolidated Fund via the ITA under Section 13 (1) of the Television Act 1964. For the first five years, from July 1964 to July 1969, the rates were nil on the first £1.5 million of revenue, 25 per cent on revenue between £1.5 million and £7.5 million, and 45 per cent above

£7.5 million. The most damaging feature of the upward revision of rates announced in the 1969 Budget at a time of ailing revenue was a reduction of £1 million in the initial 'free tranche' or 'free slice'. From July 1969 the levy became payable on revenue in excess of £500,000 through the introduction of a 7 per cent band between £500,000 and £1.5 million. The 25 per cent band stopped at £4 million, and new rates of 35 per cent between £4 million and £10 million and 47.5 per cent above £10 million were applied.

The Authority's alarm was publicly expressed by the Chairman in an address to the Scottish Centre of the Royal Television Society at Dundee on 26 September 1969:

> Since July 1964, in addition to paying the ordinary taxes – income tax, corporation tax etc, from which the BBC are happily excused – ITV has had to pay almost £103 millions in a levy on advertising revenue. I stress the word revenue because this is sometimes not clearly understood – the levy is not a tax on profits, it is a tax on income. I wish to say, in measured and deliberate terms, that the Exchequer levy on the income of Independent Television has, in the view of the Authority, been placed too high: it is more than can be fairly expected, more than can be safely borne, more than is good for programmes, more than is good for viewers, more than is good for the companies and all the people they employ, more than is good for all who wish to see the present performance of Independent Television maintained and bettered. It is beyond our capacity both to pay the levy on anything like the scale the government now demands and still do all we should be doing.

This argument was too well founded to be ignored, and in the following spring (April 1970) the Labour government relented. The effect of the reduction in the free tranche had been to hit hardest the smaller and the more vulnerable companies. This error was corrected by the extension of the nil rate to the first £2 million of revenue. The other rates were revised to become 20 per cent on the next £4 million, 35 per cent on the next £3 million, 40 per cent on the next £3 million, 45 per cent on the next £4 million and 50 per cent on amounts in excess of £16 million.

The clumsy manipulation of this blunt fiscal instrument, designed as special payments by those profiting from the privileged use of a scarce national resource, carried a number of companies to the threshold of insolvency. It helped to force Tyne Tees into the arms of Yorkshire Television, whilst Scottish Television, the most severely affected, had to be rescued through emergency measures taken on its behalf by the ITA when the company was without funds. In February 1970 it could pay neither rental nor levy nor its bills, and with debts of nearly £2 million and no surplus of income over current expenditure, a further bank loan was unobtainable.[7]

The Authority arranged for the postponement of payments due, and in the words of a later memorandum from Scottish Television, 'the Authority's understanding and support throughout the period ensured our survival. We in our turn maintained the service to the public in strict accordance with the terms of our programme contract.' The company's Managing Director, William Brown, added:

> Our own experience was perhaps the most serious of all companies' a few years ago. To give you an example, in our view, of the unfairness of the levy: in the year when my company actually made a trading loss we have paid over £750,000 in levy. That was clearly a situation where there should have been some arrangement whereby we could run the company, albeit on a very tight and modest level at that time, but which at least enabled us to run the company first and to pay the costs of running it, before paying this special taxation.[8]

This was the company whose founder, Roy Thomson, had boasted of a licence to print his own money. But what Thomson enjoyed was never a licence to print; it was a licence to save and, as the major shareholder, benefit most profitably from those savings. Now this licence had expired. Disliking Thomson's dominance, the ITA had required him to reduce his holding. He had sold at an advantageous price and left the company without reserves.

Harlech Television was another company in distress before the change in levy in April 1970. In December 1969 the board sent a delegation to the ITA's offices at 70 Brompton Road to inform the Authority of the social and political consequences if the company was forced to cut back on programme expenditure and abandon the promises which had won it the contract in Wales and the West. If the levy continued at its current level, educational and Welsh-language programmes would have to be cut, support to artists, writers and musicians in the region withdrawn, and consideration given to closing one of the company's two studio centres (probably Cardiff because Bristol had recently been re-equipped).

After the lengthy presentation of their case the Director General told them that he had not heard a single word from which he would dissent. But the great change in Independent Television's fortunes had, he said, been caused only in part by the increase in the levy. The other major factor was an erosion of the margin between expenditure and income. Thus, even if Harlech no longer had to pay an increased levy contribution (estimated at £300,000), rising costs would still present considerable difficulties, and this applied to all the companies.

Formally, under the Act, the levy was an additional rental paid by the companies to the ITA, but it was one over which the ITA had no control and of which it retained no part. Its operations were financed out of the ordinary rentals paid by the companies and in assessing them it was continually struggling to implement an even-handed policy of equalising profitability between the companies, so far as the unpredictable incidence of the levy would permit. At this time rentals were based on capital employed or, more precisely, 'capital reasonably employed'; later it became Authority policy to base them more closely on revenue.

Whereas the government enjoyed the freedom to vary the scales of the levy annually, or even more frequently if it chose, under its contracts with the companies the ITA, not anticipating the extraordinary circumstances of 1969 and 1970, had denied itself the opportunity of revising its rentals (apart from cost-of-living adjustments) more than once during the six-year contract period running from 1968 to 1974. In what was substantially a controlled economy the downward trend in advertising revenue and vagaries in the government's financial demands resulted therefore in some inequities between companies, and this placed a heavy strain on relationships between Authority and companies and among the companies themselves.

The consequences of these adverse factors between 1968 and 1970 were not at once apparent where it mattered most. 'The companies' financial problems had no visible effect during the year on the overall standard of the programmes,' reported the ITA in its Annual Report for 1969–70.

> However, it may be that the immediate effect of shortage of money is not to be found in what appears on the screen but in what does not appear. The capacity of any television service to raise its standards depends not only on the talents of its creative personnel but also on the ability at management level to underwrite fresh enterprise and experiment. Advances in television can only be nurtured in an atmosphere that permits the taking of calculated financial risks.[9]

The new levy scales which the Authority persuaded the government to accept in the spring of 1970 were designed to achieve a reduction of some £6 million in a full year. But even with this relief little more than break-even was forecast for the system in the financial year 1970–71, and a situation of continuing difficulty seemed to ministers to call for a full inquiry. The matter was therefore referred to the National Board for Prices and Incomes which, under the chairmanship of the Rt Hon. Aubrey Jones MP, undertook an intensive investigation between April and October 1970. From June, when a general election led to a change of government, the board lay under sentence of death, but it was spared to complete this assignment. *Costs and Revenues of Independent Television Companies*,[10] presented on 21 October to a new Secretary of State for Employment (Robert Carr) and a new Minister of Posts and Telecommunications (Christopher Chataway), was its swansong.

The report was valuable in publishing information little known or understood at the time. It noted an average annual increase in ITV revenue between 1965 and 1969 of about 5 per cent, followed by a drop of more than 5 per cent in the financial year which ended in July 1970. It found that, as well as inducing a loss of confidence among advertisers, ITV's declining share of the audience had coincided with a downturn in profits in the food manufacturing industry, on whose advertising ITV was heavily dependent. Blame was attached also to the companies' pricing policies. While revenue was declining by between 5 and 6 per cent, the number of minutes sold during the past year had fallen only from 87,000 to 86,000. To the NBPI this suggested faulty structuring of rate cards. Moreover, the industry's co-operative marketing agency, the British Bureau of Television Advertising, was not being permitted by the companies to operate as effectively as it ought.

The board found that the decline in revenue of £2.5 million over the past two years had been accompanied by a rise in costs of more than £11 million: costs in 1968 had been 9.5 per cent higher than in 1967; in 1969 21.4 per cent higher than in 1968. A number of special factors had contributed. A rise of nearly 25 per cent in administration expenses in 1968 included £1.3 million in redundancy payments to the staff of companies which had lost their contracts. The introduction of a fifth network contractor accounted for a third of the increase in costs in 1969. An increase of £4 million in direct programme costs was attributed to an 8 per cent extension in the permitted hours of broadcasting and to some ambitious and expensive programme-making by companies honouring promises made in their contract applications.

At the same time poor management and extravagant labour practices were held responsible for inadequate cost control. A low level of resource utilisation and a general lack of formal planning methods were found in some companies. The production resources of the industry, in studios and staff, were being used at no more than an estimated 65 to 70 per cent of capacity. The industry's occupation and wage structure needed revision to 'assist in the more flexible use of labour and the control of the rate of increase of earnings'.[11]

The Authority's evidence about the future trend in advertising revenue had been deeply pessimistic:

The rise of television advertising revenue from nothing to £100 millions in the thirteen years 1956–68 was of course a highly dramatic event in the history of advertising, but it was directly associated with a continuous rise in the number of ITV homes, and that growth has ceased. Growth associated with it has consequently ceased ... If there were some immediate recovery from the present slump, we would not be disposed to accept it as the beginning of a long-term trend.[12]

This mood prevailed on the board, which adopted a similar tone in foreseeing woes to come: 'Our projections of costs and revenues into the middle seventies, on whatever basis they are made, suggest a continuously worsening position, though without the levy profits would on average remain high in relation to profits in industry generally.'[13]

The principal remedies suggested by the board were: an increase in revenue through longer hours of broadcasting or the introduction of a second commercial service (which might, however, not produce a financial gain to the contractors); a reduction in the business risk and required level of return on capital employed through modifications in the system of awarding contracts so as to give more security of tenure and eliminate the threat of further drastic upheavals; and a reduction in the number of contractors, both central and regional, through enlarged areas and mergers, thereby reducing under-used capacity and taking advantage of economies of scale.

The report echoed a view held universally within ITV when it said that the industry's greatest requirement was a sense of stability, but the proposed means of achieving this was not calculated to appeal to those outside the system: 'We recommend that contracts should normally be renewed unless a contractor has been twice warned by the ITA that his performance was unsatisfactory.'[14]

In general, though, the report was not well received within the industry. It was thought to have failed to see some of the problems in a proper perspective. Some companies were not averse to a measure of rationalisation so long as the essential character of regional television was not lost, but – predictably – all were irritated by the sharp criticism of their business efficiency in the use of manpower and production resources and in marketing. In private 'fair comment' was admitted in some instances,[15] and, despite grumbling about the report's 'superficial' and 'inaccurate' comments, joint committees were set up to consider improvement in production methods and sales co-ordination.[16] The ITA, while favouring a full review of the system in the middle of the decade, believed that structural changes so soon after the recent mayhem could only further unsettle the industry.[17] There was general disappointment at the board's complete disregard of the social value of an admittedly costly regional system.

Even more disappointing was the fact that the board had resisted the temptation to stretch its terms of reference and state a case against what some identified as the main culprit or at least the load of straw currently breaking the camel's back: the much-hated levy. It was believed that this would have strengthened the arm of a new and well-disposed Conservative minister in persuading the Treasury to recognise the need for substantially easing this burden and reforming the whole basis on which it was assessed.

By the middle of January 1971 the Chairman of the ITA and the minister were talking privately about the prospects of securing relief from levy to the tune of £12 million in a full year. It had become apparent that in Christopher Chataway television had the most knowledgeable and sympathetic minister since the ITV service began. Unprecedentedly, he knew about the medium from the inside: from experience as a reporter for ITN in 1955–56 and a commentator for the BBC from 1956–59. Hopes of some favourable movement not only on the levy but also on extended hours of broadcasting and even on an ITV2 could now be seriously entertained.

Monday, 15 February 1971 was a red-letter day, when a government for once grasped some broadcasting nettles. In the House of Commons Chataway announced an increase of £1 in the BBC licence fee (to £7 for monochrome and £12 for colour) and a halving in the rates of the levy imposed on ITV so as to reduce the yield in a full year from £20 million to an estimated £10 million. He went on:

I am satisfied that these reductions will provide Independent Television with the resources it needs. They are also designed to enable the companies to improve the quality of their programmes and those of Independent Television News. The Authority have assured me that it is their intention to use this opportunity to secure such an improvement. The government regard the present reduction as a holding measure designed to stabilise ITV's financial position at least until 31 July 1972 – that is, until the end of the next full levy year. In the meantime, the government will review the basis on which the levy is charged.[18]

This announcement was greeted by the ITA as 'a welcome measure of essential relief'. Owing to their differing financial circumstances the relief was, as usual, shared unevenly among the companies. It saved some of the largest more than £1 million a year, and others, including the hard-pressed Scottish and Yorkshire, half a million.

The Authority's share of the relief was taken in the form of increases in rentals to yield an additional £3.5 million a year. These also were distributed unevenly, with some companies, especially the big ones, paying appreciably more, and some of the smaller contributing less. The companies acknowledged that the £6.5 million with which they would be left would 'give ITV the opportunity to maintain and extend its service to the public'.[19] Like the ITA, they warmly welcomed the promised review of the 'outmoded' levy system.

Applauded by all, this rescue operation was generally regarded as a personal triumph for the new minister, for the Treasury had much at stake in the levy. By 31 March 1971 this additional impost on ITV, levied over and above corporation tax, had brought the Exchequer's take to more than £153 million (£26 million in 1969–70; £25 million in 1970–71).[20] Chataway's announcement was said to be a 'first instalment in the campaign to restore

the industry to health' and a 'licence to make profits again'.[21] Coincident-
ally the downward trend in advertising revenue was reversed, so that the
year 1971–72 became 'one of welcome stability for the programme
companies'.[22] The NBPI's prognostication of doom proved as inaccurate
as most economic forecasts, and the government's belated rush to ITV's
sickbed with a remedial tonic came just when the crisis was passing.

Three factors in particular had contributed to the reduction in levy.
First, the government had accepted the necessity for the ITA to increase its
own rental in order to keep pace with the BBC's UHF station-building
programme during the 1970s; secondly, the NBPI report had short-
circuited what would otherwise have been a protracted financial argument;
and thirdly, as Chataway's statement of 15 February had stressed, it was
intended to provide an opportunity for Independent Television to improve
the quality of its programme service.

This last consideration set the stage for what was to be a major
preoccupation during the remainder of 1971. Higher programme standards
were the *quid pro quo* promised to the government by the Authority and
the companies, and the extent to which that promise was kept would
influence the government's intention to review the basis on which the levy
was assessed. The minister had to be persuaded that further advances
could best be stimulated by changing from a levy assessed on revenue to
one assessed on profits.

Although general improvement in such a subjective area as programme
quality might be hard to evaluate, it became evident in March that the
government expected demonstrable action under four specific headings, as
suggested by the Authority: higher programme expenditure; more high-
quality, ambitious and experimental programmes; fewer imported programmes of indifferent quality and repeats of such programmes; and a
greater readiness to schedule serious programmes at an earlier time in the
evening.[23]

Progress in these directions was facilitated by the unexpected financial
recovery. In 1971–72 the companies once again enjoyed a healthy financial
year, and the ITA was able to report:

> At the end of March 1972 there were many signs that Independent
> Television was moving into a new phase of programme development
> which seemed likely to make 1972–73 a year of advance and achieve-
> ment. If it is true that the quality of programmes is not by any means a
> simple derivation of the money that is spent on them, it is also true that a
> background of financial anxiety and restraint is not conducive to
> initiative and enterprise in planning and production.[24]

'Necessary Improvements' were specified by the Director General in a
note to the companies.[25] A wider and more adventurous range of

programming was required, with more peak-time scheduling of programmes 'of less than mass audience appeal'. The quality of educational programmes was to be raised; and there were other requirements besides better programmes. The companies were urged to accept responsibility for more training of their employees and to introduce an industry-wide pension scheme. They would be expected to complement the ITA's engineering research effort in transmission, propagation and linking by co-operating in research and development in studio equipment and programme origination.

Meanwhile argument continued to rage over the levy. During 1971 a campaign for its total abolition failed, but there were strong hopes for the early introduction of a more finely tuned mechanism. The Authority pursued its quest for an alternative system which would recognise the government's interest in securing public participation in the commercial value of the programme contracts; take account of expenditure as well as income; be immune from the need for constant change; and be capable of being properly supervised. In its representations to the minister the Authority argued strongly that the making of good programmes must be the first charge on ITV's revenue, and that thereafter money going to the government should be on the same basis as money going to shareholders: that is, profit-related.[26]

Discussions between the Authority and the ministry ran on through 1972 and 1973, when the ministry produced a scheme calculated to reward meritorious programme-making. Under this the companies would be allowed a free slice of 2.5 per cent of revenue before the application of levy and the Authority would be allowed to retain a sum equivalent to 3 per cent. Thus 0.5 per cent would be available to the Authority to reimburse companies for some of the exceptional costs incurred in making cultural programmes. But the Authority demurred, doubting whether this would lead to better programmes:

If virtuous programme performance has been made commercially attractive, there is a danger that a commercial decision not to be virtuous could become respectable. Those companies who are prepared to spend money on winning prestige, rather than audience, might dislike the suggestion that this was being done for reward; those that are not might claim that they had made a perfectly fair decision to make money in their way rather than in this way.[27]

The Director General was able to point out to the ministry a recent burst of distinguished programmes on which companies had spent more money than they could hope to recoup: ATV's Shakespeare plays and *Long Day's Journey into Night*, Granada's *State of the Nation*, LWT's *Weekend World* and *Mahler Festival*, Southern's operas from Glyndebourne, Thames's *World at War* and Yorkshire's series on the Brontës. These had been made possible by ITV's greater profitability, not by any special financial reward, and 'the worthy scheduling of them' had been more of a problem than 'the worthy making of them'.[28]

Shortage of airtime had now become more crucial than shortage of money. Cultural programmes in peak time caused more than a loss of revenue. Scheduling *Macbeth* or *Long Day's Journey into Night* before 9.30 or 10 p.m. would have meant a lost evening for the majority of viewers. In the context of a single-channel service it could be described as breaking faith with at least 90 per cent of the audience, who – whatever they were told by the wise and the good – persisted in believing that the function of a mass medium was to serve a mass public. The only solution to that dilemma was not merit allowances but a second channel.[29]

In all it took nearly eight years of argument to convince the Treasury that a levy related to profits was to be preferred to one imposed on income irrespective of profit or loss. The decision to legislate accordingly, along lines proposed by the Authority, was taken by a Conservative government towards the end of 1973 and accepted by an incoming Labour government the following year. If there had to be levy, it was now to be less of an incubus and more of a cushion, offering a fair chance that programme-making could be adequately financed even during economic blizzards.

The Independent Broadcasting Act 1974 was 'an Act to make further provision as to payments to be made to the Independent Broadcasting Authority by television programme contractors; and for purposes connected therewith'. It passed into law on 23 May and came into operation one month later. ITV was making good money again and the government had decided that it would take £35 million out of the companies' profits, currently forecast at £50 million a year, leaving them £15 million, on which corporation tax would be payable. The levy was therefore fixed at a rate of 66.7 per cent on each company's profits in excess of £250,000 or exceeding 2 per cent of net advertising revenue (whichever was the greater). Profits from the sale of programmes overseas were excluded. What were or were not relevant items of income and expenditure in calculating profit for levy purposes was to be determined by the Authority without right of appeal to the courts by the companies.

It was an occasion for modified rapture: the industry was appeased but not satisfied. The intention behind the levy had not changed, and the new mechanism was geared to siphon off profits deemed excessive as effectively as the old. If corporation tax was taken into account, the marginal impost on profits became 84 per cent. Every additional £100 of profit netted a real gain to the company of just £16. Every additional £100 of revenue expenditure incurred a real cost to the company of just £16.

Taxation at such an unhealthy level was bound to remain a contentious issue. The projections used to arrive at the rate of 66.7 per cent dated from 1973 and were already out of date when the Act was passed. In view of the risk factor, the NBPI had judged 'not less than 22 per cent' to be a fair rate of return on capital employed in the companies. This yardstick had been accepted, but by March 1975 the industry's return on capital was below that figure: there were no excess profits at all and none in sight in the foreseeable future. Yet the levy had to be paid.

In a period of high inflation the situation had altered radically in twelve months: costs had risen by 20 per cent and revenue, running at the same level as the previous year, had dropped in real value. The scale of the levy had been designed to leave the companies with a cash flow sufficient to meet the needs of capital investment and provide for an appropriate level of reserves and profit. But as a result of inflation they would together be forced to find more than £7 million of additional finance for working capital out of their own resources, and this could not come from the net profit left to them after payment of levy and corporation tax.

This inability to set aside reserves for future development during a period of inflation was only one of several serious disadvantages imposed on the industry by the levy system. If the companies obtained loans to tide them over, these carried interest at a high market rate which was not chargeable as 'relevant expenditure'. The impact of the system was erratic and haphazard as between companies. Some of the smaller ones, which inflation was driving into deficit, had derived no benefit from the switch to a profit base: thanks to the free slice they had paid no levy previously and still paid none until inflation took them over the limit. Worse, the levy worked from year to year, so that a company could one year fall well below the level of profit at which its free slice was exhausted, but was not permitted to carry the unused portion forward into future years. Thus if a company made a profit of £500,000 over a two-year period by means of two annual surpluses of £250,000 it would pay no levy, whereas the same profit achieved through recovery from a £500,000 loss one year to a £1 million profit the next would all be taken by the levy.

It was in these circumstances that the Chairman of the Authority wrote on 25 March 1975 to notify the Home Secretary of a cash shortage which would affect future programme output:

World at War, Jennie, Clayhanger, Edward VII and other such programmes need to be put into production possibly some two or three years before they appear on the screen. It is certain that if this cash shortage persists for the next few months it will prevent companies from making commitments in commissioning programmes of distinction.

The minister was reminded that the crippling impact of the levy on ITV came at a time when the BBC had, 'rightly as it seems to us', received an increase of 33 per cent in licence-fee income to counter inflation.

Under the 1974 Act the Home Secretary possessed powers to vary the rate and size of the free slice by statutory instrument, but the Authority's plea went unheeded. A letter from the new IBA Chairman in October regretted that no reply had been received to the request for modifications made in March. A negative response arrived in January 1976. The industry could only take consolation from the thought that it would have been in far worse financial straits if these 'additional payments' had still been revenue-based. The crisis passed as profits built up again. Levy payments grew with them (to £67 million by 1978–79), but whether these were too high or not high enough and whether they were collected according to the 'least worst' formula remained questions unresolved. The matter was subjected to close scrutiny and lengthy comment by successive Public Accounts Committees.

Meanwhile the Treasury – always more hostile than the Ministry of Posts and Telecommunications or the Home Office – continued to worry that the Authority had too much power in deciding what constituted profits, allowable expenses and overseas sales. Millions of pounds were involved and the Authority stood between the Treasury and the companies. Was it policing them zealously enough or leniently conniving at keeping money within the industry? The Home Office was squeezed between the Treasury and the Authority and leaned on from both sides.

Following reports by the Annan Committee, the Public Accounts Committee and the Comptroller and Auditor General in 1977, the Home Office and the Authority were obliged to embark on yet another fundamental review. 'If the view prevailed that the IBA should no longer have statutory responsibility for collecting the levy,' enquired the Home Office chillingly, 'would you consider it better to reintroduce a revenue-based system?'[30]

The IBA's answer was tetchy.[31] It was surprised at questions and points which led so directly away from what was agreed in 1973 and 1974 and was not aware of any case for a change in the present form of the levy. 'Difficulties in applying it' had been far fewer than Jeremiahs had predicted. It quoted from a speech in the House of Commons made by the Minster of State at the Home Office (Alexander Lyon) on 29 March 1974:

> Our experience over the years has been that the IBA carries out its responsibilities efficiently and well and that it has gained the confidence of public and broadcasters . . . The government can now accept that we should allow the IBA to have these wide powers in determining relevant income and relevant expenditure for the purpose of assessing profits.[32]

It was the IBA's turn to frustrate the Home Office's desire to make alterations, and exchanges were adjourned for two years until 1979 when, following the return of a Conservative government, the matter once again became a live issue. What caused concern to a cost-conscious administration were ITV's high labour and programme-making costs, to which the

BBC was loudly calling attention as part of its campaign for a higher licence fee. Early in 1980 a draft Treasury proposal advocated a half-step backwards: to a part revenue-based and part profits-based levy.

This received lukewarm support from the Home Office, while the IBA defended the *status quo* stoutly, arguing that any such change would once again identify the levy as a device which allowed revenue-raising to take priority over good programme-making. Retreat from what had become a 40:60 division of profits between companies and government towards the 1970 20:80 split might discourage new applicant groups from coming forward in response to the recent advertisements for new contracts. From 1 January 1982 the companies would be paying an additional £14 million a year in rental and £70 million in Fourth Channel subscriptions, and already the newspapers' city pages were describing the next round of franchises as licences to lose money. In the Authority's view, cost-consciousness could be protected by the normal commercial disciplines which applied to private companies required to make a profit.[33]

These arguments proved successful. No alteration in the basis of the levy was included in the 1980 and 1981 Broadcasting Acts, and the Home Secretary did not see fit to adjust the scales. The running controversy had been a destabilising factor in an industry sorely in need of stability, but the form agreed in 1974 by both Conservative and Labour governments narrowly survived until 1986.

The levy has had a dominant, if largely unacknowledged, influence on every aspect of ITV's operations. Many important decisions affecting programmes have been taken by the companies 'for levy reasons', though not the best on other grounds. There was, for example, a strong financial incentive in deploying programme-making resources to obtain levy-free income from overseas sales. But granted the political requirement of a mechanism to regulate the swing of the pendulum between excess profits and insolvency, this device performed a necessary function for which no better alternative could be found than some such form of profit-sharing between advertising-financed broadcasters and the state. Both state control and a lowering of programme standards through competition for advertising revenue were avoided, and any distortion of the system might be thought a small price to pay for Independent Television's special contribution to public funds, which by the end of 1980 had reached a total of £498 million.

3

TROUBLE AT LONDON WEEKEND

Of all the problems which vexed the reconstructed network after its public début in the summer of 1968, none was so severe or far-reaching as those which afflicted the new weekend contractor in London. The question who was to control Thames, the other new London company and largest of the contractors, caused the Authority almost as much concern and aggravation but had no wider impact. The LWT affair, by contrast, called into question most of the fundamentals of Independent Television: the methods and morals of contract awards; the role and competence of the Authority; the pattern of regionalism; inter-company relationships; and the philosophy of programming. An account of the structure and operations of these two companies will be found in Volume 4 of this history, but LWT's birthpangs belong to the story of the Authority and the network as a whole.

In writing of the award of a contract to the London Television Consortium, which was to become London Weekend Television, Lord Hill has recorded:

> It is an understatement to say that the Authority liked this application. Even allowing for the fact that promising is so much easier than performing, it was difficult to resist the thought that here was a group which would bring new thinking, fresh ideas and a lively impetus to weekend broadcasting. It had to have its chance whatever the repercussions. [1]

The appeal of the consortium lay in its conviction that there were new ways of attracting a mass audience; that ITV had been suffering from a failure of adventure in programme ideas; and that, if rightly conceived and deployed, its output could offer as complete a public service as the BBC's. Its application explained that the prime movers, Aidan Crawley and David Frost, had come together because they shared a common philosophy towards the role of television as a public service. Their first principles were respect for the creative talents of those who made programmes and for those who watched them and whose differing interests and tastes aspired to new experiences. Since it was a television company which was being planned, they saw no reason why people who knew about television, cared about television and made television should not have an effective voice in the running of the company through a substantial equity holding.[2]

No words could have been more carefully chosen to raise a cheer from members of the Authority seeking a different image from that of base commercialism. David Montagu, the merchant banker who acted as marriage-broker to bring Crawley's and Frost's originally separate consortia together, has recalled how at the end of its formal interview in May 1967 their delegation received what almost amounted to a standing ovation.[3] The application was irresistible and the contract duly awarded. The subsequent failure of the company to live up to its promises, the angry and well-publicised departure of disillusioned senior programme staff and the company's losing struggle against financial odds combined to provide material for a juicy scandal and the creation of a myth in which Creative Talent battled valiantly against Hard-faced Business in a melodrama more suited to peak-time viewing.

The failure was a failure of management. Yet the difficulties confronting the new company initially were so serious and numerous that it may be doubted to what extent they were soluble by even the most talented or hard-faced. The commitment to radical change in weekend programming ran into head-on collision with the beliefs and experience of the other major companies. Companies within the ITV system are interdependent, and for a newcomer to take his seat at the table with a publicly proclaimed brief which rubbished the programmes of the established companies on whose co-operation and goodwill it must depend was to invite disaster.

This could have been averted only by an early demonstration of the popularity of the new weekend schedules, but viewers love old favourites and do not take readily to strange fare. They expressed their preferences unmistakably by switching in large numbers to the BBC, which gratefully seized the opportunity to entice them by packing more light entertainment into its programming at weekends.

For the year to July 1968 ITV's audience share on weekdays had been 52.67 per cent; at weekends 55.83 per cent. In the early months of the new contracts (to 23 February 1969) the weekday share remained steady at 53.37 per cent while the weekend share plunged to 47.59 per cent – a rise of about 1.5 per cent in the case of Thames and a fall of about 16.5 per cent in the case of LWT. The poor reception of LWT's own programmes was exacerbated by the seven-day major companies keeping their most popular programmes for the weekday contractor after pleas to have them shown in London at the weekends had been rebuffed. They came to have more faith in the professionalism of Thames's scheduling, and competition from the BBC was less intense on weekdays.

All this coincided with the financially critical period for ITV described in the previous chapter, when an adverse economic climate depressed advertising revenue for all companies and the government chose to increase the levy and introduce a Selective Employment Tax. The report of the National Board for Prices and Incomes, published in October 1970, doubted the viability of two London companies sharing the week, and the ITA had loaded the dice in Thames's favour by miscalculating the London split, so that what was intended to be a 50:50 division of revenue turned out to be at least 60:40 in favour of the weekday contractor when the blunder was compounded by LWT's weak performance in terms of audience appeal. So far from the Net Advertising Revenue (NAR) and Net Advertising Revenue After Levy (NARAL) of the two companies being roughly equal, the first contract year showed a NAR for Thames of just over £16 million and one of just under £11 million for LWT. Yet LWT's NARAL share – the basis for all inter-company financial transactions – continued to be assessed on the assumption of the non-existent 50:50 split, and this in itself proved a heavy penalty.

Only the ablest of captains and crews could have kept afloat a ship buffeted by such rough weather from all quarters; yet LWT appeared to have them. The captains of industry and media potentates who represented the investors as non-executive directors included Lords Campbell (of Eskan), Crowther, Hartwell and Stokes, the Hon. David Astor and the Hon. David Montagu, Sir Christopher Chancellor, Sir Geoffrey Kitchen, Sir Arnold Weinstock and Evelyn de Rothschild. Aidan Crawley was the Chairman: he had been the first Editor of ITN and was political balance personified, having served first as a Labour and later as a Conservative MP. In Michael Peacock, the Managing Director, the company had secured one of the BBC's brightest stars: the first programme head of BBC2, then Controller of BBC1, and still under 40. He brought an array of talent with him from the Corporation, and other senior staff were recruited from the top echelons of Rediffusion.

The flaw – and it was a crucial one – lay in the fact that the non-executive tycoons who were vastly experienced in business matters knew little about television, while the executives who were hand-picked television professionals knew little of business. Neither Crawley nor Peacock had ever run a commercial company. There were seasoned pilots aboard, but they were not at the helm when the storm broke.

The concept that it would be managed and part-owned by its creative workers was the company's motivating ideal. But the price of non-interference is success. The executives of a well-run company in good shape can expect to be allowed to exercise effective control; but when a company runs into trouble, the board will – indeed must – step in. Under the pressure of events latent disharmonies soon surfaced at LWT. Peacock was determined to brook no interference from a board which he regarded as over-powered, over-qualified and over-numerous: cumbersome and televisionally untutored. The non-executive members thought him impossibly arrogant and became alarmed at what they saw as a company floundering from lack of entrepreneurial activity and insight, run by non-commercially-minded people who had been cosseted at the BBC and sheltered from the realities of earning a living. Peacock, they came to believe, had no qualifications to be a chief executive, yet he demanded not just editorial freedom but complete control, seeming to think that a board was no more than a source of funds, representing shareholders whose role resembled that of theatrical angels backing a play.

A strong Chairman might have retrieved the situation and forged a nexus between executives and board, but Crawley was indecisive; too nice a man for harsh measures. What then could the non-executive directors do except voice their dissatisfaction? Weinstock chose to resign and his company, GEC, disposed of its shareholding.

Peacock had plenty to contend with besides his board. Being forced to take over Rediffusion's obsolescent Wembley studios and the reluctant recruitment of suspicious and demoralised former Rediffusion staff delayed the start of programme production until May 1968 – two and a half months before the opening night. For the weekend company this was 2 August, and on that Friday evening ACTT staff allowed the new company a few seconds to announce its first programme (*We Have Ways of Making You Laugh*) and then blacked it out. On the following day the union organised a withdrawal of labour during transmission of the first *World of Sport*. The late start in May and loss of production during the national strike in August made extravagant overtime payments unavoidable if commitments for the autumn and winter schedules were to be met.

Bad feeling between Peacock and David Frost was another problem. As one of the two founders, an important shareholder and LWT's star performer, Frost was probably the single most important influence on the company's affairs. He had recruited the Deputy Managing Director (Tom Margerison), the Programme Controller (Cyril Bennett) and the Sales Director (Guy Paine). Although not a member of the board, because that would have disqualified him from appearing on the screen, he attended board meetings and his friend and business colleague, Clive Irving, was a member. The programme plans, as set out in the company's application, had been drawn up, not by Peacock (who was still Controller of BBC1 at the time), but by Frost and Irving, assisted by information from Rediffusion supplied by Bennett. Most prominently featured in these were weekly Frost programmes on successive evenings: *Frost on Friday*, *Frost on Saturday* and *Frost on Sunday*.

Peacock was not pleased when he discovered the terms of Frost's contract. For his exclusive services as a performer, for his three programmes a week for twenty weeks a year and for acting as adviser and overseas sales consultant, Frost was receiving £100,000 the first year, rising to £120,000 the second, and (if the company exercised its option) £130,000 for a third. He also operated a group of companies, using his middle name, Paradine, and these were reported to be earning him at least £600,000 in 1969.[4] Paradine Productions sold packaged programmes to LWT, to whom the big names under contract to Paradine were welcome but expensive. They included Danny La Rue, Tommy Cooper, Ronnie Barker and Ronnie Corbett. Less welcome to Peacock, in view of the possibility of a conflict of interest, was Frost's seat on the company's programme committee.

Peacock's endeavour to escape from restrictions on his freedom of manoeuvre in programming imposed by arrangements which were not of his making was generally misinterpreted as a sacrifice of ideals forced on him by greedy and assertive businessmen. In fact, it was Peacock himself, with his Programme Controller, Cyril Bennett, who took the first decision to be seen as a betrayal of contract-winning promises and gave the press its first LWT field day: the disbandment of a much-vaunted Public Affairs Unit.

This idea in the application had seemed naive and amateur to Peacock, but in order to honour the promise he and Bennett had recruited what they believed to be a high-powered team. This was designed to be capable of making an immediate and solid response to major news events breaking at the weekend. Such an eventuality, however, was unlikely to occur more than two or three times a year, so the unit was mostly engaged on researching and developing documentaries analogous to the *Sunday Times* Insight news stories, and on producing *For the Record*, a series of half-hour interview programmes.

But Peacock and Bennett soon became dismayed at the disappointing professional standards of some of those recruited and decided that most would have to go when their contracts came up for renewal. The company was embarrassed when, for example, four innocent members of the unit making a programme on race relations in Brixton behaved in such a manner as to get themselves arrested for passing cannabis. Above all, it was found that having such a unit at all was a form of over-specialisation in the programme division which inhibited flexibility in the use of staff and the development of talks and documentaries. Public Affairs were therefore judged to be better placed with Adult Education and Religion under a single executive producer.[5]

In making this explanation to the ITA in May 1969, Peacock gave an assurance that he was examining the realities of weekend scheduling in the light of experience and promised a greater effort to match the BBC's entertainment output. He did not regard the end of the Public Affairs Unit as an abdication of responsibilities; rather as a reassessment of priorities in the application of lessons learned after eight months' transmission. The priorities for the following year would be: first, the development of high-quality children's programmes; second, sports features; third, music and arts; fourth, public affairs. 'I am convinced that this is a proper interpretation of the true needs of ITV weekends,' he wrote, 'and I would be prepared to defend these priorities to anyone.'[6]

Some members of the Public Affairs Unit whose contracts had not been renewed protested publicly, and Ben Whitaker, Labour MP for Hampstead, wrote to Lord Aylestone, Chairman of the Authority, about 'the disquiet of both the public and some ITV employees at cases of apparent failure by some ITV companies to adhere to the undertakings regarding programme quality upon which they obtained their franchises'. He wanted to know whether there was any means whereby London Weekend could be made to adhere to the broad undertakings on which it tendered. 'Otherwise, obviously, such specifications are meaningless.'

Aylestone's reply (written in Sir Robert Fraser's inimitable style) was a policy statement:

Thank you for your letter of the 30th May in which you refer to the appointment of programme companies. I will try to explain.

When companies apply for appointment as programme contractors either for the first time or for a renewed period they mention, of course, their many plans for programmes, and some of these materialise while some do not. Equally, they later produce programmes, perhaps of great value, which they did not mention to the Authority at all, because the idea of producing them had not occurred to them at the time. This is the case with new and old companies alike.

In short, the Authority pays careful regard to the general character of a company's declared programme intentions, for this represents its general approach to its responsibilities, but may not attach great weight to specific proposals. In the end, in any case, the Authority offers the contract to the applicant appearing to it the best qualified.

There is nothing surprising about this, let alone disquieting. In television, programme ideas are born every day, and some that seem promising end in disappointment while others exceed expectations. In any case, the Authority does not in its contracts exact undertakings that this or that programme will be produced, or that so much drama, so much light entertainment, and so much current affairs must be provided by any particular company. The system of Independent Television does not work in that way, for the nationally distributed programmes are in the main a composite supply from five different companies, and it is the balance of the whole that is the concern and responsibility of the Authority.

But all companies have strict and binding undertakings placed on them in their contracts with the Authority, and these are much wider and more far-reaching than performance in any particular programme field.

These obligations in the matter of programme quality are two-fold.

Firstly, companies, new and old, must observe the requirements of the
Television Act, which are written into the contracts. Secondly, they
must, as the Act requires, compose their programme schedules in
consultation with the Authority and only operate schedules which have
the Authority's prior approval. In deciding whether to approve a
schedule for a given period, we must always take into account the
provision of the Act in the matter of proper balance in the programmes.

You refer to the case of London Weekend Television. Its successive
programme schedules have been duly composed in consultation with the
Authority and have had the Authority's approval from the start. The
Authority did not place in its contract any specific condition about
current affairs or any other particular types of programme. As I have
tried to explain, the Act's requirement that programmes should preserve
a proper balance is secured in a way that takes account of the needs of
the system as a whole.[7]

This did not satisfy LWT's ex-employees, some of whom were Whitaker's constituents, and the MP wrote again wondering whether there was
not a broad obligation on successful applicants to adhere to the general
principles upon which they obtained a contract. Since LWT's overall policy
had changed and been diluted, were there no means or sanctions by which
the IBA could hold the company to the implementation of its prospectus?
The Authority's letter of reply reiterated that companies were judged by
the standards of the Television Act rather than their franchise applications:

Plans are bound to be modified, not only because television programming is constantly evolving, but also because one company's plans are
inevitably affected by their relationship to the total output, needs and
preferences of the other companies in the system.[8]

What could not be contested was that, driven by necessity, the new
company was now in full retreat from Frost's proclaimed aim of leading
public taste to higher ground. Cyril Bennett was soon to introduce a new
programme schedule with a much derided apologia and home truth: 'The
first duty of a company is to survive'.

As early as October 1968, after little more than two months on the air,
Peacock had warned his board that the company was in an exposed and
vulnerable position: ratings were unsatisfactory, advertisers restive and
publicity unfavourable. Friends and allies, when desperately needed, were
few. Lew Grade, one of the most powerful figures in ITV, who had lost his
own foothold in London, initially courted the new company. But when
LWT insisted on taking over the broadcasting of the Royal Command
Variety Performance his goodwill was lost irretrievably. Peacock was not
trained in the complexities of federalism or skilled in the nuances of give-

and-take between the network companies. In the techniques and bargaining manoeuvres of programme exchange and joint scheduling he was a learner amongst old hands.

In March 1969 he told the LWT board that an ITV company must achieve financial stability through ratings high enough to make its cost-per-thousand advertising rates acceptable to advertisers, must satisfy the ITA so that its contract is renewed and must develop a sense of pride and self-respect among its staff. LWT, he admitted, had failed on all three accounts. Among the programme staff he detected too much self-indulgence, a desire to be different fro the sake of being different and 'a basic confusion and ambivalence about LWT's *raison d'être*'. The company had no roots; the workforce was not yet a team. As a cure he prescribed time, patience and stamina.[9]

In July the company was embarrassed by the unauthorised publication of its confidential contract application in the first edition of *Open Secret*, sponsored by the Free Communications Group. This released a barrage of adverse press comment.

Two months later the board finally ran out of patience and at a meeting on 19 September it resolved that 'the appointment of Mr I. M. Peacock as Managing Director be determined'. He departed the same day, battle-fatigued and bitter, with a battered reputation and a lukewarm handshake of £8,000. He had sacrificed a glittering career at the BBC where he had seemed destined for the director-generalship. He had taken the risk of resigning from the Corporation a month before the LWT consortium was awarded its contract. This was his reward.

On 9 September the Director General of the ITA had written to his Chairman, who was on holiday in Cornwall:

Aidan Crawley came to see me yesterday. I . . . was stunned to be told that Michael Peacock had just been told, on his return from leave, that he must resign forthwith and that Cyril Bennett would also depart. . . All this is the culmination of increasing trouble over the last six months between the board on the one hand and Michael and Cyril on the other, the main criticism of Michael being that, with all his talent, he is a poor adminstrator and a worse leader of men, while Cyril is just thought to be not a particularly good Programme Controller. . .I need hardly say that this will not be very good for the Authority. It will revive the whole question of the selection of programme companies, for there is little doubt in my mind that LWT owes its appointment far more to the inclusion of Michael in the group than to any other single factor. I can remember how he dominated the interview with his consistent and brilliant talk.

Peacock's resignation was the subject of front-page stories and sensational headlines in every newspaper. It was the occasion for a gleeful press attack on television: The Sickness Behind the Screen (*Sunday Times*), Uplift Stops Here (*Daily Telegraph*), Who Should Control the Communicators? (*Observer*). The *Daily Mail* predicted repercussions throughout ITV – Axe Poised over some of the Smaller ITV Companies – and there were calls for government intervention, although the *Financial Times* reported: LWT – Cabinet Move Unlikely.

'The chairman and chief executive of one of Britain's most important television corporations hid their faces from cameramen yesterday as they drove away from a meeting of their board', reported Lord Beaverbrook's *Evening Standard* on 19 September. LWT's 'discovery of the facts of business life has been an agonising – and very public – process', it continued. 'But it is debatable whether the deflowering of LWT required quite such a display of shame on the part of its chairman. The real guilty party is, of course, the Independent Television Authority.'

The verdict of a leader in *The Times* was more sober:

It would be too simple to regard Mr Michael Peacock's dismissal as Managing Director of London Weekend as a victory for the philistines of commercial television. For all the splendour of his reputation as a broadcaster, Mr Peacock was in fact running a company which was providing neither the commercial success which had been expected nor programmes of the quality that had been promised...The board of London Weekend was therefore right to conclude that changes had to be made...the person to whom authority is given must be judged on results. [10]

The furore intensified when, on the day after Peacock's departure, six senior executives in the programme division and three associates announced their resignations in protest, declaring that they were not convinced of the need for a change in management. With four others, who chose to stay, they had signed a letter to the board asking to be consulted before any irrevocable decision was taken and were angry that no such consultation had taken place. The six were Humphrey Burton (Head of Drama, Arts and Music), Frank Muir (Head of Entertainment), Doreen Stephens (Head of Children's Programmes) and executive producers Terry Hughes, Derek Granger and Joy Whitby. The three associates were Tony Garnett, James McTaggart and Kenith Trodd from Kestrel Productions. The four who stayed were John Blyton (Head of Programme Management), Eric Flackfield (Head of Programme Planning), Michael Yates (Head of Design) and Harry Rabinowitz (Head of Music Services). The two senior figures in the sports department – Jimmy Hill (Head of Sport) and John Bromley – were abroad and on their return decided to stay.

Mass meetings of staff were called – '500 Back Peacock' – and further resignations were rumoured. The Postmaster-General asked the Chairman of the ITA to come and brief him on answering Questions in the House of Commons. ACTT demanded an ITA investigation. A letter in *The Times* argued that there was an irrefutable case for a public enquiry. In the *Financial Times* of 24 September T. C. Worsley confessed himself not surprised by the 'disintegration' of LWT, 'since I prophesied from the first moment of their joining it that the particular young men who have gone would find it difficult, if not impossible, to adapt themselves to the conditions of commercial television'. With their departure the company had changed its identity: 'The London Weekend which the Authority had appointed vanished overnight. Where does that leave the Authority?'

The Authority faced this question at a meeting the next day, when the Director General gave a full account of events leading to the LWT board's decision. The Chairman reported the receipt of a letter from Humphrey Burton, who had assumed the role of leader of the dissidents, and copies of a petition signed by 800 employees of ITV companies were tabled. What began as a dispute about the suitability of one man for his job had developed into a sensation which was shaking ITV in its entirety.

The Authority agreed the wording of a press release emphasising that its first consideration was 'to ensure that the programme effort of the company is sustained and that the real promise of the past year will be fully realised in the future'. On the following day, in his speech at Dundee to the Scottish Centre of the Royal Television Society, the Authority's Chairman delivered a public response to its critics, firmly advancing a policy of non-intervention. The Authority, he declared, had no right or inclination to instruct the boards of ITV companies whom to employ or dismiss: if they were not permitted to manage their own affairs there would not be a self-respecting board left in the whole ITV system. LWT's record in programmes did not deserve the degree of criticism it had encountered: indeed LWT had been 'subjected to criticism altogether excessive in relation to its faults'. The board was as distinguished, responsible and representative as any in ITV. The Authority's responsibility was not for employment in the companies, but for the quality of ITV as a broadcasting service, and that responsibility would be discharged.

On 1 October the Chairman met the dissident six and was handed a nine-page *aide-mémoire* stating their version of events: on 8 September Peacock had returned from holiday and been asked to resign; after urgent representations Aidan Crawley had agreed to see senior programme executives at his home that evening; they had not been satisfied by his explanations for Peacock's dismissal, but he had told them that the decision was irrevocable and it would not be possible for them to see the board, which was due to meet the following afternoon.

Such was their concern that before that meeting the executives had drafted and despatched a letter to all members of the board; asked to see the Director General of the ITA (a request declined); alerted the

Postmaster-General's office; organised a mass meeting of LWT staff, which passed a motion of confidence in the existing management (more than half the staff attending at an hour's notice); arranged for Cyril Bennett, the Programme Controller, just returned from holiday, to inform the board that if it proceeded without consultation he could not guarantee programmes for the following weekend; and taken steps to arrange for voting shareholders to call for an Extraordinary General Meeting to enquire into the management of the company.

Confronted by such an energetic display of hostility, the board had stayed its hand and appointed a sub-committee. This had met the Committee of Thirteen a week later. The next day was spent in intensive lobbying and counter-lobbying of members of the board, and on the following day came the announcement of the board's adherence to its decision to remove Peacock. The Thirteen were informed that if as a matter of conscience they felt unable to stay with the company they would be released from their contracts. Those who decided to leave did so, they told the ITA, because they had lost confidence in the board.

As 'a very experienced group of television professionals, accustomed to leading large teams and responsible for annual budgets running into hundreds of thousands, even millions, of pounds', they saw Peacock as their guarantee of the company's resolve to put programmes first. The six were also holders of voting shares in the company and saw themselves as a dedicated band of programme-making part-owners until the illusion was shattered by the intervention of a board which knew little about the business of television. But what they failed to take into account were the company's dire financial straits, which were to lead to a further instalment of this unhappy saga.

At LWT a new team was appointed, with Lord Campbell as Deputy Chairman, Tom Margerison as Chief Executive (against Peacock's parting advice) and Guy Paine, the Sales Director, as Assistant Chief Executive. Stella Richman (Managing Director, London Weekend International) and Vic Gardiner (Production Controller) joined the board as executive directors. Two months later, when Cyril Bennett asked to be released from his contract, Stella Richman was appointed to the board as Controller of Programmes in his place – the first woman to hold such a position in a television company. Meanwhile shop-floor support for the dissidents evaporated.

LWT's flounderings had sent shock waves throughout the network and resulted in several attempts to rebuild the structure of Independent Television, all resisted by the Authority. An approach on behalf of ATV in 1970 was recorded by the Director General:

I had better put it on record that the leaders of ATV asked me privately last Friday what I thought would be the likely attitude of the Authority if

ATV entered into some form of association with LWT, which could include the outright purchase of LWT, followed by its operation as a fully owned ATV subsidiary.

The question was put to me without warning at a luncheon at which Lord Renwick, Mr Norman Collins, Sir Lew Grade and Mr Jack Gill were all present.

Now the truth of the matter is that ATV had by 1967 been providing London with its weekend programmes for 12 years, and one of the reasons for the appointment of LWT was the strong feeling of the Authority that the ATV weekend performance had insufficient to commend it. The Authority sought for the introduction of a greater distinction in the weekend programmes than, to judge from its record, ATV was ever likely to provide. I did not find it possible to say this in the presence of such a mixed company. It would have been too wounding to Sir Lew in the presence of his Chairman and Deputy Chairman and one of his subordinates.

Secondly, the Authority in 1967 was concerned to go over as far as possible to seven-day contracts. These were introduced in place of five-day and two-day contracts in Lancashire, Yorkshire and the Midlands, the broken week being retained only in London. Any association between ATV and LWT would mean that ATV would have a full week in one area and the weekend in another.

Thirdly, if it should prove that LWT might not always be able to carry on as an independent company, the only association or merger that would make any kind of sense, and stand a chance of giving the Authority the weekend schedules it hopes for, would be between LWT and Thames.[11]

Grades are not easily discouraged, and Sir Lew returned to the subject in August, bringing his Chairman, Lord Renwick, with him to Brompton Road to ask point-blank how the Authority would feel if ATV sought to acquire control of LWT. He said that LWT was in a rocky situation, with most of its principal shareholders willing to sell. It was not a well-conducted company. It was in the hands of amateurs. In failing to attract the size of audience needed at the weekend, it was a weakness at the heart of Independent Television. More particularly, it lacked enthusiasm for ATV and ITC programmes made with an American sale in mind, and Sir Lew was worried about a general difficulty in placing his programmes in other central areas. Granada, he told the ITA, would always take them and so would the regional companies, but Yorkshire was a poor customer and he could never be sure of LWT. For a London showing he was more interested in LWT than in Thames because his entertainment programmes were more suited to weekend viewing.

His words fell on stony ground. He was told that it would be a reversal of policy followed for fifteen years if the Authority were to agree to any one of the central companies having a financial interest in another. His suggestion that ATV and Thames might jointly assume control of LWT met with the same objection.[12]

Ever since 1954 the Authority had consistently rejected the idea of one metropolitan company with an overwhelming preponderance of audience, income, talent and clout. Nevertheless, the crisis at LWT forced an amalgamation of the two London companies to be seriously considered. An ITA Policy Meeting in July acknowledged that the network system was suffering from the effects of a divided franchise in the capital. The creation of a single dominant company would lead to greater uniformity of scheduling across the country, bringing to an end the awkward split between weekday and weekend scheduling and facilitating the provision of one overall ITV service in sport.[13]

Other arguments in favour were aired, both at this time and later during a reconsideration in 1973 when a solution to some of the network's complex financial and programme planning problems was sought. There seemed no compelling reason why all the network companies should be of a similar or comparable size. Fears of dominance from London would be seen to be exaggerated if the Authority exercised its powers to assume greater control over the system. Regulations could be imposed to ensure that the programmes of the other network companies were screened in London. The concentration of so much power might be dissipated if Thames and LWT were to operate independently under the ownership of a single holding company.

The idea of a single London company jointly owned by all the other companies, as with ITN, was dismissed on the grounds that the company serving London would then be owned and directed by companies whose first duty was to provide programmes for areas outside London. The suggestion that a single London company might be directly controlled by the Authority along the lines recommended by Pilkington was found unacceptable because it combined the disadvantages of a single company with those inherent in the Pilkington proposals and introduced the additional disadvantage that the Authority would not have the same relationship with all companies.

After his meeting with Sir Lew in August 1970 Sir Robert Fraser feared that he had not heard the last of the matter. He was due to retire before the end of the year and became anxious lest the troubles at LWT should destroy the edifice which he had taken such pains to erect. On 21 September he addressed a memorandum on the London contracts to his Chairman, his successor and his deputies and marked it explicitly 'For the Record'.

The rumour stirs again, and I have once more been asked whether I know of any moves for some form of association between Thames and LWT. I replied that I knew of none that should be given any credence. Sir Lew Grade seems to hear most of these rumours. They keep him in a chronic state of restlessness. . .

If the possibility of any form of association between Thames and LWT ever arises, it will need the most careful handling and consideration from every angle, for any proposal for a seven-day company in London would shake the whole system to its foundations.

Every one of the other four central companies, not only ATV, would feel themselves radically affected, and so would the ten regional companies. . .

As far as the central areas are concerned, it can at once be seen that the present balance of forces, cooperative and competitive as they are, would be destroyed. The income of the London company would be double that of any other in the central areas. Just as important, the metropolitan company would have the enormous advantage of London and all its unique resources behind it. This is an advantage which the present division of powers, of course, is designed to damp down. Lastly, the London company would have sole control of the choice of programmes to be shown in the London area.

It is hard to see how anything could prevent the London company becoming, as we have always said, the Titan of the system, surrounded by dependent relatives.

After all this time with the Authority, I am more than ever convinced that the variety conferred and the virility induced by its plural structure are the main reasons for its excellence and (for what it is sometimes easy to forget) its astonishing success against the combined forces of BBC1 and BBC2.

There are two main sides to this. The intense competition for prestige and standing between the central companies pours great energy into the whole system. The sheer quality of ability in the system is much greater than would be the case with fewer companies, for the arrangement in the central areas provides the opportunity for such wide employment of high talent. And each company has its own vein of talent to contribute.

As far as the regional companies go, Mr Sendall has lately been reminding me how much of its richness the system owes to their presence in it, not so much because of their occasional contributions to the network, valuable as these are when they come along, but simply because they are part of it. One can easily think of Managing Directors of quite small companies to whom we owe much.

However, let that be by the way. I am concerned mainly to say that, if the London question ever arises, it will not be a London question at all, but be regarded by all the companies as a national one affecting them all. The fact of the matter is that there would be a very large row indeed if we approached the London question, should it ever arise, simply as a London question of proper interest only to Thames and LWT.

It is perhaps worth concluding this note with some references to LWT, the past weakness of which has more than any other single factor raised the London question. At long last, I think there begins to be evidence that LWT may be about to turn the corner. I suspect that there is still, inside the company, a possibly dangerous instability, but there is no doubt in my mind that its programmes show real signs of improvement and even more signs of promise. At least for the last few weeks, its audience share has been distinctly better, and in due course, if it persists, this should begin to show in its revenue figures. And, on the question of viability, one must never forget that LWT has an annual surplus of revenue over expenditure of some £2.55 millions and that it is not its fault that the whole of this disappears in the levy.[14]

Fraser's statement to Sir Lew Grade that the Authority would not permit an association between two central companies had been carefully phrased. It was about to become public knowledge that an association between a central company and a regional company had received the Authority's, albeit reluctant, blessing. The announcement of the linking of Yorkshire and Tyne Tees under the Trident banner provoked another central company into action. Denis Forman of Granada floated the idea of an association between Granada and LWT, only to be told firmly by Brian Young, the new Director General, that in no circumstances would two of the five be allowed to merge.[15]

The next approach came in February of the following year from Ward Thomas of Yorkshire, who astonished Young by suggesting a single London company. When reminded that the Authority's insistence on a London split was due in part to maintaining a balance among the five, he replied that Yorkshire, Granada and ATV would be prepared to merge to preserve the balance.[16] But on the very next day Grade was proposing a different re-structuring to give him back his foothold in London: Thames to merge with Southern; LWT to merge with ATV and Anglia; and Granada with Trident.

These manoeuvrings illustrate the extent of the impact of LWT's under-performance on the intricate mechanism and interdependent well-being of Independent Television. All foundered on the rock of the ITA's determin-ation to preserve the established system. But, with those options closed and LWT still in deep financial trouble, the Authority found itself left with a scarcely more palatable alternative.

On 13 July 1970 Fraser lunched with the Australian publisher, Rupert Murdoch, and his legal adviser, Lord Goodman. Murdoch had been in touch with LWT's major shareholders and estimated that 63 per cent of LWT shares were for sale. What, he wanted to know, would be the ITA's attitude if he decided to purchase them? Fraser replied cautiously that the

Authority preferred companies to continue throughout their contract period without change of ownership; it exercised no control over non-voting shares, but transfers of voting shares required its prior approval; the Authority was also averse to control by a single shareholder, whether an individual or a company. Murdoch then supposed that no proposal to acquire more than a 25 per cent interest would be approved and Fraser replied that the first question was whether the Authority would approve the purchase of any voting shares at all.[17]

In November the Authority consented to Murdoch's News of the World Organisation acquiring the 7.5 per cent of voting shares held by Sir Arnold Weinstock's General Electric Company. In the same operation 16 per cent of the non-voting shares were acquired from GEC and others as well as the 11 per cent of the loan capital held by GEC. Murdoch became a non-executive director of the company in Weinstock's place.

The debt which LWT and the whole of Independent Television owes to Rupert Murdoch has been overshadowed by his alleged ineligibility and unsuitability to guide the destiny of an ITV company on the grounds that he was the foreign proprietor of sex-and-sensation newspapers (the *News of the World* and the *Sun*) and had a reputation for rough business methods. The Television Act disqualified individuals who were 'not ordinarily resident in the United Kingdom, the Isle of Man or the Channel Islands' from being programme contractors and prohibited newspaper shareholdings 'leading to results which are contrary to the public interest'.[18]

Towards the end of 1970 LWT had run through its authorised and fully-paid share capital of £1,500,000 non-voting and £15,000 voting shares. Twenty-one per cent of its £3,090,000 loan stock had been called and the remainder was being effectively used to secure bank overdrafts totalling nearly £2.5 million. The cost of the new studio centre being built on the South Bank was £5.5 million, £4 million of which was being provided by National Coal Board Nominees. In December a one-for-three rights issue was therefore proposed, which Murdoch agreed to underwrite through his News of the World Organisation.

At this point he represented only one among a number of powerful institutional shareholders which included The Bowater Paper Corporation, the Imperial Tobacco Company's Pension Trust, Pearl Assurance, Lombard Banking, Samuel Montagu, The Observer, the Daily Telegraph, Economist Newspapers and The London Co-operative Society. None had sufficient faith in the company's future to risk additional investment; nor did the executive directors or senior staff who between them still held 36.6 per cent of the voting and 28.1 per cent of the non-voting shares. There was a slump in television and throughout the economy generally, and the institutional investors had financial problems of their own. Murdoch suggested to Lord Hartwell of the *Daily Telegraph* that they put up the money jointly, but Hartwell declined. An Australian television group expressed interest, but a substantial investment by an overseas company was not acceptable. Without Murdoch's money, Margerison told Anthony Pragnell of the ITA, LWT would have no alternative but to merge with Thames.

'I am inclining fairly strongly to the view,' Pragnell reported to his Chairman, 'that, although the Authority might prefer not to see this particular change in LWT, we are faced, as a result of the financial pressures affecting ITV as a whole, with a situation where there is no alternative open to us which we can properly press upon the company.'[19]

In LWT's circumstances, the injection of funds from almost any source was welcome. All members of the board agreed to Murdoch's underwriting of the rights issue. Among senior executives who were consulted by Margerison there was general approval. In contributing, single-handed, the £505,000 needed to ensure the company's survival, and thus becoming the owner of 35 per cent of the non-voting shares, Murdoch was not, of course, acting as a philanthropist. He was a shrewd investor whose farsightedness put the timid natives to shame. It was a situation reminiscent of the 1950s when Roy Thomson, an elderly businessman from Canada, was hard put to find a single Scotsman or Scottish institution canny enough to risk money in what was to become the financial bonanza of Scottish Television. It offered, too, a preview of the 1980s when TV-am, with a galaxy of stars and programme ambitions remarkably similar to LWT's, collapsed under a wealth of talent and dearth of business aptitude and had to be rescued by Packer money from Australia and made financially viable by the importation of an Australian chief executive.

On 21 December it was announced that Murdoch had agreed to attend meetings of the executive directors and to devote a part of his time to the affairs of the company. He was doing what no other non-executive director had done before: taking his coat off, going down to the studios and finding out what was really going on. On 31 December the new issue of shares was formally approved by the ITA. Noting that this could safeguard LWT's continuance as an independent company, that Murdoch himself could be a revitalising influence, that he would still have less than 10 per cent of the votes, and that the *News of the World* (although then UK-owned) had already controlled another ITV company (TWW) without objection, it found itself with no choice.[20]

On 20 January 1971 Stella Richman reported to the LWT board on the completion of her first year as Programme Controller. It had been 'a bad year and disappointing in almost all aspects', and she itemised fourteen problems facing the company, all of which she described as 'currently insurmountable'. Among them were the constrictions of the weekend franchise, the paucity of non-peak viewing time (for experimenting with new programmes and new ideas in programming), the number of unprofitable religious and adult educational programmes mandated by the Authority, and the NARAL system of cost-sharing among the companies.[21]

Faced with an emergency and such despair, the Dirty Digger (as his denigrators in the company nicknamed him) did not hesitate to fire. Stella Richman left the company before the month was over. Guy Paine had preceded her. On 18 February Tom Margerison's resignation as chief executive was announced and Murdoch became chairman of an executive committee formed to run the company. This reorganisation was announced by Aidan Crawley, who was himself to survive only a further three weeks before being kicked upstairs into the functionless post of President. LWT's suffering once more erupted into the public arena, and alarm bells rang again in Brompton Road.

The day after Margerison's ejection Pragnell sent a letter by hand seeking the urgent advice of the Authority's solicitors on the contractual position: 'I feel sure that the Authority would take the view that the changes announced yesterday, which seem to put Mr Murdoch in executive charge of the company, would have meant that LWT would not have got the contract had they been in operation before the contract was entered into'.[22] In the Commons a full enquiry was demanded, but the minister (Christopher Chataway) replied that it was not for him to say whether changes in shareholdings were fundamental enough to warrant withdrawal of contracts. That lay within the powers granted to the ITA.

The press hailed this new twist in the story as another scandal, now featuring not only a rival medium but also a rival newspaper proprietor. 'Lost Weekend – Chapter Three', ran a headline in the *Evening Standard*. Murdoch parried with a press statement that he subscribed to the programme philosophy of the original consortium: what had been wrong was not the intent but the execution. In a personal letter to Brian Young four days earlier he had similarly attributed LWT's troubles not to 'unrealism or unattainable ideals' but to 'very bad management and a company structure which led to nobody being really accountable'. Clive Irving, who had left the board at the end of 1970, joined the fray with an attack in the correspondence columns of *The Times*, claiming that Murdoch had fired Margerison without reference to the full board and had drawn up the company's new programme schedule without reference to the programme staff. 'The company's initial philosophy was too ambitious,' he allowed, but 'it is one thing to concede that. It is another to make undiluted commercial expediency the alternative.'[23]

On 23 February Brian Young wrote a 'horns of our dilemma' Note for members of the Authority, setting out the arguments for and against action. On the one hand, the company was indisputably a different one; Murdoch indisputably held the main executive power; and 'the fact that Murdoch's newspapers are what admirers call popular and detractors call vulgar increases the discrepancy between Peacock's promise and Murdoch's method'. On the other hand, the company was likely to become more effective under Murdoch's leadership; sixteen employees had left, but a thousand remained and did not deserve a second total upheaval in three years; the shareholders would get a raw deal; and the only practical alternative was a merger with another ITV company.[24]

The Authority met on the morning of 25 February, and a letter was despatched by hand to LWT the same afternoon stating it to be the view of the Authority that 'the changes announced in your company's statement to the press of 18 February constitute an event which entitles the Authority to determine the programme contract under Clause 1(2) of Part I thereof'. While reserving the Authority's right to do so, the letter did not, however, give the company notice of the termination. Instead it invited the Chairman and members of the LWT board to prepare a submission and meet the Authority for an interview and discussion about, *inter alia*, the company's future plans for its management and programme provision.

This decision relied heavily on the advice of the Authority's solicitors that the changes announced did indeed 'constitute an event affecting the nature and characteristics of LWT which, if it had occurred before the signing of the programme contract, would have induced the Authority to refrain from entering the contract and which entitled the Authority to determine the programme contract under Clause 1(2)'. The Authority duly and formally so resolved but were influenced against taking immediate action by a number of considerations, including the points made by Young in his memorandum. How certain was it that the Authority would not have awarded the franchise to the reconstituted LWT? Not only had LWT changed, but so had the Authority, which now had a different Chairman and Director General and several new members, none of whom had been involved in the original decision to appoint the company.

The legal advice, moreover, struck a note of qualification and caution. Termination of the contract would amount to a repudiation of it. If based on a misinterpretation of the wording of the Act, it would entitle LWT to claim damages from the Authority and the damages, related to the loss suffered by LWT in being unable to continue as programme contractor during the remainder of the period of its contract, might be substantial. The Authority was therefore advised to obtain leading counsel's opinion before reaching a decision to terminate. In all these circumstances the ingenious device of re-interviewing suggested by the Director General was an acceptable compromise.

After the meeting a press notice was issued announcing the Authority's decision to allow LWT six weeks in which to prepare a new submission and choose (after consultation with the ITA) a new Managing Director and a new Programme Controller. In the meantime it made public the Authority's concern about the readiness and ability of the company's management to pursue the aims which were set out when LWT obtained its contract in 1967, and about the proper division of responsibility between non-executive members of the board and those exercising day-to-day management of the company.

In retrospect it is difficult to fault the Authority's suspension of judgment as the most sensible option open to it. At the time its decision was generally well received, although taken in the face of a vigorous 'sack LWT' campaign in the press. The restraint displayed by the Authority had been severely criticised by a variety of media pundits and represented as at best pusillanimity and at worst a clear dereliction of its duty to terminate the contract of what had become a Murdoch company. 'Americans who grumble about the feebleness of their FCC can now stop grumbling,' wrote Bernard Levin in the *New York Herald Tribune*: 'Britain's ITA is even feebler.'[25] From the left had come demands that the Authority should exercise its powers to take over LWT and run it as a standard-setter for the whole of ITV. Strong emotions had been whipped up and the Authority's credibility was said to be as much at stake as LWT's future.

Early in March Murdoch returned from Australia and paid a call at 70 Brompton Road to accuse the Authority of character assassination. In an outburst lasting for an hour and a half he demanded to be told categorically whether the Authority welcomed him or disapproved of him. Before leaving he handed over a letter which stated that he had honoured every arrangement made by him with vendors, shareholders, directors, and above all the Authority, and required the Authority's co-operation in repairing the gratuitous damage inflicted on his good name.

Murdoch had cause for anger. Engaged on rescuing a company slithering toward bankruptcy, he was being publicly pilloried as a man unfit to have control of a television company in Britain. But while there can be little doubt that he was the buccaneering kind of entrepreneur not favoured by the Authority during the Young years, it had been scrupulous in not personalising the dispute and was well able to reassure him in a mollifying letter deploring some of the newspaper comments and enclosing copies of its own press statements.[26] In reply Murdoch accepted the assurance. He denied that he had ever intended to play an executive role in LWT, but declared his intention of taking a very active interest in the running of the company in support of the new Chairman and Chief Executive.

This last was John Freeman, the retiring British ambassador in Washington, who had been named as prospective Deputy Chairman in the London Television Consortium's application. He arrived from America like a *deus ex machina* to end the drama and guide the company to a peaceful and prosperous future. A relieved Aylestone broke the news of the appointment to Chataway on 9 March, and the minister expressed his own relief in reply. The initiative in what the Chairman of Scottish Television described as an inspired move not only for LWT but for the industry as a whole was taken by Murdoch. Freeman was also a friend and former *New Statesman* colleague of Lord Campbell, the Deputy Chairman. But most of the credit belonged to David Frost and his talent for persuasion. It was also David Frost who had originally recruited Freeman to the consortium, and this was not the first attempt to persuade him to join as Chairman. For this purpose Peacock had, one weekend, flown on a desperate and unsuccessful mission to Delhi, where Freeman was then the British High Commissioner.

On 3 March a searching questionnaire prepared by officers of the Authority had been sent to the company. It covered all aspects of the company's structure and operations: financial, managerial, creative and technical. On the crucial matter of programme content it included the Authority's analysis of 'serious' programmes in LWT's schedules: during the first few weeks of transmission these had represented 31.6 per cent of the whole and 30.2 per cent of peak hours (although August 1968 was recognised not to be a typical month). By January 1969 the corresponding figures were 29 per cent and 20 per cent; by March 1971 they had fallen to 23.8 per cent and 20 per cent.

On 6 April the Authority convened for a special meeting to consider the company's answers and agree on arrangements for 22 April, the day of decision, when it was scheduled to hold a preliminary meeting at 10.30 a.m., a meeting with the LWT delegation at 11, and another meeting of its own at 2.30 p.m. The company was, in effect, being forced into a full reapplication for its contract.

For the meeting at 11 LWT fielded a team of ten. Headed by Freeman, it included Murdoch, Campbell and the two board members who were leading merchant bankers: David Montagu and Evelyn de Rothschild. They were thanked for their full written answers and it was noted that the executive committee to which the Authority had taken exception had been disbanded. The absence of a Programme Controller on the delegation was commented on and Freeman replied that he would be recommending an appointment to his board the following day. Although no name was mentioned, it had already been made known to the Chairman in confidence: Cyril Bennett, who had resigned in January 1970, would be returning to the company in May.

Freeman, the newcomer, handled the company's side of the interview and chose to answer most of the questions himself. The arguments which he marshalled and presented were authoritative and convincing, and no one was left in any doubt that he had already assumed full executive control of the company's affairs. In the press he had been quoted as saying: 'I would be happier if the company did not have the language of the original submission hanging round its neck. But I would be ashamed if we did not have its ideals behind us.'[27] As his opening statement expressed it:

London Weekend Television still believes in the ideals it presented to the Authority in its original application.

That said, it would be pointless to deny from a position of hindsight that we underestimated some of the difficulties we should face and that some of the detailed commitments we made in good faith at the time would probably not have been made – or accepted by the ITA – if either party had fully appreciated the consequences of the limitations within which we were to work. . .

Our setbacks have derived from administrative, executive and commercial shortcomings, for which both Board and Management must accept a due share of responsibility, not from creative weakness, and we wish to affirm that any shortfall in our performance has not come about because we have at any time abandoned our essential objectives. Indeed we can claim to have fulfilled many and attempted most of the intentions set out in our original application. In comedy we have achieved some striking and original successes. In drama and in our programmes dealing with religion and the arts, or designed for children, we have been nothing if not adventurous. Although sport was mentioned only briefly in our application we consider it to be of the highest importance to Independent Television as a whole and we are proud of the leadership which LWT has shown in the process of eroding the BBC's formerly dominating position. . . Technically, and particularly in the introduction of colour, we can claim with confidence that our achievement has not been surpassed anywhere. Given the solution of our management problems – and this we believe is now in sight – there is no reason why LWT should not now realise its full creative potential.[28]

The Authority's satisfaction with the company's response was made public the following day in an announcement that LWT's franchise now had the same security as that of other programme companies throughout the current contract period. The management arrangements were now 'of a kind to inspire confidence'. Discussions with the company about programmes had ranged widely, and it had been agreed that the original concept, of an adventurous approach to the planning of weekend programmes, was still a valid one. The need for securing a firm financial base, for close cooperation with other companies and for good popular programmes had been fully accepted on both sides; so had the requirement to cater for a variety of interests in addition to serving the majority.

The Authority concluded its statement with an expression of pleasure that uncertainty about the future of London Weekend Television had been removed. Six years later the Annan Committee, while unjustly allowing the Authority no credit for the happy ending and making misleading comments on the affair,[29] was to acknowledge that 'LWT under Mr John Freeman's management finally came to match in performance what its application had promised'.[30]

4

PARLIAMENTARY INVESTIGATION

By the late 1960s, despite lingering prejudice and resentments, experience of the competing performances of the two television services had brought about a general acceptance of a channel financed by advertising: 'The battles having been fought and the dust having settled . . . this is the system the country has . . . This combined system is to the benefit of the country,' said Edward Heath.[1] But this acceptance was always dependent upon the exercise of satisfactory controls, and the source, nature, extent and purpose of such controls became a central issue of discussion among that minority of the British public, in and out of parliament, who engaged in public statement and argument about the broadcasting institutions.

In April 1972 Tony Benn, who had been Postmaster-General between 1964 and 1966, emphasised that the debate about broadcasting in Britain had moved to a new level. The question was no longer simply how to control what programme providers – and advertisers – were doing. It was rather a matter of deciding on the proper status and role in society of an increasingly powerful mass medium. Television was fast becoming for most people their main – almost exclusive – source of information and vicarious experience. It had supplanted older information media and was likely to become a more potent social force than any of them: a fact to which parliament must awaken, and legislate accordingly.[2]

He was not alone in fearing that the power of television might challenge even the sovereignty of parliament itself, but in seeking to establish greater accountability and a greater degree of parliamentary control over broadcasters, politicians risked impairment of the equally cherished democratic principle of the freedom of the Fourth Estate – the information media – from political supervision and control. Some, nevertheless, had become convinced that the limited lives and periodic renewals of the Television Act and the BBC Charter, following wide-ranging public enquiries on the models of Beveridge or Pilkington, were not enough. Still less were they satisfied with the sparsely rationed parliamentary time available for debating the ITA and BBC annual reports. They looked for some other constitutional device whereby, without prejudicing freedom of thought and expression, broadcasting bodies could be made more clearly responsible and responsive to the public they were appointed to serve – and to that public's democratically elected representatives.

The Select Committee on Nationalised Industries (SCNI) was devised by the House of Commons as a means of enabling MPs to circumvent the rule that Questions could not be asked in the House about the day-to-day operations of nationalised industries. In July 1968 a special report from the Committee had recommended the inclusion within its terms of reference of a number of other bodies under government control. These included the Bank of England, the Horse Race Totalisator Board, Letchworth Garden City Corporation and the Independent Television Authority.

In evidence to the committee, Roy Mason, the Postmaster-General and minister responsible for broadcasting, doubted whether the ITA could properly be described as an industry and pointed to the convention that the broadcasting organisations were already subject to full periodic reviews. Sir William Armstrong, Joint Permanent Secretary of the Treasury, saw an advantage in having the ITA and the BBC examined by one and the same committee (the Public Accounts Committee) and virtue in leaving those appointed to run broadcasting free from interference between major enquiries. A wary ITA requested its Chairman to inform the Postmaster-General of its reluctance to appear before the SCNI. This he did; and there the matter rested until a new minister (John Stonehouse) decided that the specific exclusion of the ITA from the list of new candidates for examination would be inappropriate. Meanwhile another major investigation by the government had been planned.

On 14 May 1970 Stonehouse announced that he had invited Lord Annan, Provost of University College, London, to head an independent committee of enquiry into the long-term future of broadcasting following the expiry of the Act and the Charter in 1976. No other members were appointed to the committee for the time being, and it was understood that written evidence was not likely to be required before the autumn.

The ITA and the BBC, when consulted earlier in the year, had both informed the minister that they were not persuaded of the need for an enquiry at such an early date. If one had to be held, the broadcasting authorities thought it would be soon enough in five years' time. The Prime Minister, Harold Wilson, was believed to be not unsympathetic to this point of view, but the argument put forward by ministry officials prevailed: that on general constitutional grounds an early enquiry by an independent committee was necessary in order to demonstrate that the broadcasting authorities were accountable, and must be seen to be accountable, to the public. Current concern about the impact of television on society proved the decisive factor.

It so happened, however, that the accountability of government to the electorate took precedence. The dissolution of parliament took place the very next day, and in the general election which followed, Labour's expectations, together with those of almost all the opinion pollsters, were confounded. A Conservative government led by Edward Heath was returned to power with a parliamentary majority of 31, and on 23 July in a written reply to a parliamentary question Christopher Chataway, the new Minister of Posts and Telecommunications, stated the government's reservations about the value of launching another major enquiry at that time. Instead he intended to invite his Television Advisory Committee (TAC) to identify and study the main technical developments which his predecessor had adduced as a justification for the enquiry. When TAC had reported early in 1971 he would consider again whether a full-scale enquiry by an independent committee was really necessary or desirable.

Heath and Chataway were well aware of the disruptive effects of committees of enquiry on the work of the broadcasting bodies, and when the TAC report was received the proposed committee of enquiry continued to be adjourned *sine die*. It came as no surprise therefore that the earlier proposal for an investigation of the ITA by the Select Committee on Nationalised Industries should be revived on the introduction of a Sound Broadcasting Bill destined to extend the responsibilities of the Authority into commercial radio. Sub-committee B, chaired by Labour MP Russell Kerr, was accordingly instructed 'to examine the Reports and Accounts of the Independent Television Authority'.

The broadest of interpretations was placed on this simple formula, and the extent of the committee's intentions was revealed in a letter dated 29 July 1971 in which the Clerk to the Committee instructed the ITA to furnish it with a general paper setting out the duties of the Authority and how it discharged them. The committee, he added for good measure, had expressed a particular interest in the following subjects: allocation of contracts and securing adherence to their conditions; relations with the public; staff relations; and political objectivity.

In Anthony Smith's account of subsequent proceedings:

The top brass of the world of commercial television ... as well as an assortment of prominent pressure groupers of the media world shuffled in group by group and were publicly grilled by a Select Committee ... and it is possible to see, through fumbling weeks of evasion and enquiry, rather more of the structure and the sustaining assumptions of Independent Television than have ever been revealed before.[3]

Those grilled included Sir John Eden, the new Minister of Posts and Telecommunications, and his senior civil servants; the ITA Chairman and his top executives; the Chairman and two members of the Authority's General Advisory Council; spokesmen for the broadcasting trade unions; advertisers' representatives; Free Communications Groupers; assorted

MPs; and directors of four of the ITV companies. Members of the committee visited the United States to study broadcasting institutions there, and information was collected on other overseas systems. Professor Hilde Himmelweit was appointed as an expert consultant.

The ministry's evidence contained a summary of the minister's powers over the ITA, which revealed that government restrictions on the freedom of broadcasting were far from negligible. The means of rigorous control were readily available to any government which chose to use them. Powers under the Wireless Telegraphy Acts and the Television Act enabled him to control all frequency allocations and technical developments, including the siting of transmitters and establishment of new stations, changes in the definition standards from 405 to 625 lines, the choice of the television colour system and the timing of its introduction. He had powers to regulate the amount of broadcasting time, to require the broadcasting of government announcements, and indeed to require the Authority to refrain from broadcasting 'any matter or classes of matter' (a power not exercised since the withdrawal of prohibitions under 'the 14-day rule' in December 1956). He could prevent exclusivity in the broadcasting of sporting or other events of national importance (such as coronations, Cup Finals and the Oxford and Cambridge Boat Race), and he enjoyed comprehensive powers to control the broadcasting of advertisements and powers (never exercised) over the award of contracts to companies with newspaper shareholdings, which might be contrary to the public interest. He appointed, and could dismiss, all the members of the Authority. Their remuneration was decided by him. He directed the form and content of their annual report and accounts and approved their auditors. He had powers to require the Authority to provide additional payments, 'commonly called the levy', and to control the Authority's capital expenditure, its reserve fund and the disposal of its surpluses.

Some of the ITV companies were understandably reticent in responding to an invitation to comment on the Authority's performance of its duties. Anglia wrote that it greatly appreciated the opportunity but preferred not to take advantage of it. Westward did not consider that it had any useful observations, far less criticisms, to make of the Authority. Granada recorded its opinion that the Authority discharged its duties under the Act fairly and firmly and that press criticism of its lack of control over programme schedules betrayed an ignorance of the facts. For ATV, Sir Lew Grade was more forthcoming. He gave both written and oral evidence, complaining that the Authority interfered too much in the compilation of programme schedules: its mandatory decisions, for example, often prevented ITV from providing alternative programming to the BBC. He also seized the opportunity to state a case against fixed-term contracts. The other companies to welcome this public platform were Thames (represented by Howard Thomas and Brian Tesler), HTV (by

Anthony Gorard and Patrick Dromgoole) and Scottish (by William Brown).

From the outset members of the committee displayed a highly critical attitude. They were not easily going to be persuaded that the Authority was exercising its statutory powers to serve and protect the public interest in the way it should. Equally apparent was the committee's disapproval of the government's cancellation of the proposed Annan Committee which would have been able to give a more thorough public airing of the problems. Those were seen as not only the shortcomings of the present system and how it might be improved, the implications of new technical developments, especially in cable and domestic video recording, and the destination of a fourth channel for which the ITA had recently made proposals; but also, more fundamentally, the whole future role of the broadcast mass media in British society.

So witnesses were pressed on how far the Authority might play a more 'positive' role in relation both to programme output and the overall pattern of schedules instead of acting as the companies' 'partner and friend'. Complaints from Sir Lew at being bullied into schedule changes and the more measured statements from Thomas, Tesler, Gorard and Brown about the firm but reasonable way in which they were from time to time overruled made little impact: such incidents were evidence of negative rather than positive control. Questioning explored how far the Authority could participate in the very early stages of schedule planning so as, for example, to afford a wider network showing for programmes of merit from the smaller regional companies and, when it deemed necessary, to commission from outside sources or even itself produce such programmes as would ensure a sufficiently wide range and balance of subject matter to serve significant minority audience interests. Lawyers' explanations of the strictly limited and contingent nature of the Authority's existing statutory powers in this last respect were ignored.

Nor was the committee deterred by witnesses who pointed to the logically anomalous position of a statutory public control body which itself participated in the provision of what it was required to supervise and censor. Instead attention was focused on the recent crisis in LWT as an occasion on which the Authority might well, it was implied, have terminated a contract and itself assumed programme responsibilities. This in its turn offered an opportunity for expressions of scepticism about the Authority's competence to undertake such responsibilities. Surprise was voiced that, in contrast to the relatively large staff engaged in technical research and development, there seemed to be little or no similar activity on the programme side. The small amount of audience research undertaken by the Authority seemingly assessed reaction to what was provided instead of discovering what needs and interests remained to be served or even, perhaps, awakened.

Much attention was directed by the committee towards a range of topics concerning the ITA's relations with the outside world. Was it, for example, sufficiently well known to the man in the street? Should it not be at pains to be more open in its decision-making, especially when programme companies were awarded or deprived of their contracts? Should not the professional workers in the industry be given the right, through their unions, to discuss with the Authority matters of general programme policy? The MPs did not seem at all happy with the General Advisory Council as a two-way channel for information between Authority and general public. It was, they thought, neither representative nor independent. The recently created Complaints Review Board was a small step in the right direction, but it too was not independent. Perhaps the whole question of public accountability might best be served by the establishment of an independent National Broadcasting Commission to represent the public interest *vis-à-vis* both Independent Television and the BBC.

The committee's first hearing of oral evidence was held on 8 December 1971 and its last on 26 April 1972. During this five-month period the ITA team underwent four uncomfortable sessions totalling nearly eight hours of cross-examination. John Golding (Labour), for example, complained of 'a considerable amount of rubbish' on ITV.[4] He believed that ITV was failing to provide for the serious interests of working-class people and failing to hold the balance between classes in society.[5] Sir Henry d'Avigdor-Goldsmid (Conservative) believed that the problem the committee was investigating was 'the direct commercial clash between the quality programme and the popular programme'.[6]

Phillip Whitehead, the Labour MP who was ITV's most persistent critic, told the committee in evidence that the ITV companies 'know that they can only survive ultimately by profit maximisation, audience maximisation, by producing a good return within the limited period of their contract on the capital invested'.[7] He wanted the Television Act amended to strengthen the ITA's powers. At present, he told the committee, 'it is extremely difficult, if not impractical, for ITA to require a programme to happen'.[8] He wanted the Authority to recruit a new section of programme originators, quite separate from its regulators. This would have enabled it to take over from LWT when the company fell down on its promises.

Michael Peacock, LWT's lost leader, called in evidence, expressed reservations about Sir Henry's analysis and Whitehead's claim that the companies' survival depended on maximising profits and audiences:

> I am not sure if I would go all the way with this rather easy polarisation between quality of progammes on the one hand and being at odds with the urge to maximise profits on the other. The two can co-exist . . . There is no inherent reason why programmes of quality cannot be combined with reasonable profitability within a commercial system.[9]

The Select Committee's report, published on 1 August 1972, together with minutes of its proceedings, minutes of evidence and appendices, ran to nearly 500 pages and, as if that were not enough, contained the recommendation that a full-ranging enquiry be set up at the earliest opportunity to consider the structure of broadcasting after 1976.

The report noted that the role and function of the Authority derived from the Television Act 1964 which consolidated the 1954 and 1963 Acts and remained in force until 31 July 1976; after which, in default of further legislation, the Authority (renamed the Independent Broadcasting Authority under the Sound Broadcasting Act of 1972) would cease to exist. Under the 1964 Act it had explicit powers to direct programme contractors over the drawing up of schedules and the deletion or exclusion of individual programmes. It was thus alone statutorily responsible for the whole of ITV's programme output. But now, in addition to the powers exercisable over the Authority by the Minister of Posts and Telecommunications (since the abolition of the office of Postmaster-General under the Post Office Act 1969), the committee recommended that consideration be given to the establishment of an independent commission to safeguard the interests of the general public.

'The Authority's transmission service, technical competence and its engineering programme was found to be of high quality; as was its news service and indeed the quality of individual programmes of all kinds'.[10] But this handsome accolade and the finding that the Authority was well established and running a popular service were buried beneath a mountain of disparagement.

The committee did not favour the structure of the two-tier system, nor the Authority's definition of its role within it, any more than Pilkington had. It saw the major ITV companies as too dominant and wanted a more interventionist Authority and more programme opportunities for the smaller companies. Big was bad, whether in companies, audiences or profits. The committee judged that:

> Even though there has been a real improvement in the service compared with 1960, the overall output of programmes falls below what might have been achieved. Just as the Pilkington Committee expressed criticisms of the Authority's interpretation of its duties *vis-à-vis* the more generally phrased 1954 Act, so Your Committee are critical (as were the Prices and Incomes Board) of the Authority's interpretation of its more clearly defined responsibilities under the 1964 Act. The Authority appears to have been more influenced by the requirements of the individual companies, especially of the major companies, than by the requirements of the public which should be its first concern.[11]

The differences between the commitments and obligations of the five central comapnies and those of the ten regionals, as specified in their respective contracts, were not appreciated. It was judged wrong that the Big Five should horsetrade over network programme slots while the smaller companies had to fight their way in. On this point the committee made four recommendation: that major companies, like the smaller ones, should make programmes of local interest; that guaranteed network times be made available to the regional companies during prime viewing time; that once the smaller companies had proved their competence they should be in an identical position to the major companies so far as networking was concerned; and that funds should be allotted by the Authority to the smaller companies to enable them to make high-quality programmes. Made aware that, owing to its regionalism, there was already over-investment in studios and equipment within the ITV system, the committee suggested official or unofficial liaison by which the major companies would make such facilities available to the smaller companies.[12]

The Authority's arrangements for programme and audience research were yet another cause of censure. Its research was condemned as small in scope, with results not widely enough disseminated. It revealed 'an obsession with scheduling, and the manipulation of audiences that goes with it, to the exclusion of other socially more important areas of research'.[13] New areas of programme content should be researched in order to identify present and future audience needs.

More and worse was to follow. There were found to be too few serious programmes in peak time (less than 30 per cent) and too much foreign material (in excess of 14 per cent). The Authority had even refused to make the networking of current affairs programmes at the weekend mandatory despite LWT setting up a special unit. The Authority was wrong in dismissing criticism of its programme output as 'an over-intellectual attitude' inappropriate to a broadcasting service for the general public. The committee was concerned not 'about the showing of esoteric programmes but about freeing a service to produce its best creative output rather than repeat the safe formulae which previously persuaded a mass audience not to switch off'.[14]

The Authority, it was suggested, should have a broader, more representative membership. It should strengthen its programme staff, institute a Programme Planning Board and itself commission programmes. In times of crisis, such as had occurred with LWT, it should be prepared and able to take over the provision of a full service itself. Much stricter control should be exercised over the regular programme schedules, especially in peak hours, in order to meet the Act's requirements for balance, having regard to the time of day and day of the week. The Authority had taken this to mean showing what most people wanted to see when most people were able to view, but to the committee it meant showing more programmes of minority interest in peak hours.

This was not the only disagreement in interpretation. The Authority was thought to have been too rigid in enforcing the Act's requirements relating to due impartiality and offence to public feeling. Too lax in the control of balance, it had been too zealous and inflexible in not permitting greater freedom for the exposure of experimental programming and heterodox views.

The committee recommended that, because of the sense of uncertainty and instability created for those working in ITV by the periodic termination and re-award of contracts, consideration should be given to the device of rolling contracts proposed for the independent radio stations. But this should not be so used as to give sitting tenants a perpetual advertising monopoly and preclude infusions of new blood.

The Authority was urged to be generally much less secretive and more prepared to make fuller information generally available about its decisions, whether relating to franchise awards, fixing of company rentals or rulings on specific or general programme policy matters. Only thus could it be held to be properly accountable in the discharge of its public responsibilities. The claims which won Harlech and LWT their contracts had never been officially published: warnings of a substantial reorganisation should have been given to TWW and Rediffusion and generally throughout the industry. The committee advised the Authority to study the methods used in other countries, especially in Canada, where public hearings were held when applications were invited for contracts.

There are several salient features about this report. First, the weight of the attack was excessive and therefore counter-productive. Secondly, the charges were levelled specifically (and sometimes, it seemed, maliciously) against the Authority, and to a lesser extent the major companies, rather than against the system and operations of ITV as a whole. Indeed the committee appears to have seen itself as coming to the assistance of programme-makers and the smaller companies in an attempt to redirect a misguided Authority towards its proper path.

Thirdly, the report was strongly Pilkingtonian in flavour. Under the heading 'The Requirements of a Good Broadcasting Service', for example, it quoted with approval a key passage from the Pilkington report: 'It is by no means obvious that a vast audience watching television all the evening will derive a greater sum of enjoyment from it than will several small audiences each of which watches for part of the evening only. For the first may barely tolerate what it sees; while the second might enjoy it intensely.'[15] This the committee swallowed whole, failing to note that it is one of those valuable and unexceptionable statements which, owing to the careful use of 'may' and 'might', are of little practical relevance because the opposite is at least equally true: 'It is by no means obvious that a vast audience . . . will *not* derive . . .'

But, whatever the rights and wrongs, the Authority had been publicly chastised. To one company managing director who offered consolation Brian Young wrote philosophically:

> Naturally one's first feeling was of indignation at the misunderstandings and prejudices and sheer ill will that emerged from the report. However ... the function of these committees is to carp, however loosely and unjustly, and we must simply accept that they may here and there be on to a good point that we should ponder further. [16]

The *Observer* too thought that the report suggested a lack of understanding of the problems of television: 'the most obvious of these is that you cannot make people like what they don't like'. [17] To the leader-writer, who identified lack of public knowledge about these problems as the most impressive of the report's complaints, Young replied privately with the quotation, 'I can give the honourable gentleman the facts; I cannot give him understanding.' 'The IBA and four of the companies,' he continued, 'devoted really rather a lot of time last winter to explaining to this group of back-benchers some of the facts of television life. We clearly failed to get very much through to them; but the firm prejudices with which they entered on the enquiry are evident from many of the questions.' [18]

Most of the press, however, found justice in at least some of the criticisms. In the *Daily Telegraph* Sean Day-Lewis remarked that this most comprehensive rebuke to the Authority, in spite of being in parts amateurish, self-contradictory and financially unrealistic, had made several useful points: its shortcomings would not justify easy dismissal of demands for more minority programmes in peak time and a stronger scheduling role, better programme research and greater public accountability from the Authority. [19] Each in its own idiom, *The Times* and the *Daily Mirror* delivered the same verdict on the Authority: 'The Sleeping Cerberus of Television' [20] and 'Probe MPs Pan the Sleeping Watchdogs of Telly'. [21]

Lord Aylestone, the Authority's Chairman, called the report scathing and partial, while the tone of reactions among the companies was one of incredulous indignation. If the MPs owed their position to a majority of votes from the electorate, how could a majority preference for ITV programmes by those same electors be so lightly overlooked? Sir Lew Grade's response was in character. He hired a BOAC Jumbo jet to fly seventy American journalists to Britain to see 'the finest television in the world', starting with two episodes of ATV's *The Strauss Family*, 'its most expensive production yet'. 'BOAC has renamed the Jumbo "the Sir Lew Grade" for the occasion,' reported the *Financial Times*. [22]

In its published response the Authority was in part smoothly circumspect, in part withering; steering a zigzag course between courteous condescension to the ill-informed and resentful irritation at palpable naiveties. It found, it said, much common ground with the Select Committee, particularly over the paramount importance of programme output. It agreed that the chief successes of television were programmes which extend the limits of viewers' awareness. It would be giving further consideration to increasing its involvement in scheduling, although it had reservations about a revision of the Television Act to enable it to move into programme production or commissioning. It accepted the need for the potentialities of the regional companies to be used in the best interests of the service.

Then, rapping knuckles in retaliation, it highlighted some errors: 'The report treats balance and quality together. They should be distinguished ... A well-balanced schedule can consist of programmes each of poor quality; and programmes all of high quality can constitute an ill-balanced one.' The Authority, it went on, did not (as the committee seemed to) regard 'quality' as the opposite of 'popular': 'popular programmes need not be of lesser quality than those designed for a minority.' As for solving the problem of balance, that was largely dependent on overcoming the handicaps of a single-channel service, but good balance was 'not a simple question of putting less popular items into peak times at the expense of more popular ones. Balance at peak times ... will continue to be different from the balance over the programmes as a whole, and the committee here appears to have misunderstood what is statutorily required.'[23]

Another misunderstanding by the committee had led to a contradiction:

Quality cannot be prescribed; it grows out of the creative talent to which the system has access. So, in the Authority's view, quality is best achieved in a situation where the programme companies and those who work for them have the freedom (in the committee's words) to produce their best creative output. A theme of the committee's report is that this freedom would be increased by being reduced [by the Authority taking greater control].[24]

A counter-attack launched at the heart of the committee's most cherished belief stressed the benefits and implications of the policy and practices of plurality to which the Authority was committed. Taken to its logical conclusion, the argument that the Authority should assume greater powers could lead to its becoming like the BBC, but financed by advertising.

Anything less is unlikely to satisfy those criticisms of ITV which, though concentrating on the detailed interpretation of the Television Act, are based on a distaste for any profit motive in broadcasting: but it could well give rise to other criticisms, of monopoly, reduced access, and lack of response to public attitudes; and it would be at variance with parliament's decision that Independent Television should be a system of private enterprise under public control.[25]

The Authority saw its task as ensuring that the best possible programmes were produced by the companies and did not believe that this would be best achieved by itself prescribing programmes. It quarrelled with the committee's statement, 'it is inherent in the system that the companies should be the producers of programmes while the Authority is responsible for maintaining standards and quality,' arguing that 'standards and quality are not separable items which can be superimposed on an alien product'. Certainly the final responsibility was the Authority's, but 'under the Television Act, the duty of securing programme standards and quality has to be extended, through the programme contracts, by the Authority to the companies'.[26]

Programme production and the regulation of the production of others were identified by the Authority as distinct functions which could not be satisfactorily combined. Different staffing and resources were required, and 'one cannot both hold the ring and enter the ring'. As for scheduling, the Authority had itself proposed a Programme Planning Board in its submission for ITV2, but was not convinced of its value in the case of a single channel.

To the committee's stricture that those making the vital decisions about programmes were few and unrepresentative, the Authority replied by pointing out that they were many and varied. All fifteen companies were equally represented on the Authority's Programme Policy Committee and on the companies' own Network Programme Committee. Each company had a board of directors and programme staff and advisers who had sometimes formal, sometimes informal, discussions with the Authority and its staff at various levels. Responsibility for programme output was, in fact, diffuse and, with so many committees and so many different agreements between the various parties, might even be thought cumbersome. As for the regional companies, a wider role would become possible following the extension of broadcasting hours recently granted after representations by the Authority, and even more so when there was an ITV2 for which the Authority had long been pressing.

It was conceded that more effort could be made to provide the general public with information about the Authority's policies and planning and research activities. But a wide range of information about ITV was already published in Authority and company annual reports, the annual hand-books, the series of ITV Notes and other occasional publications and press releases. The Authority would certainly be considering the further de-velopment of its research programme, but it was wrong to assume that its current preoccupation was mainly with audiences' tastes and preferences. The *potential* needs of viewers, proposed by the Select Committee as a subject for research, were 'more a matter for debate and discussion of values and objectives'.[27]

The committee's long-considered recommendations, which appeared to reflect Labour Party thinking and attitudes rather than those of parliament generally, found little favour with the Conservative government, which brushed them aside in some terse Observations.[28] The minister, Sir John Eden, told the House that, in the light of his Technical Advisory Committee's advice that new technical developments were unlikely to have a significant impact on broadcasting until the early 1980s, 'the government consider that there is no cause in the present decade to alter the basic structure of broadcasting in this country'.[29]

The government therefore proposed the extension of the Television and Sound Broadcasting Acts and the BBC Charter until 1981 without any substantial change. Separate consideration would be given to the question of a fourth television channel. The government did not think it desirable to establish a single consumer organisation for all communication media, but it would discuss with the Authority and the BBC ways in which viewers' interests could best be represented. It was not prepared to legislate in order to give the Authority unrestricted powers to commission or undertake programme production on its own behalf.

The Select Committee report and the government's Observations were debated in the House of Commons on 3 May 1973. Strictly speaking, the debate was not on the report itself. It was on an Opposition amendment deploring the government's refusal to implement the committee's main recommendations, in particular the minister's 'cursory dismissal' of the proposal for a full and early enquiry into the future of broadcasting. It was therefore that much easier for the House to widen the discussion to include topics which such an enquiry would presumably cover, and the minister was unsuccessful in his attempt to narrow the debate to the Select Committee's terms of reference.

In introducing the debate Russell Kerr (Labour) was at pains to emphasise that the report was unanimous, and thus dictated by the facts, not by political preconceptions. He complained that the IBA's official response was patronising, misrepresented the committee and was used as an excuse for making yet another plea for the fourth channel. 'The unofficial reply,' he correctly surmised with some satisfaction, 'is no doubt unprintable.' But, he reassured the House, 'like the Lord, the Select Committee desireth not the death of a sinner, but rather that he should turn from his wickedness and live.'

Sir Paul Bryan (Conservative), a director of Granada Television, rebuked the committee for concerning itself with the machinery of ITV rather than the programmes themselves: satisfying the viewer was what mattered and that was what ITV was doing. Evelyn King (Conservative) lamented that in his lifetime the power of the House of Commons had sunk and the power of the media risen; he feared for the future of democracy. The suggestion, made by other Conservative members, that the simplest way to mitigate the risks of excessive concentration of media power was to have more channels, including an early allocation of the fourth channel to ITV, was disputed by the Opposition.

The debate ranged at random over various broadcasting topics and the affairs of the Authority occupied a relatively minor part of the six-hour session. John Golding (Labour) insisted: 'We must go back to Pilkington and understand that we shall not have strong regulations until our regulatory body, the IBA, itself sells advertising time,' but, as Tom King (Conservative) pointed out, the presence of no more than twenty members in the chamber did not help the honourable member in advancing his view that there was overwhelming interest and concern throughout the country for a wide-ranging enquiry. The future of broadcasting had attracted fewer members than a debate on water resources the previous day.

John Grant (Labour) wound up for the Opposition with the declaration that 'the future of broadcasting in this country is far too important to leave to the government of a commercially sponsored party'. He repeated the pledge given by earlier Opposition speakers: on returning to power a future Labour government would reinstate the Annan Committee without delay.

In spite of the diversity of views expressed in debate, the vote was strictly on party lines, giving the government a majority against the amendment of 119 to 97. Included in that majority were three Conservative members of the Select Committee, the rejection of whose unanimous report and recommendations their votes helped to endorse.

5

THE FRUITS OF
DERESTRICTION

In Volume 2 of this history it is recorded that the collective talent of Independent Television in the 1960s was formidable,[1] but that its best years of achievement lay ahead.[2] In the 1970s it not only became more formidable; its sights were set higher. Initially, progress was arrested by the traumas of 1968 and the recession of 1969 and 1970, but thereafter ITV's programme service to the public made some impressive creative and technical advances in both scale and breadth.

Three events rescued the system from the constraints of 1968–70. Alleviation of the levy happily coincided with an upturn in advertising revenue in 1971, and early in the following year came a long-sought act of liberation. On 19 January 1972 Christopher Chataway, the Minister of Posts and Telecommunications, announced that he would no longer exercise the powers of control which he possessed under Section 17 of the Television Act 1964. It was the end of government restriction on hours of broadcasting.

Since 1955, in a campaign to secure the relaxation of this irksome control, the ITA had made representations to no fewer than seven Postmasters-General and two Ministers of Posts. Grudgingly conceded minor easements, while welcome in themselves, had complicated rather than simplified the regulations, and in 1965 Lord Hill, on behalf of the Authority, had pressed for total abolition. His successor, Lord Aylestone, in a letter to Edward Short at the General Post Office written on 1 January 1968, described this restriction as 'quite the most persistent bone of contention over the years between the Authority and your Department'. He asked for an increase in the basic weekly maximum from 50 hours to 57 and in the daily maximum from 8 hours to 9, and was granted 53½ and 8½. He asked for an increase in the annual allowance for live outside broadcasts from 400 hours to 500 and was granted 450. These increases took effect from July 1968, and further small concessions were made in January 1971. With additional allowances for special categories of programming – adult education, schools broadcasting and religion – and for ministerial and party political broadcasts, each ITV station was then permitted to broadcast for a total of 3,330 hours a year. The rules drawn up by the Post Office and issued to the ITA in February 1969 illustrate the tightness and detail of government restriction on broadcasting time during the 1950s and 1960s. They are set out in Appendix C.

The return of a Conservative government in June 1970 had raised hopes that this form of rationing would be brought to an end. Twelve months earlier Professor E. G. Wedell and Brian Blake had completed *Hours of Television Broadcasting*, the report of a study undertaken at the request of the then Labour Postmaster-General, John Stonehouse. The authors' approach to relaxation or abolition of the rules was based on the assumptions that there should be no general increase in the BBC licence fee and that the government would be anxious not to encourage consumer spending and not to increase capital expenditure in the public sector, of which ITV was deemed to form a part.

The force of these assumptions overrode the report's recommendation that hours should be extended, with broadcasts permitted between 9.15 a.m. and 12.30 p.m. daily. No action resulted therefore, but in a letter in October 1970 Aylestone drew the attention of Chataway to the Wedell-Blake conclusion: 'There can be little doubt that the hours rules, as they at present stand, circumscribe administratively the power and the responsibility that parliament intends the governing bodies to exercise.' He expressed his confidence that the Authority could safely be left to regulate ITV's output 'in ways which are both socially desirable and economically defensible'.[3]

The obstacle was the BBC. Regulation of the hours of broadcasting served to protect the Corporation from competition by disallowing extra time which ITV could finance out of additional revenue but which the BBC could not afford without an increase in the licence fee. There had, however, been no parity in broadcasting hours between the two organisations since the launch of BBC2. In 1971 the BBC with its two channels was transmitting for an average of 125 hours a week while ITV was restricted to about 75. The ITV objective was 100 hours; which would do no more than reduce the handicap under which it was labouring in attempting to provide a full alternative service. Yet in August the BBC sent the minister a paper arguing persuasively that BBC1 would need to match any extra hours on ITV and that this must mean a higher licence fee.

But unless the BBC was to remain privileged and cosseted at the viewer's expense the case for either unrestricted hours or a second ITV channel was unanswerable. In conceding the former the minister appeased the strong BBC lobby by announcing at the same time that he was not persuaded that the time had come to allocate what ITV desired most: the fourth channel. Derestriction was a consolation prize.

The announcement of an immediate and total freeing of hours was so unexpected that it threw the television services into a state of confusion.

Most surprising was the absence of strings: there were no reservations, no qualifications, no conditions, not even a prescribed starting-date to enable the two contenders to get off the mark together. Other ministers at other times would have required a phased development by both organisations acting in consultation with one another and ministry officials. Some intimation of the possibility of an announcement had reached the ITA earlier in the month and a tentative discussion had taken place at a meeting of the Standing Consultative Committee on 12 January, but there had been no detailed programme planning and the ITV system was quite unprepared for such a heady dose of freedom.

Although press forecasts of 'Round-the-Clock TV' were premature by a couple of decades, the minister's statement 'was received with consternation at the BBC and with delight by Independent Television', as the *Daily Telegraph* reported. Bernard Sendall, the ITA's Deputy Director General (Programme Services), met Huw Wheldon, the BBC's Director of Television, and David Attenborough, Controller of BBC1 and 2, on 21 January and found them anxious, feeling that they had been wrong-footed by the minister. They enquired about ITV's plans for breakfast-time television and, when told that this would be considered by the companies and the ITA, warned that, although they had no wish to introduce such a service, if it was introduced by even one company in one region the BBC would follow suit over its entire network. In the event network breakfast-time television had to await the award of a separate contract by the Authority in 1980. ITV's *Good Morning Britain* opened on 1 February 1983 and the BBC had duly launched a competitive service two weeks earlier.

The 'great step forward'[4] into unlimited hours was taken cautiously. The Authority intended to proceed in an orderly manner by means of a special ITV working party under its own chairmanship. It had shouldered the task of making the case for derestriction and was determined to ensure that promises which it had made to the minister were honoured. It did not want the companies to get the idea that Chataway had set alight a bonfire of controls. Its own requirements relating to such matters as the amount of deployment of foreign material, the use of feature films and the 35 per cent proportion of serious programming would all be affected by an extension of transmissions forward into the daytime and backward into the small hours, but they were all to be strictly maintained: adapted where appropriate, but not eroded.

Adult education and religious and Welsh language programmes too would continue to be protected by order of the Authority. Agreement was reached with the BBC that arrangements for the religious 'closed period' on Sundays would be unchanged.[5] Hitherto the special programme categories had enjoyed the advantage of 'off the ration' time. Henceforward they would be competing with all other programmes for their share of airtime, as would live outside broadcasts.

To the companies it seemed that bureaucratic controls relinquished by the government were being too readily assumed by the Authority and that broadcasters were in danger of being denied much of the freedom which the minister had intended. Their restiveness found expression in a powerful deputation headed by John Freeman, the recently arrived statesman. The other members, all company principals, were Aubrey Buxton (Anglia), Howard Thomas (Thames), Ward Thomas (Yorkshire) and David Wilson (Southern). At a meeting with the Chairman and Director General of the Authority on 1 March 1972 they complained of an authoritarian attitude by the ITA and an increasing tendency to intervene in the companies' affairs. Action taken by the Authority following the announcement of derestriction was seen as an example of a general trend. They protested that edicts had been issued without consultation and that the Authority was, in effect, taking over the function of programme planning from the Programme Controllers Group. These criticisms were not accepted by the ITA, which saw derestriction as a special case.

There were differences between and within the companies too. Breakfast-time television did not form part of the ITA's plans, more because of the difficulties which it would create in the manning of transmitters than from any objection in principle. Most company chiefs were also unenthusiastic, but the two Thomases were keen to experiment and took a more bullish view. They were Thomases without doubts, who welcomed ITV moving into a higher gear with the major companies, not the ITA, in the lead.

Throughout the spring and summer of 1972 the repercussions of Chataway's announcement reverberated through the system. There was hardly a committee whose proceedings were not affected. At the ITA these included the Authority itself and its Programme Schedule Committee, the General Advisory Council and the Educational and Religious Advisory Committees; among joint bodies, the Programme Policy Committee and the Standing Consultative Committee; and within the inter-company structure the Network Planning Committee and virtually all its subcommittees, from the Network Operational Sub-committee to the Network Children's Sub-committee.

A new committee was charged with the responsibility of avoiding chaos in a scramble by frustrated regional companies ambitious for network exposure. This was the Inter-Company Production Committee (ICPC), which met under the chairmanship of David Plowright of Granada to agree on the supply of network programmes made by the regionals for the new daytime hours. In the light of this agreement the Programme Controllers Group representing the major companies had to evolve a new-look network schedule, operative from the autumn and acceptable to all the company managements and the ITA.

The trade unions made their presence felt and flexed their muscles. The

ITA held a special meeting with their Television Safeguards Committee to air the implications of derestriction in relation to the quota of foreign material and the number of feature films and programme repeats. Company managements experienced difficult negotiations over the rostering of staff for extended hours of transmission, and in September and October industrial action interrupted late-night programmes and the early morning transmission of programmes for schools. Disruption of the latter was especially damaging in the context of ITV's delicate relationship with the educational system and authorities. Plans for an extension of Friday and Saturday night schedules until 2 a.m. had to be abandoned in the face of union insistence on uneconomic manning levels and golden payments for unsocial hours.

All this discussion, argument and sometimes outright conflict exerted a severe strain on the intricate machinery through which the ITV programme network was conducted, but what emerged was generally appetising and commendable. The extra hours which were provided brought the total to more than the hundred hours a week which had been the Authority's target. The number of fully networked programmes increased by 40 per cent and the number of networked programmes supplied by the regional companies by 70 per cent.[6]

All the benefits which had been dangled before the minister by the Authority,[7] and which had influenced him in overruling his officials, were achieved: a regular daytime service for housewives, shift-workers, retired people, the house-bound and toddlers; experimental programmes which would not otherwise have been scheduled; and a wider scope for creative talent in the regional companies. Within a few months audiences at lunchtime were numbering more than 4 million, rising to more than 5 million by 4 p.m.[8]

One valuable product of derestriction was ITN's *First Report* at 12.40 p.m., which later became *News at One*: an informed and authoritative presentation of the news, with Robert Kee as both newscaster and interviewer. Despite the experience of five successful years of *News at Ten*, it was believed by some company managements that this would obstruct the build-up of the daytime audience and should be kept to an earlier and less prominent time-slot, but, as the Authority was pleased to record:

> *First Report* soon acquired a prestige and an impact of its own. Cabinet Ministers, MPs, trade-unionists and leaders in industry and commerce were quick to recognise its importance and showed themselves keen to accept invitations to be interviewed. Audience appreciation research showed that *First Report* pleased viewers as much as its parent *News at Ten* and the overall popularity of the lunch-time news amply justified the Authority's decision that it should be transmitted at 12.40 p.m. rather than at noon.[9]

Another new arrival which, against many expectations, succeeded in establishing itself as an ITV institution was LWT's *Weekend World* on Sunday mornings. This was first transmitted in September 1972 at 11 a.m. Its success of esteem owed much to the skill and personality of its first presenter and chairman, Peter Jay. Wisely, it did not attempt to attract a popular following, but excerpts were often shown to much larger audiences in subsequent news bulletins and it was widely quoted on radio and in the newspapers. Subsequently transmitted at 11.30 a.m., *Weekend World* eventually found a regular slot at noon.

The Authority's review of the programme year 1972–73 was prefaced with a broadside against those who were wilfully failing to recognise ITV's progressive improvement in range and quality:

> Readers of the following facts. . . may well find difficulty in reconciling the story they tell with the largely fictitious image of Independent Television conjured up by some of its critics. Regrettably, it is not merely the hysterical fringe of television baiters who are apt to regard the ITV service as little more than a trivial round of undemanding entertainment. Those who so freely criticise along these lines betray a deplorable ignorance of what the service contains.[10]

In the following year this point was rubbed home in an account of ITV's news, current affairs and documentary coverage. This drew attention to the manner in which ITN met the challenge of reporting the Yom Kippur War in October 1973:

> Forty-eight hours after the start of the fighting ITN was able to carry two reports by satellite from war zones in Israel, providing film that was shown on the same day in seventy-five other countries at a time when rival news organisations were still travelling to the areas of fighting.[11]

Among other programmes selected for special mention by the Authority were ITV's extensive coverage of the general election held in February 1974; Jonathan Dimbleby's remarkable *This Week* film for Thames on the then almost unreported famine in Ethiopia; *The Year of the Torturer*, a Granada *World in Action* film; *Weekend World's* elucidation of the Watergate Affair; Thames's *World at War*, a 26-part history of the Second World War acclaimed as a classic among television documentaries; and Granada's *State of the Nation*, which in five hours on three successive evenings examined the working of parliament and the need for reform.

There were documentaries on the problems of the deaf (Thames's *Sunday and Monday in Silence*), on scientology (ATV's *Thank you Ron*), on local farming and fishery (Yorkshire's *Children of Eskdale* and *The Linehams of Fosdyke*), on the teen-aged Russian Olympic gymnast Olga

Korbut (Granada) and on *The Forbidden Desert of the Danakil* – an Anglia *Survival* programme which won an international award for the best contribution to the cause of anti-pollution and the preservation of nature. In Thames's *Something to Say* Bryan Magee conducted debates between two opposing spokesmen on subjects ranging from foreign aid, through psychiatry and crime, to the future of mankind.

All this was the ambitious face of people's television, intent on feeding its viewers with knowledge of the world around them and on stretching their minds to comprehend it.

New documentary and factual programmes in the afternoons introduced by courtesy of derestriction included regular magazines for the housewife and the house-bound at 2.30 p.m. on weekdays and, from a variety of smaller companies, a series of networked programmes, *About Britain*, which demonstrated some of the benefits and versatility of the regional system. At lunch-time, after the news, light entertainment was provided, much of it locally originated. There were also four new weekday afternoon drama serials or series, two of which became so popular that they ran on into the 1980s: *Emmerdale Farm*, a serial of country life from Yorkshire Television, and Granada's *Crown Court*, a series of three-day serials portraying the process of law and exploring current social issues.

Expansion of pre-school programming was undertaken by four separate units located in different companies. Thames produced *Rainbow*, ATV *Inigo Pipkin*, Yorkshire *Mr Trimble* and Granada *Hickory House*. These programmes were transmitted just after midday on weekdays, the companies taking it in turns to present them sequentially over a year. In time they won the confidence of the specialists and popularity among children and their parents. Planned in 1971 (well before derestriction was announced), they had grown out of a desire to compensate for government failure to implement the 1944 Education Act's promise of nursery school education. They were pioneered in three areas by experimental screenings of the American series *Sesame Street*, and a research study into the reactions of parents and children had then confirmed that there was demand for a comparable native product. *Rainbow* won an award for the best programme for young children in 1975.

Fears that adult education programming would dwindle without the special advantage of 'off the ration' time proved groundless. Instead there was growth. A working party representing Authority, advisers and companies agreed on the adoption of a new definition of adult education based on one used by the European Broadcasting Union.[12] All regions began to transmit at least three hours of formal, 'validated' adult education programmes each week. In addition, and over and above the regular feature and current affairs programmes, a new genre took shape, combining entertainment with instruction. Examples were *A Place in the Country* (Thames), a networked series about National Trust properties shown in most areas at 10.30 p.m., and two series networked in the afternoon: *Generation 3* (Westward) for retired people and *The Splendour Falls* (HTV) about the Welsh coastline. Such programmes, whether 'validated' or not, were educational in the widest sense: they motivated viewers 'into taking part, into doing something, into following up'.[13]

The momentum of these advances made in the wake of derestriction was maintained in subsequent years. Despite the internal stresses and bickering, ITV's response to Chataway's act of faith may be judged to have vindicated the nature of a system in which the interplay of forces generated by private enterprise worked with a public authority which conscientiously exercised responsibilities well beyond the merely regulatory. Viewers were fortunate, for no similar benefit in derestricted hours was granted to the customers of shops and pubs. The two gaps in the Independent Television service which remained unfilled until the 1980s – one by ITV's own decision and the other by that of successive governments – were breakfast-time television (apart from a brief experiment in the Yorkshire area) and a second channel working side by side with the first to complement and amplify the service.

There was, however, a small hiccup which began on 17 December 1973 and, with a brief interlude over Christmas and on New Year's Eve, lasted until 8 February 1974. Like the three-day working week imposed at the time, it was one of the measures considered necessary by the Heath government to effect economies in the consumption of fuel during an industrial dispute involving miners and other workers. By government direction no television was permitted after 10.30 p.m.; although radio was not similarly affected. The order was conveyed to the Authority in a terse communication citing Section 21 of the Independent Broadcasting Authority Act 1973, in which the minister had retained discretionary control over broadcasting hours. It was signed by a civil servant humorously misdescribing himself as 'Gentlemen, Your Obedient Servant'.[14]

The Central Electricity Generating Board then found that it could not tolerate the abrupt termination of all three television services at the same time because this created a surge in demand for electricity as other appliances were switched on. ITV and BBC were therefore prevailed upon to take it in turns to close down at 10.20 p.m. and 10.30 p.m. respectively on alternate days. The subsequent claim that this highly unpopular measure had indeed achieved a significant economy met with some scepticism. The public was never convinced of the need for it, and the resulting resentment may have tipped the balance against the Conservative government in the general election which followed.

ACTT, the technicians' trade union, claimed that ITV's hastily rejigged programme schedules relied heavily on the showing of old feature films and repeat programmes, thereby reducing the transmission of original television material and threatening the employment of its freelance membership. The charge that cuts were applied mainly to worthwhile, home-produced material was also made by the playwright Robert Bolt in a letter to *The Times*,[15] but there was no difficulty in demonstrating that it was ill-founded.[16] While the enforced early close-down caused more problems for ITV than for the BBC with its two channels, the Programme Controllers Group had contrived to avoid deferring most network drama and documentaries. There was a reduction of around 13 per cent in hours occupied by repeats and films acquired from outside. The most serious home-produced casualty was the Sunday play.

As Lord Aylestone reported to the minister on 23 January:

> One way and another it has been possible to replace earlier in the day some of the valuable local programmes and network documentary material which is normally accommodated after *News at Ten*. Even so, *News at Ten* itself has had to sustain reductions in length, and many valuable programmes of information and education, which in our system are often scheduled on a local basis, have had to be held over.[17]

In contrast to the BBC, the television curfew cost ITV dear in loss of revenue, but advertising had been buoyant for three years. Aided by levy relief, it had lifted Independent Television out of the financial doldrums and provided the resources for a larger supply of more ambitious programmes. As in the late 1950s and early 1960s, the most expensive category of programming flowered most luxuriantly, and the spearhead of success was drama, now particularly drama in series and serial form. In 1971 came the launch of *Upstairs, Downstairs* (LWT), a saga which was to run with enormous success for six years and sixty-eight episodes. Jane Austen's *Persuasion* in five parts, also in 1971, was followed in 1972 by the *Country Matters* anthology (both from Granada) and by *The Strauss Family*, a series of eight plays about the composers from ATV. In 1973 came *The Brontës of Haworth* (Yorkshire) and *Sam* (Granada); in 1974 *Jennie* (Thames) and *South Riding* (Yorkshire); in 1975 *Edward the Seventh* (ATV) and *The Stars Look Down* (Granada); and in 1976 *Clayhanger* (ATV).

It was Chancellor Barber's boom which made 1971, 1972 and 1973 years of unforeseen prosperity for Independent Television despite an economic climate of rising costs, price control and high taxation, but in 1974 and 1975 the shadows of hyperinflation gathered and the financial outlook darkened once more.

Such ups and downs were all the more disconcerting in an industry unable to relate expenditure to income within a normal accounting period. Whatever the circumstances certain programmes had to be made, and to an acceptable standard. How then could programme budgets be geared to rise and fall in harmony with revenue and profits if commitments to expensive dramas and documentaries had to be made at a time when forecasts of sales revenue could be no more than guesstimates? Too much euphoria might precipitate a financial crisis; over-caution could expose a company to accusations of under-spending. The companies always seemed either too rich or too poor in a manner which perplexed city analysts and made investment advisers wary despite the profits of earlier years. Those crying woe were accused of crying wolf.

During the decade to mid-1974 programme costs almost trebled, rising much more steeply than advertising revenue, which almost doubled. At constant prices (adjusted to take account of inflation) the industry's profits fluctuated from £19 million in 1964–65 to £4.8 million in 1969–70, reaching a peak of £20.5 million in 1973–74.[18] Thus, in real terms, progress in programme-making was not matched by any increase in revenue or improvement in profitability.

Owing to the usual time-lag, the financial anxieties of 1974 and 1975 were not reflected in programmes seen by viewers during those years. 1973–74 was hailed by the Authority as a vintage year for documentaries and 1974–75 as overall probably the most successful programme year since Independent Television began. After giving due credit to the producers and their creative teams the Authority commented in its Annual Report: 'Administrators and Authorities, having accepted that they do not drive the train themselves, must also accept that unless they lay the right kind of track they will make it impossible for even the most talented and professional people to get moving and take the public somewhere worth-while.' Money, time, freedom, encouragement and expectation were identified as the pre-conditions of distinguished programme-making.[19]

Money was the ingredient causing most concern in June 1974 when the Authority unveiled its intentions for the period 1976–79. The six-year contracts operative from 1968 had already been extended for two years while the government was making up its mind about the future of broadcasting. The Conservatives had intended the existing system to continue until 1981, and this would have permitted a new contract period of five years from 1976, but the new Labour administration was not prepared to guarantee the system – or indeed the Authority – life beyond 1979. This abbreviation of the period to three years effectively precluded the possibility of introducing a system of rolling contracts and made the intended advertising of contracts for competition in 1975 unrealistic.

Instead, the Authority decided to review the present contractors' performance and formally appraise their composition, management, financial approach, programming and operational procedures; the initial conclusions of the review to be communicated to each company in the autumn of 1974. In the light of the review, of its research into the opinions of the public in each region and of each company's ability and willingness to take any remedial measures required, it would decide early in 1975 whether special limitations or restrictions should be attached to the contracts to be awarded from 1976.

This procedure was followed, and the final appraisal of each company was published in the Authority's Annual Report for 1974–75 (after some diplomatic editing). The companies had little liking for being reported on as though schoolboys and were unenthusiastic about an extension of contracts whose financial viability was in doubt. Toward the end of 1974 their representatives sought meetings with the Chairman of the Authority to discuss their future in the light of forecasts which suggested that, under prevailing conditions, the system could not survive beyond 1976 and the five smallest companies not until then. A realistic estimate for 1975–76 predicted static revenue and an increase of 20 per cent in costs, so that the companies wondered whether any contract which the Authority intended to offer for 1976–79 would be attractive enough to be acceptable. Managing directors were going to have to convince their boards and shareholders that there was some benefit in making losses or minimal surpluses with no immediate prospect of better times and no more than the possibility of a new contract in 1979.

At a meeting held in February 1975 it was reported that the companies jointly had lost a million pounds in January, although February might break even and the loss might be recovered in March. Their financial adviser from Peat, Marwick told the Authority that there must be a long-term future, otherwise investors would no longer put money into the system: companies could not be expected to carry on business with little or no return in hard times, only to face the penal taxation of 'additional payments' when good times returned.[20]

On 11 March there was discussion of the terms of a letter which the Authority proposed to send to the Home Secretary formally requesting a review of the levy. To strengthen the argument, the companies were asked for information about the benefits of diversification: there had been criticism that television profits were being spent on extraneous activities and the Treasury might maintain that some companies were in cash-flow difficulties because they had diversified. To assist the smaller companies, the Authority wanted agreement on a reduction in the prices for network programmes paid by Westward, Ulster and Grampian – a concession already made to Border and Channel. It was promised that the Authority would exercise particular care over its own expenditure. Rentals, which were subject to cost-of-living increases, were already £2.5 million below the contractual level and would continue to be kept down. It was conceded that engineering costs were high, but they were said to compare favourably with the BBC's. The companies were assured that members of the Authority fully appreciated their difficulties.

Diversification was a delicate and sore subject. HTV confessed to disappointments, but blamed the state of the economy. Southern said that it had kept its money in reserve within the system, only to have it eroded by inflation. Granada had diversified out of cinemas into television and pointed out that no one had complained about diversification then. ATV made the point that if its cash flow deteriorated the resources of the rest of the group would be available to keep the television company afloat.

On 20 March representatives of the companies were invited to a meeting of the full Authority. Those who attended were Aubrey Buxton of Anglia (the current Chairman of the companies' association, ITCA), John Freeman of LWT, Denis Forman of Granada, Bruce Gyngell of ATV, Anthony Gorard of HTV and Alex Mair of Grampian. Asked to demonstrate the financial problems which confronted the industry in programme terms, Forman replied that the present cash shortage would not have a significant effect on programmes until the autumn of 1977 (two and a half years later). Gyngell explained that a major drama series took eighteen months to two years to prepare for transmission. He pointed out that such series tied up a lot of money at high interest rates for long periods. If the companies had foreseen the gravity of the current situation he felt sure that they would never have embarked on some of the very expensive series currently in production. In order to retain enough money in the system to finance future programme-making, the companies' representatives urged the Authority to press the government to allow 10 per cent of each company's advertising revenue to be levy-free.[21]

The Authority's letter, as already mentioned in Chapter 2, went to the Home Secretary (Roy Jenkins) on 25 March:

> During the past five months the Authority has been engaged in detailed discussion with the television programme companies about the financial problems we were facing. We are, obviously, trying to minimise the damage to programmes; but we have had to announce some curtailment of the full programme service that we have been providing over the past two or three years, and this has provoked some disappointment, mostly from housebound viewers who have come to value a vein of programming which we introduced in the afternoons following derestriction. We have no immediate plans for further economies in programming, but even if the level of revenue is sustained and the rate of cost inflation does not increase we can only hold the line for a limited time. It is clear that if our financial forecasts for the coming year are borne out we shall, sooner or later, have to take further action to reduce the cost of the programme service.

The letter went on to refer to the alarming prospects for an essentially self-supporting service at a time when the licence-financed BBC had been granted a 33 per cent increase in income to offset the effects of inflation: 'As inflation begins to bite, we expect that the companies together will have to provide over £7 million of additional finance for working capital from their own resources.' The scale of a levy which was intended to siphon off excess profits would, it was feared, discourage necessary investment during the three-year standstill period from July 1976 to July 1979 and prevent ITV from maintaining the service required of it by parliament.

This plea of poverty from the supposedly rich fell on the ears of a Labour government deafened by cries of distress from all quarters. In the midst of a national economic crisis with the rate of inflation running at 25 per cent per annum, and with Lord Annan's resurrected committee of enquiry into the future of broadcasting now at work, there was every excuse for the Home Secretary to make no positive response. His inaction was prudent, for this was, as it proved, a financial stutter, not the feared plunge into an abyss. All the companies accepted the three-year renewal of their contracts, and prosperity soon returned to the system.

The evolution in ITV programming between 1970 and 1975 was a significant development. Between 1970–71 and 1974–75 the total weekly hours of transmission for the average ITV company rose from 71½ to 98. Within this total the balance swung away from entertainment towards programmes defined as 'serious'. The wide field categorised as entertainment covered variety shows and one-off spectaculars; situation comedies; chat programmes; pop and musical shows; talent shows such as *Opportunity Knocks* and *New Faces*; and quizzes and competitions (*Sale of the Century, Celebrity Squares, Looks Familiar, University Challenge*). But, contrary to general belief, these kinds of popular programmes played a decreasing part in the schedules.

By contrast, the amount of serious programming increased both absolutely and relatively. Hours rose from just over 23 to just over 34 a week – an increase of 572 a year – and the proportion went up from 33 to 35 per cent. Programme categories defined as serious were news, current affairs, documentaries, arts, religion, education, and information programmes for children. Drama was extra: however earnest or classical, plays were not categorised as 'serious'.

Critics continued to assert, however, that the number of serious programmes transmitted in peak time was very small; that the serious-minded viewer was expected to be either a shift-worker or an insomniac. But in fact the 'serious' element in the main viewing period of 6 to 10.30 p.m. amounted to 30 per cent, with the half-hour *News at Ten* being consistently rated as among the most watched programmes on any channel.

The other important trend during this period fittingly moved what had been designed as a regional system towards a greater degree of regionalisation, and this move gathered pace in the years that followed. As a result of pressure from the Authority the provision of news, information and entertainment catering specifically for regional needs and preferences was enlarged by up to 25 per cent in the case of the regional companies and, in the case of the major companies, from 4 or 5 hours a week to 6 or 7. In Yorkshire, for example, a range of local programmes extended through *Calendar News*, *Calendar Sunday* (on local parliamentary issues) and *Calendar Profile* (an interview with a local personality) to *Calendar Kids* for children. In London Thames covered local sport in *Sportscene*, commerce and industry in *Time for Business* and social action in *Help*, while LWT catered for local current affairs in *The London Programme*, for teenagers in *The London Weekend Show* and for children in *Our Show*; and there were corresponding programmes in the ATV and Granada areas.

Making the productions of one region available throughout the whole country was another aspect of regionalism, and the output of the regional companies on the network registered a dramatic upturn from between 160 and 165 hours to 420. Anglia contributed drama, *Sale of the Century* and the *Survival* series on wildlife; Southern provided children's programmes and operas from Glyndebourne; from HTV came children's programmes and Christmas fare; from Scottish came Hogmanay and *The Prime of Miss Jean Brodie*. Even Border, the smallest mainland company, achieved national coverage for a long-running series: *Mr and Mrs*.[22]

6

THE AUTHORITY

The reign of the first Director General ended in October 1970 on the retirement of Sir Robert Fraser, architect-in-chief and master builder of Independent Television. The philosophy and framework of ITV were his achievements: its pluralism, its regional structure, its two-tier system of Authority and companies. He was a socialist intellectual who came to believe that a monopoly such as the BBC had enjoyed was 'an enemy of the free spirit of man' and that free-enterprise television was as valuable to the preservation of a democratic society as a free and independent press. The debt which the Authority owed him for his conduct of its affairs as head of its permanent staff for its first sixteen years, it found 'difficult to put into words'.[1]

Born an Australian of Scottish stock, Fraser was educated at Melbourne University and the London School of Economics, where he fell under the influence of Hugh Dalton and Harold Laski. In the general election of 1935 he stood as a Labour candidate, but was unsuccessful. During the Second World War he served in the Ministry of Information and was appointed the first Director General when it became the Central Office of Information.

In 1954 responsibility for the appointment of a first Director General of the new Independent Television Authority rested with its first Chairman, Sir Kenneth (later Lord) Clark, who is reported to have received 332 applications.[2] 'The Treasury always hopes to get rid of some troublesome claimants, in particular disgruntled members of the armed forces,' he recorded in his autobiography:[3] 'They gave me the choice of nine admirals, seventeen generals and six air-marshals.' Disenchanted after interviewing these and others, Clark chose Fraser, whom he had known at the Ministry of Information. Fraser had applied late for the job, but he seemed to be what Clark was looking for: an experienced and 'innocent' communicator. The choice incurred the displeasure both of the Conservative government, outraged at the appointment of a socialist, and of the Labour opposition, no less outraged at the appointment of a renegade comrade.

Though intellectuals themselves, Clark and Fraser were of one mind in believing that, to establish itself, ITV must initially concentrate on mass entertainment and attract a popular press, not a *Daily Telegraph* or *Guardian*, audience. Unlike the BBC, ITV had to please to live: 'We have no income which goes on just the same no matter what the public thinks of us.'[4] After the financial crisis of 1956 Fraser took unconcealed pride in ITV's growing prosperity, in the large audiences it entertained and in the advertisements which made this possible. It was thus on his head that the wrath of Pilkington fell. He was accused of being in the companies' pockets in not insisting on more balanced and serious programming when the system could afford it. His principle that the companies were partners to be trusted and not merely agents to be instructed attracted censure as well-meaning but misconceived.[5]

Fraser was a warm-hearted, unassuming man who never stood on dignity and displayed none of the flamboyance of a Bernstein or a Grade. They were constantly in the public eye as the tycoons of ITV, and it was a point of honour with him never to upstage them or criticise them or their companies in public. But appearances were deceptive. He was very much in charge, although asserting himself without raising his voice and by taking a firm grip on elbows rather than throats. When his leadership was usurped, he was patient and suffered Lord Hill's rumbustious chairmanship and demands for more centralised control with good humour and wise advice.

Fraser was blessed with exceptional clarity of vision, but his deeply felt philosophy was perhaps too often and too bluntly expressed. Independent Television was created by Act of Parliament as a public service but, he was fond of pointing out, since it received no public subsidy or grant, 'like the newspapers it must support itself: and rightly so. . .If viewers do not freely choose to watch our programmes, we are out of business: and once again, rightly so. We rest, as governments and newspapers and manufacturers rest, on public consent.'[6]

Because the income of a state-financed broadcasting organisation does not fluctuate in accordance with public demand for its programmes, Fraser argued, the BBC's income, though it might be deserved, was not earned. 'The performance and standards of such an organisation may be detached from the preferences of the ordinary viewer, for he has no means of enforcing a change in the service by withdrawing his custom.'[7] This was the situation which had changed in Britain with the coming of ITV in 1955.

Fraser did not refrain from criticising critics:

> If you decide to have a system of people's television, then people's television you must expect it to be, and it will reflect their likes and dislikes, their tastes and aversions, what they can comprehend and what is beyond them. Every person of common sense knows that people of superior mental constitution are bound to find much of television intellectually beneath them. If such innately fortunate people cannot

realise this gently and considerately and with good manners, if in their hearts they despise popular pleasures and interests, then of course they will be angrily dissatisfied with television. But it is not really television with which they are dissatisfied. It is with people.[8]

When at a farewell staff party he looked back over his years in broadcasting, Fraser identified one strong and persistent quality in the Authority: a capacity for organisation and innovation – for original thought and original solutions. In its engineering and administration, in its regionalism and control of advertising, above all in its programming, 'ITV trod in no one's footsteps'. After thirty years of broadcasting according to one pattern, a new kind of service had been created, more varied, more interesting and closer to people. 'We began as an entertainment service, and clearly our instant acceptance by viewers was a sign that one was wanted. But then, before too long, we began to widen and deepen. We added school programmes, and adult education programmes, and religious programmes. We broadened our children's programmes. We began that exploration and expansion in the field of news and current affairs that now finds its best-known expression in News at Ten.' The retiring Director General's valedictory advice was: 'Never let Independent Television stand still. Try fresh things.'[9]

After his retirement Fraser served as Chairman of ITN from 1971 to 1974 and continued living quietly in London until his death in January 1985 at the age of 80. He had constructed and presided over a system which brought pleasure and the sight and stimulus of wider horizons to almost the entire nation. Yet he received no public recognition for this achievement: his OBE and knighthood had been rewards for his work at the COI. While others in ITV were awarded peerages and other honours, Fraser, the modest master-mind (like Norman Collins, ITV's founding father), remained conspicuously unhonoured. Nor did he choose to bequeath an apologia, deflecting would-be publishers of his memoirs with the excuse of a poor memory.

Brian Young, his successor, was a man of stature, both physically and intellectually. His appointment was as unlikely as Fraser's, and like Fraser (and John Whitney after him) he had not applied for the job initially. John Freeman had declined to be considered, and Young's name was added to a short list of five culled from a field of nearly a hundred applicants. The front runner was Sir Michael Cary, a senior civil servant. The challengers were Robin Day and three internal candidates. Casually approached for names by a member of the Authority, Young had suggested Michael Swann (later Chairman of the BBC) and then became interested himself. He knew nothing of the business of entertainment or journalism, but viewed the job as an important and exciting form of public service. For Independent Television this was the most crucial appointment since Fraser's.

The son and grandson of colonial governors, Young had been educated at Eton, where he was a King's scholar, and at King's College, Cambridge, where he was a scholar, took first-class honours in both parts of the classical tripos, won the Porson Prize and half-blues for athletics and Eton fives and played the clarinet. He returned to Eton as an assistant master before being appointed headmaster of Charterhouse at the age of 29. At the time of his appointment as the second Director General of the ITA, he was Director of the Nuffield Foundation, a grant-giving charity specialising in scientific and social research and in education.

Young was not someone readily associated with a television set, let alone one tuned to Channel 3. Almost inevitably, the news of his appointment was greeted with the journalistic cliché that this was the most extraordinary appointment since Caligula made his horse a consul. At a party a few weeks later the journalist concerned was greeted by 'a placid and smiling man' with the words: 'May I introduce myself? I am Caligula's horse.'[10]

In Victorian times Young's natural progression from Charterhouse would have been back to Eton as headmaster or Cambridge as Master of a college, followed (since he would have been in holy orders) by elevation to the bench of bishops. Instead he stayed with ITV for twelve years (1970–82) until retiring at the age of 60, when he continued to serve the causes in which he believed by becoming a member of the Arts Council and Chairman of Christian Aid. He was knighted in 1976 for his work in ITV.

The mould had been formed by Fraser the free marketeer. Young was recruited to exercise a more interventionist leadership and he was well qualified for the task of high-minded expansion. For him television was an opportunity to enrich the quality of the lives of ordinary people. It could, for instance, open windows of unforeseen enjoyment by introducing opera to the masses. He sought to encourage any programme – new, risky, ambitious – which no one concerned only with profit would make. But if his head was in the clouds his feet were mostly on the ground. While himself a believer in magic casements rather than bread and circuses, he supported ITV's role as a provider of popular entertainment and was fully aware of the network's financial dependence on popular appeal.

Young's main achievement lay in the growth of minority programming, and this was made possible by more broadcasting hours, changes in the levy and a second ITV channel, for all of which he lobbied with perseverance. His retirement in November 1982 came ten days after the launch of Channel 4, for which he had fought long and hard. Under his supervision Independent Local Radio was born (with thirty-eight stations launched between 1972 and 1982) and the entire ITV system re-engineered.

His vision of television as an educational medium in the broadest sense was expressed in the IBA's evidence to the Annan Committee,[11] and in a number of speeches and lectures:

Television has a unique power to focus many of the conflicts and difficulties of our society on a small screen; and it has this power just because it takes all of us into so many new and different situations that our own circle could not provide. It is a crowding together of possible experiences and sensations in a mixture which some believe is too heady for the viewer to cope with.[12]

Aiming high was his objective: 'The same service that puts out cheap and popular programmes must also put out programmes which are expensive and ambitious.'[13]

At other times Young could be almost as populist as Fraser:

I think that the disdain for what people like in the ordinary way is an unattractive feature of television critics; they generally speak as though what they don't want to see is tripe, and they do not see that in fact there is a mixture of, if you like, wooing the viewer and leading the viewer into something else.'[14]

His own approach to the serious business of being the public-service conscience of commercial broadcasting was often pursued in a far from serious manner. He had an aptitude for light verse, witty speech-making and marginal comment. 'Wrong' he commented in the margin of a letter from a viewer who presumed to inform him: 'You will be interested to read the enclosed.' One ITV gathering was entertained by the following:

At meetings and talks by an effort of will
I always contrive to keep perfectly still;
For it needs but a word of annoyance or pity
And wham there I am on another committee.

Second only to the influence of the Director General was that of the Chairman of the Authority. The job was part-time and Chairman and Director General had to work in partnership. They performed most effectively when their qualifications and talents were complementary, and in this respect Young and his first chairman were admirably suited, for they had little in common. Lord Aylestone had been a senior politician. His political antennae were well tuned and he was experienced in picking his way through the thickets of Westminster and Whitehall. What was more, he actually liked game shows, leg shows and the rest of ITV's popular entertainment. To be a natural ITV viewer was a rare – indeed perhaps unique – attribute in a member of the Authority, and certainly not an accusation which could be levelled against Young.

As Bert Bowden, Aylestone had been a Labour MP for more than twenty years from 1945: in 1954 he had opposed the introduction of advertising-financed competition to the BBC and voted against the Bill to

establish an Independent Television Authority. Of upright military bearing
(after war service as a Special Constable and in the RAF police), he made
his name as Chief Opposition Whip from 1955 to 1964. Then, in Harold
Wilson's cabinets, he became Lord President of the Council and Leader of
the House of Commons (1964–66) and Secretary of State for Commonwealth Affairs (1966–67). In 1967 he was rewarded with a life peerage and
appointment to the chairmanship of the ITA. The circumstances of his
appointment are of unusual interest because he succeeded Lord Hill,
whose transfer to the BBC was a ploy in the Prime Minister's feud with that
organisation: 'Charlie Hill has already cleaned up ITV, and he'll do the
same to BBC.'[15]

Aylestone has recollected the event as follows:

In April or May 1967 I decided to leave the government mainly for
political but partly for domestic reasons. I saw the Prime Minister
privately and told him of my intentions and that if he was agreeable the
actual timing of my resignation would be in his hands as I had no
intention of rocking the boat. The Prime Minister with characteristic
kindness agreed and said: 'Leave it to me, for you must have something
to do.' We did not discuss the position again until 15 June 1967 when at a
cabinet meeting the Prime Minister received a note informing him of the
death of Lord Normanbrook (Chairman of the BBC), which having read
he passed to me, having himself written on the note: 'BBC??' I replied
on the same note: 'Speak to you after the meeting'. When the cabinet
meeting broke up there was no opportunity to speak privately with the
PM. However, a few days later we did speak briefly and he observed:
'You don't seem very keen on the BBC and of course Charles Hill has
only a short while to go before his term of office expires with the ITA.
Anyhow I will speak with you again.' The idea of sending Charles Hill to
the BBC and myself to ITA was no doubt already in his mind, which he
confirmed to me early in July 1967. Before the end of July and the
summer recess the Postmaster-General (Edward Short) was made aware
of the PM's proposed changes and told me so. During August I was
staying in Penarth, South Wales when I was told by the local police that
the Prime Minister wished to speak to me from the Scilly Isles where he
was staying. For security reasons I spoke from the Penarth police station
and the Prime Minister from the coastguard station of the Scillies. This
was now final confirmation, and he gave me the date of the official
announcement of the change he proposed to make. I therefore became
Chairman of the ITA with effect from 1 September 1967.

My reason for preferring the ITA's company and not that of the BBC
was mainly the result of dealing with both broadcasting bodies during the
nine years when I was Chief Opposition Whip and the two years as
Leader of the House of Commons in government. During those two

years I was Chairman of the Cabinet Committee on Broadcasting. I did feel I could much more easily work with Robert Fraser and the ITA than with Hugh Carleton Greene and the BBC.[16]

Cabinet divisions over Rhodesia were responsible for Aylestone's departure from government and he found life at the ITA more congenial. He was already on Bob and Betty terms with Fraser and his wife (another former Labour candidate), and he and Fraser were both on good terms with the Prime Minister. It was Wilson's practice to keep in touch with his appointee and the affairs of ITV by means of pre-arranged lunchtime telephone calls, when he would start by saying how pleased he was at the way things were going and then raise just one or two points. . .[17]

Aylestone's gift in politics for making many friends and few enemies did not desert him in broadcasting. He was quiet and unassertive; stolid, straightforward and unflappable. No intellectual, no philosopher, he was good at conducting meetings and working to an agreed brief. The troubles which he experienced during his three-year stretch in harness with Fraser – a period of almost incessant difficulty and strain – were for the most part either legacies from the contract decisions made by the Authority under Hill's chairmanship or fierce tussles with a former colleague, Jack Diamond, Chief Secretary to the Treasury, over ITV's financial problems.

To Aylestone working with the companies was a pleasure, and in a role reversal after the Hill/Fraser regime the new Chairman was sometimes inclined to take their side against the new Director General's determination that the will of the Authority as interpreted by its staff should prevail. In July 1973 his term of office was extended by a Conservative government for eighteen months – a tribute to his political impartiality – and he finally retired on 31 March 1975 after seven and a half years in the chair. For his services he was created a Companion of Honour.

Lady Plowden, his successor, had been a governor and Vice-Chairman of the BBC since 1970. Kindly, high-principled and dedicated to good causes, she held strong convictions, as befitted the daughter of an admiral. Her main area of interest lay in education. As Chairman of the Central Advisory Council for Education (England) she gave her name to the Plowden Report on primary schools (1966). She had been a co-opted member of the Education Committee of the ILEA and Vice-Chairman of the ILEA Schools Sub-committee and a member of the Houghton Inquiry into the Pay of Teachers. She was Chairman of the Governors of the Philippa Fawcett College of Education and President of the Pre-School Playgroups Association. In 1976 she became Chairman of the Voluntary Organisations Liaison Council for Under Fives, and in 1980 President of the National Institute of Adult Education.

The new Chairman's arrival at the IBA headquarters at 70 Brompton Road in April 1975 would have been greeted with greater enthusiasm within ITV if this intense interest in education had not duplicated that of the Director General. The Plowden/Young team which ruled until her retirement at the end of 1980 was formidably worthy but unbalanced: Independent Television had a headmaster and a headmistress at the same time. By some they were seen apprehensively as missionaries come to convert the native ITV viewer from the worship of frolicsome light entertainment to the true god of solemn adult education.

Bridget Plowden was nonetheless respected as a conscientious and energetic Chairman and a robust guardian of ITV's interests in the difficult Annan and post-Annan years. But some decisions in the 1980 contract awards for which she was held primarily responsible attracted heavy criticism, and the succession of Lord Thomson of Monifieth (who had served as Deputy Chairman for eleven months) was welcomed. Like Bert Aylestone, George Thomson had been a Labour MP for twenty years and become a Cabinet minister. He was an affable but shrewd Scot, an experienced trouble-shooter and a 'consensus' chairman: a man of moderation who, like Aylestone, came to prefer the SDP to the Labour Party.

The ordinary members of the Authority, on whom, no less than on the chairman, rested formal responsibility for ITV's obedience to the will of parliament, were appointed initially for varying periods of three, four or five years. During the 1970s most served for five years and some for longer. The system was not rigid: one farewell luncheon for a member was interrupted by a telephone call from the Home Office extending his term of service. The composition of the membership illustrates the strengths and weaknesses of the long-established British cult of the disinterested outsider, whereby a group of reputable but unrepresentative individuals, advised by professionals, make key decisions in what they judge to be the public interest.

Of the sixty-two members of the Authority appointed between 1954 and 1980, 25 per cent (sixteen) came from the world of education. These included six professors (two in electrical engineering and one in physics) and three headmistresses (of independent schools). The other substantial categories were businessmen (twelve), trade unionists (ten) and politicians and their widows (nine). Five came from the voluntary services, four were former civil servants, two had worked for the BBC, and there were one lawyer, one clergyman, one librarian and one film critic. Their names, occupations and terms of office are listed in Appendix B.

Would a significant minority with experience of broadcasting or a stronger element with experience of commerce and advertising, journalism and the arts, have come amiss? The particular precedent for reliance on the detached good sense of the Great and the Good was to be found in the BBC, whose governors were similar people similarly appointed (governors by the Queen in Council i.e. the Prime Minister; members by the Postmaster-General, later the Home Secretary, subject to the Prime

Minister's approval). Indeed it seemed as though the members of the Authority and the governors of the BBC could have changed places *en masse* at any time without any perceptible effect on either organisation.

After considering a number of proposals for a different system the Annan Committee, in 1977, recommended that members and governors should continue to be appointed as individuals, not representatives, and normally for a five-year term. It hoped for a wider selection which would leaven the Great and the Good with some of the Lesser and Better.[18]

In mid-1968 the Authority numbered thirteen, nine of whom were due to retire during the following twelve months. Hill had thought thirteen too many and Aylestone agreed. He suggested to the minister, Roy Mason, a drop in numbers to eleven and the desirability of rephasing lengths of service to allow for the steady arrival and departure of two or three members a year, so that the Authority would receive its infusions of new blood without loss of continuity. If two three-year terms were extended to five and two five-year terms to six, only three new appointments would be needed.[19]

A month later the GPO reported that Stormont was agreeable to the reappointment of the member with Northern Ireland as his special responsibility and that the Treasury saw no objection to the other reappointments, but comments were still awaited from the Scottish and Welsh offices about the members specially concerned with Scotland and Wales. As soon as these were received, the Postmaster-General intended to consult with the Prime Minister.[20] These special geographical arrangements were an essential ingredient in the system, but there was no formal – and often no informal – representation of the North-East, the North-West, the Midlands and other English regions.

The next step was to take soundings about suitable new members: specifically, someone from the trade unions to replace Sir Vincent Tewson, someone with technical knowledge to succeed Sir Owen Saunders, and a businessman to take the place of Sir Patrick Hamilton. The Authority's store of financial know-how had not been replenished since the earlier departure of Sir John Carmichael and Sir Sydney Caine. Short lists were drawn up, candidates were approached and those selected declined the invitation. To busy and important people the attractions were not great. In addition to meetings of the full Authority, usually twice a month, there were at that time six committees: Establishments, Finance, Policy, Programme Schedule, Television Gallery and Television Fund, generating an amplitude of papers to be studied. The Chairman, who was expected, not to be full-time, but to make the affairs of the Authority his main interest, was paid £5,000 a year; the Deputy Chairman £2,000; and the other members £1,000. The Chairman enjoyed also an entertainment allowance of £500 a year and the Deputy Chairman £250, while the other nine members divided £150 between them. By April 1980 rates of pay had risen to: Chairman £19,048, Deputy Chairman £5,000 and other members £2,500. In real terms all except the Chairman were even more poorly remunerated than in 1968.

In September 1969, when all the proposed changes had been effected, the Authority was constituted as follows: Lord Aylestone (Chairman), Sir Ronald Gould (Deputy Chairman), Mary Adams, David Gilliland (Northern Ireland), Dr W. Macfarlane Gray (Scotland), Sir Frederick Hayday, Stephen Keynes, Professor J.M. Meek, Baroness Plummer, Baroness Sharp and Sir Ben Bowen Thomas (Wales). This body included two former civil servants, two serving trade union officials, two university teachers (one a scientist), one former BBC executive and one merchant banker, but no businessman experienced in management. Lady Plummer, the ennobled widow of a Labour MP, was replaced in 1971 by Lady Macleod, the ennobled widow of a Conservative MP; these two and, earlier, Baroness Burton, a former Labour MP, were the only overtly party political appointments among ordinary members. In 1972 financial and management expertise was strengthened by the appointment as Deputy Chairman of Christopher Bland, a management consultant who had been Chairman of the Bow Group. Professor Meek, an electrical engineer at Liverpool University, was replaced in 1974 by Professor Ring, a physicist from Imperial College, London, who had been a successful performer in ABC's adult education programmes and an outstanding Chairman of the Authority's General Advisory Council.

Subsequent changes followed the established pattern. The trade unions continued to be well represented. Businessmen such as Alex Page (Metal Box), Tony Purssell (Guinness) and George Russell (British Alcan) were recruited with difficulty and were sometimes too busy to attend meetings. The dominant element remained educational; Dr Tom Carbery (Scotland), Professor Huw Morris-Jones (Wales) and Mary Warnock (a former headmistress and Oxford philosopher) joining Lady Plowden and Professor Ring. From August 1979 numbers were increased from eleven to twelve.

Throughout the 1970s the Authority was generally a united and harmonious body, with differences amicably settled. Fewer members had lessened the likelihood of factions and pressure of work was eased with the abolition of three of the committees instituted by Lord Hill – Establishments, Finance and Programme Schedule – when it was found that discussion was being duplicated. Joint responsibility was a binding factor. All programme schedules had to be approved at meetings of the full Authority, and the BBC tradition that programmes were never viewed in advance by the governors was regarded as an abrogation of responsibility, Lord Aylestone remarking that he who carried the can had a right to feel the weight of it.[21]

At the BBC Lady Plowden had found that the governors had, in practice, virtually no power and no responsibility. Their function was to tell the staff how well they were doing and to defend them when attacked.[22] At the IBA, by contrast, she and other intellectually gifted and strong-minded members like Baroness Sharp, Mary Warnock and Lady Anglesey, although steered by staff, exercised a real influence over issues of importance in policy, in the showing or not showing of controversial programmes and in the selection of television and radio companies.

Baroness Sharp, for example, formerly Permanent Secretary at the Ministry of Housing and the redoubtable antagonist of her erratic minister, Richard Crossman, held strong views on all subjects. At one Authority meeting she made the mistake of holding a box of matches in her hand while banging the table to emphasise her opposition to what was being proposed. The matches caught fire and she had to be hurried from the room to receive first-aid treatment for burns. This did not deter her from returning to resume thumping the table with her bandaged hand.

Dr Carbery (Scotland) was another articulate and influential member, serving for nine and a half years in all, and Christopher Bland made an incisive Deputy Chairman during a term of seven years. Exceptionally, both spent time familiarising themselves with the work of the staff, and it was not helpful that the services of these knowledgeable members were lost in 1979 just when, in the award of new contracts, the Authority was preparing to make its biggest decisions for more than a decade.

There were no executive members of the Authority. The chief executive was the Director General, who headed a staff of more than a thousand. In 1971 the total was 1107, 70 per cent of whom worked in Engineering.[23] The other divisions were Programme Services (which included advertising control and audience research), Administrative Services, Finance and Information.

The quality and, by this period, the know-how of the senior staff were outstanding. Bernard Sendall, the Deputy Director General (Programme Services) from 1955 to 1977, had served with Fraser in the Central Office of Information. He was a civil servant with an uncommon talent for finding ways through or round apparently intractable difficulties, thus contriving that systems and relationships ran smoothly. Gentle but persuasive, he was held in almost universal esteem and affection. Evidence of the depth of his knowledge and understanding of the workings of Independent Television is to be found in his authorship of the first two volumes of this history, written after his retirement.

In May 1977 he was succeeded in the restyled post of Director of Television by Colin Shaw, who had been with the BBC since 1959, latterly as Secretary (1969–72) and Chief Secretary (1972–77). Shaw's experience in broadcasting matters proved invaluable in the major tasks of briefing the Authority to make new contract awards in 1980 and in formulating the structure of the fourth channel and the policies it was to pursue. In programme judgments he was a liberal, and he believed that the less closely his staff were involved in production the better they would be able to assess programmes offered by the companies for transmission.

The Deputy Director General (Administrative Services) from 1961–1977 and the sole Deputy Director General from Bernard Sendall's retirement until his own in 1983 was Anthony Pragnell, who had joined the ITA from the Post Ofice in 1954. He had a calm, rational, good-humoured approach to keeping the wheels of administration turning without friction. After wartime service as a navigator in RAF bomber command he could face the flak of the Authority's enemies with equanimity. Over a period of service of nearly thirty years, he became much more than an administrator. He was involved in all important legal matters and contributed to policy-making. Between the Authority and the companies he often formed a bridge of understanding.

Howard Steele and Tom Robson, successive Directors of Engineering, Tony Curbishley, Tony Brook and Roy Downham, successive Directors of Finance, Barney Keelan and Barbara Hosking, successive Heads of Information, and Bryan Rook, Secretary to the Authority, were other leading members of staff who ensured that the Authority was served by an infrastructure of sound management and professional expertise. Senior programme staff included Joseph Weltman and David Glencross: Glencross was to succeed Weltman as Head of Programme Services and then Colin Shaw as Director of Television. Among its backroom experts the Authority possessed a one-man think-tank in Kenneth Blyth, the Director General's chief assistant.

John Thompson, the first Director of Radio, was appointed in 1972 when Independent Local Radio became (unsought) another of the Authority's responsibilities and a conspicuous distraction from the business of television. Under the Sound Broadcasting Act effective from 12 July 1972 the Authority, which had borne the title of Independent Television Authority since July 1954, was renamed Independent Broadcasting Authority. In the following year the Television Act 1964 and the Sound Broadcasting Act 1972 were consolidated in the Independent Broadcasting Authority Act 1973. Under the terms of the 1964 and 1973 Acts the functions of the Authority were due to terminate on 31 July 1976, but in 1974 the 1973 Act was amended by two further IBA Acts. The first introduced a levy based on profits instead of turnover,[24] while the second extended the Authority's life until 31 July 1979 to allow the Annan Committee time to complete its deliberations on television's future. Further Acts were to follow in 1978 (extending the Authority's life until 31 December 1981), in 1979 (empowering the Authority to undertake engineering work for the fourth channel), in 1980 (empowering the Authority to provide a second television service and extending its life until 31 December 1996) and in 1981 (consolidating the Acts of 1973, 1974, 1978 and 1980).

To supplement its headquarters staff the Authority employed ten National and Regional Officers to cover the fourteen ITV areas. They were located in Belfast, Glasgow, Cardiff, Norwich, Birmingham, Newcastle, Manchester, Southampton, Plymouth and Leeds. There were three shared areas: the two Scottish regions; North-East England and The Borders; and South-West England and the Channel Islands; and until the 1980s it was thought unnecessary to have a Regional Officer in London. The task of these officers was to expound and interpret policies at meetings with the general public and local dignitaries and to keep the Authority informed about opinion in the region. Their function was partly executive and partly ambassadorial. They were the Authority's eyes, ears and mouth in the area and were expected to keep in touch with and keep an eye on the regional contractor. By some companies they were regarded with suspicion and kept at arm's length; others treated them almost as company board members.

Advisory bodies involved specialist representatives and members of the general public in the Authority's decision-making procedures. Among its twenty-four or twenty-five members the General Advisory Council, established in 1964 as 'unpaid antennae', contained both. Appointed by the Authority, they were selected from different parts of the country, from different age groups and from different walks of life. In 1978 the function of the GAC was explained as follows in a memorandum to a House of Commons Select Committee:

The valuable service which members of the Council can perform is primarily to express their own views (tempered by the views of other people with whom they are in contact). It may well turn out then that there is a consensus in the Council on a particular subject, however contentious it may have appeared to begin with; but, even if there is disagreement, the Authority and its professional advisers will be grateful for the expression of a range of views ... The position of the Council might in some respects be compared with that of a jury, on which twelve good men and true are not expected to be legal experts, still less to mirror public opinion exactly, but to use their common sense and reach their own conclusions on the evidence.[25]

There were also advisory National Committees for Scotland, Wales and Northern Ireland, which met regularly under the chairmanship of the members charged with making the interests of those countries their special care. Advertising, education, charitable appeals and religion were other specialist areas where the Authority and its staff took no awkward decisions before inviting and considering the advice of experts through its Advertising Advisory Committee or Medical Advisory Panel; its Educational Advisory Council or Schools Committee or Adult Education

Committee; its Central Appeals Advisory Committee or Scottish Appeals Advisory Committee; its Central Religious Advisory Committee or Panel of Religious Advisers.

In October 1971, when talk of a Broadcasting Council was in the air, the Authority instituted a Complaints Review Board under the chairmanship of its Deputy Chairman. The other members were the Deputy Director General (Administration), the Chairman of the General Advisory Council and one other GAC member. This board was empowered to investigate in depth any complaint relating to the content of programmes transmitted or prepared for transmission. Its terms of reference required it to review reports of complaints received and investigated by Authority staff and to consider complaints itself in cases when the complainant remained dissatisfied. Its findings were to be reported to the Authority for appropriate action.

Although membership was deliberately chosen from outside the programme division, the board could not be, and was not, described as an independent review body. An internal Programme Complaint Review Commission established by the BBC also came under attack for that reason, but the Authority insisted that, unlike the BBC, it had by Act of Parliament a purely controlling and supervisory function, separated from programme production, and could therefore investigate complaints about programmes objectively. This it did, and full accounts of the board's investigations, like those of the activities of the General Advisory Council, were published each year in the Authority's Annual Report. By 1980, when Mary Warnock took over the chair from Christopher Bland, members of the GAC were in a majority of three to two. But the Annan Committee held that the broadcasting organisations' own complaints bodies did not 'command public confidence'[26] and its recommendation for an independent Broadcasting Complaints Commission was accepted by the government.

The public viewed programmes made and/or presented by the companies and knew little of the Authority, but individuals who chose to write to the Chairman or Director General could be sure of a personal reply. The Authority prided itself on not resorting to the BBC system of diverting personally addressed letters to a correspondence department. It was the Director General himself who would write to explain to a correspondent that the Authority did not itself make programmes 'and so we must regretfully decline your invitation to film your home in Alfreton'.[27]

But it had not been the ordinary ITV viewer who formed the membership of, or gave evidence to, the Pilkington Committee. However popular, a television system under public control could not survive in the long term without the approbation of the educated, middle-class opinion-former, and this was a danger to which the Authority was once again alerted by the hostility of the report of the Select Committee on Nationalised Industries in 1972.

Commercial PR is inconsistent with the dignity of a statutory body, but during the 1970s the Authority, like other official organisations, grew increasingly aware of the value of building a favourable public image. If the public perception of its role was not to be distorted by political adversaries, sensational press reports and other misrepresentation, there was a need to disseminate more information, more widely, about its policies and plans. From 1972–73 its Annual Reports were therefore expanded to publicise its views and arguments on important issues. That year's report printed its observations to the Minister of Posts and Telecommunications on proposals for a Broadcasting Council; affidavits by its Chairman to the Appeal Court on ATV's Warhol programme;[28] and a statement of policy on television programmes for Wales. The next year's report contained plans for Independent Television during the period 1976–79; a further submission on ITV2; the Chairman's statement to the minister on communication with the public; and an account of the Authority's policy on diversification by the ITV companies. The 1974–75 report published the Authority's appraisals of the companies' performances before the renewal of contracts from 1976 to 1979.

In the summer of 1974 a quarterly IBA magazine, *Independent Broadcasting*, was launched as a forum for the dicussion of broadcasting issues. This was not a commercial publication:. some 10,000 copies were printed and distributed free of charge. In 1975 a series of IBA lectures was inaugurated, the texts being publicised in the quarterly. On a more popular level the annual handbook was rewritten and redesigned as a glossy paperback containing basic information about ITV and ILR services. Renamed *Television & Radio*, it had a print order of 35,000, was extensively promoted by means of on-screen announcements and sold at a low price. Other publications, issued free, included *ITV Education News* (114,000 copies distributed to schools), *ITV for Colleges*, *IBA Technical Review* and a range of leaflets and other *ad hoc* publications supplying the public and the trade and industry with engineering information. The Engineering Information Service at Crawley Court dealt with some 20,000 written and telephoned enquiries a year; questions about programmes and other non-technical matters were answered by Information staff at Brompton Road.

The Television Gallery on the first floor of 70 Brompton Road, formally opened on 25 September 1968 by Earl Mountbatten of Burma, President of the Society of Film and Television Arts, was a unique permanent exhibition illustrating and explaining the development of the medium. The result of three years of planning and building, it covered the whole story of television, beginning with its history and international aspects and continuing through its organisation and functions to methods of programme-making and audience research techniques. Although admission was restricted to small guided parties, it received more than 5,000 visitors during the first six months. Afterwards it was continually updated and in 1974 expanded to include radio and become the Broadcasting Gallery.

Relationships with the companies as a body were formally conducted through the Programme Policy Committee (PPC), which was chaired by the Chairman of the Authority and attended by the fifteen company chief executives, members of the Programme Controllers Group, the editor of ITN and officers of the Authority, and through the Standing Consultative Committee (SCC), which was chaired by the Director General, attended by company chief executives and relevant Authority staff, and dealt with matters not relating to programmes. It was through these committees that the Authority exercised its routine control over policy and general supervision of the composition of the network programme schedules, whose detailed planning was the responsibility of the central companies' Programme Controllers Group. Day-to-day dealings between the Authority's programme staff and the companies were designed to ensure that the Authority's final approval of schedules did not become merely a rubber-stamping formality.

The tool most effectively employed by Authority staff to bring aid and encouragement to programme-makers (as opposed to managements) and to realise the objective of stimulating more ambitious programme-making was the Consultation. Authority Consultations were held on an average twice a year for the purpose of exchanging information, discussing problems, disseminating ideas and pondering on future possibilities. It was of their essence that they were open-ended, leaving echoes and unanswered questions. A special audience research survey was normally undertaken in advance, and the conference programme of addresses and discussion groups with convenors was carefully structured. Producers and creative staff were the main participants. The Director General usually presided and ended proceedings with a summing-up. Afterwards a full report was distributed within the Authority and the companies and to those who had attended.

At a Consultation with regional programme-makers in Newcastle in November 1968 Fraser took the opportunity to reaffirm Independent Television's formative principles of dispersion and pluralisation stemming from the Authority's belief in the value of the separate regional communities in Britain: it was, he emphasised, the primary function of the central companies to produce network programmes and that of the regional companies to produce local programmes, and that distinction was fundamental. Another Consultation on regionalism was held at Stirling in January 1975, when the agenda for more than a hundred IBA and company staff included Reporting the Region, Involving the Region, Entertaining the Region, and Regional Scheduling and Programme Planning. By then the fundamental distinction had come to be modified. A quota for the production of programmes of local interest had been prescribed for each of the central companies.

Issues concerning the special classes of religious and educational programming were regularly and frequently debated at Consultations
attended by the specialist advisers. Other chosen subjects sometimes
related to areas of acknowledged weakness such as children's programmes,
discussed in 1965, 1973 and 1981. A conference on Sex Education for
Adults in 1973 decisively supported the view of the GAC that ITV should
fill a gap by providing programmes on this subject. During a Consultation
on comedy and light entertainment at Sheffield in October 1975 Denis
Norden contributed the view that the generality of ITV comedy was like
soda water when it ought to be like wine – all sparkle and no taste. A
Consultation on drama at St Andrews in July 1979 agonised over the single
play.

Business between Authority and government was conducted through the
ministers and ministries responsible for broadcasting. Their names, from
the birth of ITV to 1980, are set out in Appendix A. The Post Office was
considered insensitive and obtuse on all except the technical aspects of
broadcasting. On the extinction of the ministry of Posts and Telecommunications after a life of four and a half years, responsibility passed at the end
of March 1974 to the Home Office. There, as only one among numerous
responsibilities in a huge department of state, broadcasting sometimes
seemed to serve as a temporary respite for civil servants on the move
between the prison service and Northern Ireland.

Whether this was the appropriate ministry to have charge of television
and radio was a question often raised, but in practice it brought important
benefits to broadcasters. The Home Secretary was invariably a leading
member of the Cabinet, and a Roy Jenkins or a William Whitelaw had the
necessary seniority to protect broadcasting interests from interference by
colleagues. The Home Office, moreover, traditionally governed at arm's
length. It saw its function as legislative; its duty as policy-making. Beyond
that it did not seek to intrude on those whom it appointed, whether police
authorities or broadcasting authorities. It was staffed accordingly: in 1978,
for example, the total complement of Broadcasting Department, Home
Office (BDHO) was no more than thirty-two. Those in charge, though,
were normally of the high calibre and wide experience to be expected in a
large ministry, and Assistant Under-Secretaries such as Dennis Trevelyan
and Shirley Littler, however professionally self-effacing, exercised great
influence over broadcasting policies.

The most frequently advocated alternative to the Home Office was the
Department of Trade and Industry, whose minister was less powerful and
whose preoccupations were primarily commercial, or – what was most
dreaded by broadcasters – a media ministry with no function except that of
constant attention to the affairs of television, radio and the press.

Authority relationships outside the United Kingdom were fostered in
Europe through the European Broadcasting Union (EBU) of which ITV

became an associate member in 1956 and a full member in 1960. The Union was, and is, 'a professional association of broadcasting organisations whose objects are to promote cooperation between its members and with the broadcasting organisations of the entire world, and to represent the interests of its members in the programme, legal, technical and other fields'.[29] Because members were required to have responsibility both for the transmission facilities and for the programme supply of a national service, ITV's membership was a joint one in the name of ITA/ITCA, later changed to United Kingdom Independent Broadcasting (UKIB). This represented a major innovation for the EBU on two grounds: ITV was the first broadcasting system from a large country to be entirely self-supporting from advertising, and it was in competition with an existing member, the BBC, which had been a prime mover in the formation of the Union in 1950.

In the event these factors caused few problems, and ITV played a full part in EBU affairs through the technical, legal and programme committees. It was, however, precluded from attendance at the Administrative Council except by special invitation. This was the top executive body, where membership was limited to one member organisation from any one country. Only in 1982 were the statutes changed to permit UK membership of the Council in the names of the BBC and UKIB jointly.

Operationally, membership of the EBU was of particular value to ITN through its participation in the daily news exchanges organised through the EBU control centre in Brussels. This gave ITN access to European news material and the opportunity to provide, in exchange, material from its own bulletins. Through the EBU the ITV system also gained access to the Eurovision network for bilateral or multilateral programme movements in or out of the UK, a facility of central importance in the coverage of major sporting and other events in Europe. The EBU also negotiated collectively on behalf of its members the rights for showing international sporting events – the Olympics, the World Football Cup, the European Football Championship.

International acceptance of Independent Television as a national broadcasting organisation of equivalent status to the BBC was assisted by a growing recognition of advertising as a legitimate – and, indeed, often a necessary – form of financial support for public service broadcasting. ITV's experience as a pioneer in this field proved of great value in the councils of the EBU.

Although increasingly respected in its dealings with government and other bodies, in the exercise of its duties the Authority became unavoidably embroiled in controversy and attracted little public understanding, sympathy or thanks. It was a standing Aunt Sally ready-made for attack by one interest or another. If it stopped a controversial programme, it was denounced as a censor; if it allowed one to be shown, it was denounced for opening the floodgates of licence. To some it was a needlessly interfering busybody; by others it was scorned as a regulatory body too feeble, in the words of one Member of Parliament, to knock the skin of a rice pudding.

Important decisions affecting the delicate balance between freedom and control, and on the bestowal and removal of valuable contracts, were made by men and women appointed through the democratic process but transparently unrepresentative. Even a Home Secretary of the day thought some of their contract award decisions unfair.[30] Yet whenever reform was demanded the flaws in the system were seen on close examination to be less serious than those in the alternatives proposed. The conscientiousness of the Authority in interpreting and realising the will of parliament was an important, possibly decisive, influence in saving broadcasting from the fate suffered by, amongst others, education, local government and the economy during the upheavals of the 1970s – in an apt irishism, improvement for the worse.

7

EDITORIAL CONTROL OR CENSORSHIP

The broadcasting services in the United Kingdom have always been subject to controls and restraints additional to those which they share with other forms of public communication. From the early days of radio, well before the term 'mass media' became current jargon, it was accepted that for a public message system so widespread, so accessible and potentially there-fore so influential, something more was needed beyond the limitations imposed by the laws governing libel and obscenity, contempt of court and national security.

At the time of Independent Television's creation the Reithian image of broadcasting as a national institution devoted to the public enlightenment, uplift and culture still prevailed. But a service wholly dependent on revenue from advertising would, it was feared, tie the medium to the chariot wheels of commerce; market forces might play too great a role in the shaping of programme policies. The Independent Television Authority was therefore set up as an independent statutory body charged with responsibility for ensuring that programme providers complied with an array of controls and requirements going a good way beyond those which publishers and communicators in other fields had to observe.

It was the duty of the Authority to supply 'a public service for dis-seminating information, education and entertainment'; to ensure that the programmes which it broadcast maintained 'a high general standard in all respects' and 'a proper balance and wide range in their subject-matter'; and to secure a wide showing for 'programmes of merit'.[1]

Its ground rules for programme scheduling were accordingly strict. A third of every schedule had to be devoted to 'serious' programmes; there was to be no bunching of entertainment programmes; all programmes transmitted before 9 p.m. had to be suitable for audiences which included children; foreign programmes were restricted to 14 per cent of weekly output; each company was required to transmit a minimum number of hours of programming 'calculated to appeal specially to the tastes and outlooks' of the public in its area; a schedule was not considered properly balanced unless it contained, on every weekday, at least an hour of programmes made for and addressed to children; 'suitable quantities' of school, adult education and religious programmes were required, and these would not be so classified unless submitted at the planning stage for examination, discussion and approval by the Authority's specialist advisory bodies.

In addition the transmission – and in most cases the timing – of specific programmes was mandatory on all companies. The principal programmes so mandated in 1971, for example, were: *News at Ten*; one weekday play; one weekend play; two weekly current affairs programmes (*World in Action* on Mondays at 8 p.m. and *This Week* on Thursdays at 9.30 p.m.); a special documentary programme at least 39 times a year (13 of which had to be shown before 10 p.m.); a weekend arts special (such as *Aquarius* or Lord Clark's *Modern Painters*) to be shown before 10.40 p.m.; and *World of Sport* on Saturday afternoons. On rare occasions the Authority was prepared to consider allowing an individual company to opt out of a mandated programme, subject to there being a good local reason for offering an alternative and to the substitute programme being of equivalent character and quality.

The powers granted to the Authority to prescribe the nature, proportions and timing of different classes of programme constituted its main editorial function, but it was required by parliament to enforce other requirements. Advertisements had to be tightly rationed and supervised; they had to be clearly seen as separate from, and unconnected with, the programmes. Programme sponsorship was banned. Accuracy and 'due impartiality' had to be observed in news and the discussion of current public issues. The Authority had a right and duty to satisfy itself whether any programme or programme item was likely to incite to crime or public disorder, or to be offensive to public feeling, decency or good taste.

By virtue of all these powers the Authority became not merely a printer or publisher. It regularly made decisions about the content of programmes and the balance of programming 'in much the same way as does an editor about the contents of his newspaper'.[2] The mechanics of day-to-day control were partly formal, partly informal. For example, a synopsis of every play and every episode of every series and serial had to be submitted to the Authority's programme staff, who would then raise with the producing company any problems which they foresaw. The companies would in turn raise any problems which they themselves anticipated. Since the early 1960s divisions between the creative and the watchdog functions had become blurred in a dialogue between people with common aims and interests. Self-censorship was practised by writers, directors and producers, and each company operated its own internal channels of control, even though the decisions which emerged could be overturned by Authority staff or members.

Descriptions of such exchanges formed the substance of so-called

Intervention Reports which were presented to Authority members for formal noting at their meetings. For the most part these were records of judgments made and decisions already taken. On exceptional occasions, when no agreement had been reached, a special report would be submitted with an Authority staff recommendation. This could lead to the involvement of company senior management and the viewing of the programme by members. Viewings by the full Authority were normally discouraged by the Director General and his programme staff, but sometimes recommended and requested. Between 1971 and 1977 such previews averaged two a year.[3]

These private interventions, whether by Authority staff or members themselves, were judged to be more effective than public criticism, which was usually reserved for contract renewal time or published assessments of company performances. Unlike the Press Council, the Authority was part of the system and its power to influence programme output for the better depended more on securing the co-operation of programme-makers than on castigating them. But were its influence, its powers of persuasion, its sanctions short of franchise removal strong enough, or should it take a larger part in the programme scheduling process, for which it had ultimate responsibility?

Those were the questions which the Director General urged members to consider towards the end of 1972. Three times during that year (over its first statement about derestriction, over the timing of the lunchtime news, and over the Christmas schedules) the Authority's right to have its way had been asserted in the face of deputations and other protests suggesting that the companies knew best and that scheduling interventions by the Authority were blows to their professional pride.

The companies could – and did – argue that their schedules were drawn up by people with considerable experience who were in touch with programme production and knew what was available. They were emphatic that ITV's fortunes depended very largely on scheduling competitively against a rival who had the advantage of two channels. But the Authority saw reluctance to lose audiences at any time as a weakness, and it was becoming dissatisfied with a system of scheduling by five people of equal status on a basis of fair shares for all regardless of merit.[4]

The collective programme-planning machinery, as developed piecemeal during the 1960s, was the Programme Policy Committee (attended by representatives of the Authority and all the companies), the Programme Schedule Committee (five members of the Authority), the Network Planning Committee (all the companies), the Network Programme Secretariat (ITCA) and the Programme Controllers Group (the five central companies only). It was this last body which drafted the network schedules and which was expected to consult regularly with the Authority's programme staff so that the Authority would have advance information on what was being planned. All this was the outcome of an Authority decision in 1967 not to appoint its own Director of Network Planning and establish its own Network Programme Secretariat.

In 1972 matters had been brought to a head by the late arrival of the Christmas and winter schedules at Brompton Road (with the implication that Authority approval could be taken for granted) and the Authority's subsequent insistence on the last-minute insertion of a Glyndebourne performance of Verdi's *Macbeth* from Southern Television, to be shown at 9.30 p.m. on Wednesday, 27 December. The companies' anger at this intervention was great and not concealed from the press. This in turn angered the Authority, which believed that its command of ITV would be threatened if it was appearing to acquiesce in the five central companies having *de facto* control of the schedules, with its own staff unable to do much more than warn them of the danger points and the Authority itself only able to prescribe certain requirements of general application and make occasional late changes, sometimes against heavy resistance.

During the year the Authority's Deputy Director General (Programme Services) and Head of Programme Services – Bernard Sendall and Joseph Weltman – had joined the Programme Controllers Group for part of their meetings on six occasions. Now the Authority demanded a closer and more systematic relationship and decided to strengthen its programme staff by the appointment of a specialist scheduling officer.[5]

The companies took alarm at the prospect of any deeper intrusion into their professional work and any further impediment to successful competition with the BBC. A large measure of freedom to determine the scheduling of the programmes which they produced and acquired was essential to their operations and they saw no good reason for Authority interference provided that the existing rules and contractual obligations were observed. John Freeman put their case at a meeting of the Authority to which the chief executives of the five central companies were summoned on 3 January 1973. He argued that the scheduling of an 'unpopular' programme at a 'popular' time was not of itself good broadcasting. He conceded that there were exceptions like *This Week*, but minority programmes like *Aquarius* or *The Scientists* were just not suitable for showing in peak hours. The companies, he said, could see no automatic virtue in scheduling programmes which the majority did not want to see at a time when that majority wanted to view television.

The five had come armed with two peace-making proposals. First, they promised to produce and schedule at favourable hours six programmes of distinction – described as 'television events' – every year. When asked for examples, Sir Lew Grade announced ATV's plans for productions of *A Long Day's Journey into Night* (with Laurence Olivier), *The Merchant of Venice* (with Laurence Olivier and Joan Plowright), The Royal Shakespeare Company's *Antony and Cleopatra*, and *Carmen* from the Royal

Opera House. The second proposal was for the Authority's Deputy Director General (Programme Services) to join the Programme Controllers Group as a permanent member in his own right, so that in future the Authority would have full access to all the group's plans and deliberations.[6]

Agreement was reached on these new arrangements and they were confirmed at a PPC meeting the following month. In addition Frank Copplestone, a former ITA officer, was appointed Director of Programme Planning within ITCA's Network Secretariat and Chairman of the Programme Controllers Group.[7] This met the Authority's view that network scheduling required one or two minds undistracted by other preoccupations and not committed to particular companies. In April, after studying a memorandum from the companies on Christmas schedules, the Authority told them bluntly that equating 'a good Christmas schedule' with an hour-by-hour attempt to get a larger audience than BBC1 was not the right approach.[8] This formally concluded the controversy, which was not the crude conflict of views sometimes supposed: for the companies were well aware that ITV had public service obligations and the Authority was well aware that the main thrust of ITV was to make what was popular good and what was good popular.[9]

What then was the outcome? Sendall's role in the Programme Controllers Committee (and the one which he was best suited to play) was to guide the five controllers in the formulation and presentation of their proposed schedules in such a way as to anticipate and avert future collisions. 'One-off programmes of distinction' became a regular ingredient in the schedules. To qualify as 'special events' these productions had to conform to the requirements that they should be in some respect exceptional in quality and content, should usually be of more than customary length and should be transmitted at, or near, peak viewing time. They included Glyndebourne operas from Southern, ATV's Shakespeare plays, Royal Ballet performances from ATV and Thames, plays by Christopher Fry about the Brontës (YTV), LWT's *Death of Adolf Hitler* and some ambitious documentaries: *The State of the Nation* series (Granada), YTV's *Johnny Go Home* and *General Strike*, LWT's *Akenfield* and ATV's *Opium Warlords*. It may be doubted whether such a harvest of prestigious creativity would have enriched ITV's schedules, even at a time of anticipating new legislation and campaigning for a second channel, but for the Authority's action over the 1972 Christmas schedule and determination that the service must move forward and not simply try to repeat past successes.

But eight years later members were once again complaining about predictable and unambitious schedules and facing the limitations on their powers of control:

There are almost always some programmes in the schedules which set members' teeth on edge because they seem unworthy or shoddy or too obviously possessing a vulgar streak. *Crossroads* and *Stars on Sunday* had long runs in this role, and children's programmes like *Tiswas* and *Fun Factory* are now the stars or villains of the piece. It is never easy to see how outside pressures can make these programmes better. What they need is sensitive producers who, without losing whatever common touch they may have, can eliminate the incompetence or crudity which jars on some of their viewers. Such programmes cannot well be 'banned'. The Authority can refuse to accept another series, or cut down the amount which it will accept (as it did with *Crossroads* and *Stars on Sunday*). Moreover, it can, and does, send a stream of comment and criticism towards the management, who are sometimes themselves equally helpless but occasionally also uncaring. Where we are certain that our dislike of a programme is not the result of a narrow aesthetic judgment or a class judgment, but that the programme is substandard or even demeaning, we might more often say firmly that unless it is improved it will not again be accepted for the schedules. But those who love it will be deeply pained; and, had we actually insisted on *Crossroads* being scrapped altogether, there would have been quite a sharp tussle with the company about our powers to do this, and a very loud cry from many people (including particularly the elderly and the lonely) who, for all its faults, find it a very dear companion.[10]

These were broad issues, but the time of Authority staff was more often devoted to the detail of specific instances of possible breaches of one or other section of the Act.

In parliament's words, nothing shall be included in any programme which 'could reasonably be taken to state, suggest or imply that any part of any programme... has been supplied or suggested by any advertiser; and, except as an advertisement, nothing shall be included in any programme broadcast by the Authority which could reasonably be supposed to have been included therein in return for payment or other valuable consideration'.[11]

There is scanty evidence of any improper practice under this heading. Cases recorded were either indadvertent or trivial, mostly concerned with the use of brand names. This became a problem when in November 1972 Thames's *Good Afternoon* programme embarked on a series of consumer advice items. In the early 1970s presenters of programmes with prizes, such as *The Golden Shot*, were told to moderate their enthusiasm when describing the prizes and, if possible, to avoid mentioning the brand names. The emergence of the T-shirt as a medium of communication caused difficulty. The producer of Yorkshire Television's local news magazine *Calendar* was forced to admit an error when in November 1973 the heavyweight boxer Richard Dunn was interviewed with the name of his employers, J. Gibson Scaffolding Company Ltd, displayed across his chest.

A similar incident occurred in February 1977 when, in the Tyne Tees programme *Be a Sport with Brendan Foster*, the shot-putter Geoff Capes presented the viewer with a beefy front bearing the legend Dewhurst the Butchers. In August 1980 Judith Chalmers earned a reproof when her description of the outfits worn by competitors in the *Miss United Kingdom* contest contained the information that they had all been chosen at Richard Shops.

In this context growth in the commercial sponsorship of sporting and other events was a continuing worry to ITV and BBC alike when covering events which were not, to a significant extent, under the control of the programme-makers. The elimination of 'an undue element of advertisement'[12] in outside broadcasts was not easy. Sponsorship in the 1970s rapidly came to cover a wide range of sporting and cultural events from horse racing, athletic meetings, association football and cricket Test matches to drama festivals, opera and art exhibitions, and there can be little doubt that in some cases the belief and expectation that the sponsored event was likely to attract television cameras proved a powerful incentive to generosity. In every case the Authority attempted to interpret the Act by applying three main criteria. Would the event have taken place anyhow, irrespective of the likely presence of the cameras or the sponsor's support? Did it already command wide public interest? Could it be held to represent, or include, an unacceptable level of publicity for the sponsor? In most cases none of these questions could be answered with an unequivocal 'yes' or 'no', and the rulings of the broadcasting authorities inevitably appeared at times self-contradictory.

Billboards around sports grounds were contentious. The BBC was opposed to any advertisements; the companies objected to advertisements for which they received no payment; the Authority had to determine whether or not the advertisements were 'undue'. The BBC made a stand aganst the display of condom advertisements on the bodywork of racing cars; the IBA refused to permit display advertising for Biostrath at a tennis tournament because its medical advisers had vetoed acceptance of commercials for the product. But both services regularly screened major public sporting occasions sponsored by tobacco manufacturers, who prudently avoided any direct allusion to cigarettes, advertisements for which on ITV had been outlawed by ministerial directive in 1965.

An appearance of sponsorship sometimes occurred when programme-makers were afforded facilities of recognisable pecuniary value in the course of coverage of matters related to the activities of an official body or public institution. When in November 1974 Scottish Television produced a documentary *Why Defence?*, the question arose how far NATO had (albeit indirectly) contributed to the programme costs. Camera crews had been given free flights in RAF aircraft to and from West Germany; some of the film used had been obtained at no charge from the Ministry of Defence. The conclusion reached between company and Authority staff was that, so long as it was demonstrably clear that no conditions had been attached to the help given, and that the selection, editing and presentation of all the material shown had been at the untrammelled discretion and editorial responsibility of the company, then no breach of the Act could reasonably be alleged.

But in January 1978 a YTV *Calendar* programme ended its first half with a trailer announcement that the second half would be devoted to a discussion of the Health Education Council's *Keep Fit* campaign. A commercial break with a paid advertisement for the Council's general work followed. Then, after the break, came a discussion of the campaign by four studio guests. However worthy the cause, the appearance of sponsorship was blatant, and the producer was taken to task.

'Opinion advertising' was another problem. The Authority rejected an approach to Ulster Television by the Stormont government in November 1971 with the aim of buying commercial time for a series of slides designed to discredit the IRA. A slide with the caption 'The IRA's contribution to employment' was to be followed by one showing a destroyed factory; another with the legend 'The IRA's final solution to the housing problem' was to be followed by a picture of a graveyard. These violated the legal requirement of 'due impartiality' in 'matters of political or industrial controversy or relating to current public policy'.[13]

Another difficulty under this rule arose over the semantics of a series of paid public service announcements from the Central Office of Information about housing legislation. The scripts had been written in popular, non-legalistic language to help viewers who were tenants to a clearer understanding of their rights under the 1972 Housing Finance Act. The phrase 'fair rent laws' was used. Immediately after the first transmission in autumn 1972 Anthony Crosland, the Opposition spokesman on housing, complained that the words 'fair rent laws' constituted a tendentious political statement and were therefore in breach of the Act. Legal advice was taken and it was decided that, by themselves, the words 'fair rent' were acceptable, being part of the current vocabulary of housing legislation. On the other hand, to describe the legislation itself as the 'fair rent laws' could well be construed as underwriting government policy and therefore not politically impartial. No further transmissions were permitted until the scripts had been revised.

Behind the legislation covering such cases lurked the persistent fear of abuse of power. Gaining privileged access to the television screen through the purchase of airtime or the granting of favours in order to promote a cause was a development which opponents of commercial broadcasting had apprehensively predicted. The principle at stake and the controls to be applied were the same whether the cause was commercial or political; good or bad.

In May 1977, in the religious magazine programme *Saints Alive*, ATV planned a weekly feature on the plight of prisoners of conscience in overseas countries. The information was to be supplied by Amnesty International. Viewers who wished to add their protest would be invited to write courteously to the head of state in question. Consulted by the producer, Authority staff expressed doubts about a religious programme which might seem a form of advertising directed to a political end. As a result the invitation was altered, and viewers were given a neutral address in the United Kingdom to which to write for more information.

The proclamation of an article of faith could also be a matter of public controversy. In January 1978 during a series of Scottish Television epilogues under the generic title *Late Call* Monsignor Brendan Murphy devoted one broadcast to an exposition of Roman Catholic teaching on abortion. This provoked some strongly worded complaints and, although the Authority's Scottish Officer had been consulted in advance and raised no objection, STV's senior management decided to avoid such incidents in future, mindful that they were serving an area in Glasgow where religious differences readily took on political overtones.

Sometimes bureaucratic niceties took precedence over humanitarian considerations. Rightly anticipating a strong public reaction to the harrowing scenes presented in a *This Week* programme in October 1973 on famine in Ethiopia, the producer proposed to end the programme with an announcement that Oxfam was co-ordinating all the charitable relief work. But the Authority's officers felt that this would turn the programme into a thirty-minute charitable appeal which had not been previously considered by the Central Appeals Advisory Committee. They therefore refused his request, ruling that if the Disasters Emergency Committee (which included Oxfam) wished to broadcast an appeal on behalf of the famine victims it should approach the Authority through the normal process.

Immediately the *This Week* transmission ended, company switchboards were swamped with calls from viewers asking how they could help. A compromise was then agreed which permitted the broadcast, just before *News at Ten* that evening, of a brief announcement of Oxfam's role. During the following week a request by the Disasters Emergency Committee for airtime for an appeal on behalf of the Ethiopian famine victims was promptly granted.

Less insensitivity was shown six years later when *Year Zero*, an ATV documentary about famine in Cambodia, was broadcast on 30 October 1979. Although a public response to the film was expected, no arrangement was made for a Disasters Emergency Committee appeal. Instead it was agreed after discussion between the DEC and the Central Appeals Advisory Committee that donations could be invited via an ATV Post Office box number. Through this £1 million were received.

The provision in successive Acts prohibiting programme appearances by

Authority members and staff and directors and officers of ITV and ILR companies expressing their personal opinions on matters of current public policy was much resented. It frequently silenced those well qualified to contribute authoritatively to a public debate.

In Thames's historical documentary series *The Day Before Yesterday* (1970), for example, the producer was unable to include interviews with several eminent public figures about politically important events of which they had first-hand knowledge. One of these was Lady Gaitskell, who was also barred, in January 1972, from appearing in a *Today* programme on the Indo-Pakistan war. The disqualification in each case was her non-executive membership of the board of Yorkshire Television. In January 1975 the director of Age Concern was barred from appearing in an item on housing for the elderly in STV's local *Housecall* series because he happened to be a director of Radio Forth. In June 1976 Granada's *Reports Action* was reproved for including Lord Winstanley, a regular contributor but also a director of Manchester's Piccadilly Radio, in a discussion about candidates for the Liberal Party leadership. In November 1979 ITN was similarly rebuked for using a recorded telephone conversation about the Blunt affair with Sir John Colville because he was a director of Thames Valley Radio.

The Authority and the companies were at one in their irritation over this unnecessary legal restraint, which on occasions merely resulted in ITV personalities making their contributions to a public debate through BBC channels. On other occasions defensible methods of circumvention were employed, as in January 1976 when the Authority's Head of Advertising Control was invited to appear in LWT's *The London Programme* in order to explain why some government-sponsored advertisements had been rejected. He was not able to appear, but his explanation of the Authority's position was read out.

This restraint was eventually relaxed in the Broadcasting Act 1981, which excluded from ITV programmes any expression of opinion on matters of current public policy (other than broadcasting) by the Authority, its Fourth Channel subsidiary or its contractors.[14] Statements on such matters by members, directors and officers in their personal capacity were no longer prohibited.

The main bulwark against abuse of 'media power' lay ready-made in the insistence of the law on due impartiality. Yet it was hardly possible to define and identify 'impartiality' to the general satisfaction, and the qualification implied by the word 'due' introduced a sizeable additional element of subjective judgment. Legally, due impartiality on the part of those providing ITV programmes was what the Authority deemed to be duly impartial, but its judgments were inevitably, and rightly, open to challenge. 'Due' enabled broadcasters to come down on the side of civilised moral standards. They were not required to be impartial about murder or racial discrimination. They were expected to share the assumptions of the parliamentary democracy in which they operated and not give equal time and weight to those who would destroy it.[15] Nor were programmes required to be indifferent or bland. 'Due' was taken to signify a balance of objective presentation within a single programme or programme series.

In August 1968 Authority staff gave a ruling on the dueness of the impartiality of an LWT film portrait of the actress Melina Mercouri. She expressed violently hostile views about the regime of the Greek Colonels in her native land, and transmission was justified on the basis that, although it might be held that the programme was lacking in impartiality, it was defensible on the grounds that a true picture of Miss Mercouri's personality could not possibly be presented without including an account of her attitude and self-justification as a voluntary political exile.

But in the same month a less tolerant view was taken of a Thames *This Week* report on a recent tour of the British Lions rugby team in South Africa. The South African teams, it claimed, had been encouraged to use all methods, however brutal or unfair, in order to vindicate a political and social system which was widely condemned. The company was told that the programme could not be defended, and Thames's Managing Director wrote a letter of regret to the South African ambassador.

The Anglia documentary *Red under the Bed*, produced by Brian Connell in November 1973, created a storm. It was an investigation into the systematic penetration of the British trade union movement by the Communist Party. Anchorman for the programme was Woodrow Wyatt, a right-wing Labour politician, broadcaster and journalist well known for his anti-communist views. His contribution was to present and link a series of extracts from interviews with leading communists, trade union officials, industrialists and journalists. These extracts were clearly chosen as evidence to support the case that the Communist Party had the consistent aim of fomenting strikes and industrial unrest, and that it did so by getting its members and fellow travellers into positions of influence in the unions. Reference was made to the notorious falsification of votes in the 1959 election for the post of general secretary of the ETU, as well as to the more recent instances of violence by flying pickets.

Authority staff were invited to a preview. They expressed concern over the extent to which Wyatt, as presenter, had been allowed to voice his own opinions, and over the editing of interviews. Had all those whose revealing statements were featured been made aware of the nature and purpose of the programme to which they were contributing? Were the used extracts fairly representative of what had been said or had they been deliberately taken out of their qualifying context in order to support an *a priori* case? The latter could all too readily be suspected if the programme presenter in his linking commentary sought to air his own views rather than allow the evidence to speak for itself. Changes in Wyatt's contribution were requested and agreed with the producer. In a letter to *The Times*[16] Wyatt complained that this was an act of censorship, but had he been one of those interviewed, and not the programme presenter, his opinionated statements and inferences would have been left untouched.

He was not the only presenter of programmes to find compliance with this requirement of the Act unduly restrictive. In June 1977 the IBA's Director of Television was instructed to write to the Director of Programmes at Thames to let him be in no doubt about the strength of feeling among members of the Authority about the intrusion of the personal views of the presenter/reporter, Jonathan Dimbleby, into *This Week* programmes.[17] The most serious accusation of bias had come from Iain Sproat, a Conservative MP, who complained to the Chairman of the Authority about a *This Week* programme, broadcast on 20 January 1977, on the abuse of social security benefits. Making charges which reflected the kind of anger felt by some participants in other ITV and BBC current affairs programmes during this period, he wrote at length and in detail, concluding:

In short, the Thames programme:
(a) was totally different from what I had been led to believe would take place;
(b) distorted the point of my speech and campaign about social security;
(c) gave me no opportunity to reply;
(d) cast offensive doubts on my speaking the truth; and was generally an attempted hatchet job on myself;
(e) failed to make any objective or impartial investigation;
(f) was a vehicle for the personal views of the reporter, Mr Dimbleby. The closing minutes of the programme was surely straight, personal editorialising by Mr Dimbleby in a way that should be totally unacceptable – whatever his views happened to be.[18]

The IBA's reply did not accept that Sproat had been deliberately misled, nor that his views had been distorted, but it did accept that Dimbleby's closing remarks had been out of place.[19] It offered the complainant the opportunity of appearing in another *This Week* programme on the same subject, to which the company agreed but from which it subsequently excused itself. Dimbleby was far from contrite, believing that, by pointing to the heavier losses suffered by the Treasury from tax-dodging, he had legitimately set allegations of social security scrounging in context.[20]

The problem of reporting on apartheid was a running source of worry and, at times, equivocation. The Authority consistently pressed for the inclusion of statements in explanation and possibly even justification of the South African government's policies. Early in 1978 an ATV series, *The South African Experience*, occasioned a lengthy discussion at the Programme Policy Committee, when the Authority was particularly exercised by the fact that those working for progress in South Africa were seldom heard in programmes. ATV's Programme Controller pointed out that the company's intention that those very people should be heard in the last programme in the series had been frustrated by their arrest.[21]

At the same meeting William Brown of Scottish Television drew attention to a programme on Argentina which his company was preparing in advance of the forthcoming World Cup football competition. What was one to do, he asked, when representatives of a government turned down an invitation to take part in a programme. The producer had wished to include an official reply to charges of brutal political repression made in a report from Amnesty International. In the event the programme – *The Other Side of the Peso* – was previewed by the Authority's Scottish Officer and transmitted in April 1978 with no balancing response since no one could be found to refute the indictment.

It was another Amnesty International report which led the *Sunday Times* to headline a leading article in June of that year 'TV's Redundant Censors' and inform its readers that 'the Independent Broadcasting Authority is one of the biggest menaces to free communication now at work in this country'.[22] Amnesty had investigated allegations of ill-treatment of suspects held for questioning by the Royal Ulster Constabulary at the Castlereagh Interrogation Centre and, by and large, found the charges proven. The *This Week* team had in preparation a programme discussing the report, whose contents had already been leaked and widely featured in Irish and British newspapers.

The Authority insisted nonetheless that transmission be delayed until the report was published, arguing an important difference between the leaking of selected items as news reports and mounting a public debate on a controversial issue before all the material on which the debate would be based was available. Thames's ACTT members protested at this postponement by refusing to put out a substitute programme, and the company was off the air for twenty-seven minutes.

At Thames it was thought that the Authority had simply lost patience with continual attacks by *This Week* journalists on the work of the security forces in Northern Ireland (this would have been the third critical programme in succession), and that the stated reason was no more than a pretext for suppression.[23] At Brompton Road the Authority's own programme staff were deeply upset by its decision. They believed that a detailed statement in the House of Commons by Roy Mason, Minister for Northern Ireland, on the afternoon of the day of intended trasmission changed the position, even though the report had still not been published, and that a programme that evening responding to the ministerial statement was not only justifiable but perhaps even desirable.[24] By that time, however, the members' decision had been taken and made public.

The Authority's discomfiture under attack from all sides was compounded when the BBC's *Nationwide* programme broadcast excerpts from the unshown *This Week* programme, cheekily introducing them with the words: 'The programme the IBA bosses won't let you see.' Nor did it gain much comfort from a letter from the BBC's Chairman to *The Times* in which he approved the use of the extracts while expressing regret over these words.[25]

Jeremy Isaacs, Thames's Director of Programmes, had authorised the handing over of six minutes of the *This Week* programme material to the rival channel, and this act of defiance was symptomatic of the irreconcilable differences between current affairs journalists and any regulatory body, however liberal-minded. An enforced postponement was a 'ban' or a 'gag', and the Authority had suffered a similar drubbing from the press in 1974 when *Cudlipp's Crusade*, a strongly flavoured political programme, was found acceptable but delayed from September to October to avoid the period of a general election.

This was the decade of Watergate, and the mood of the time favoured a journalistic ethos which bestowed more glamour and acclaim on investigative journalism and committed reporting than on the due impartiality which parliament required the Authority to impose on the journalists of Independent Television. The case against members of the broadcasting organisations making programmes which expressed their own commitments was forcefully argued by Lord Hill in the House of Lords debate on the Annan Report,[26] but his dictum that the great publicity apparatuses of ITV and BBC should not be used to promote the personal views of staff, who had not been appointed for the wisdom of their opinions, was not generally accepted in current affairs departments.

Within ITV the most expert practitioners of the adventurous committed programme were to be found in Granada's *World in Action* team. Volume 2 of this history has recorded how no single ITV series was the occasion of more discussion between Authority and company,[27] and this dialogue of the 1960s continued unabated through the 1970s. The Authority alternated between sorrow and anger. Granada, valiant for truth, remained aggressively unrepentant. But both parties strove to avoid a revival of earlier acerbity.

In May 1969 *World in Action* made a filmed report of conditions in the southern Sudan where a rebel provisional government had been set up by dissidents. It recorded a number of unsubstantiated charges against the central government in Khartoum, and no opportunity was given to government spokesmen to comment. The programme commentary did, however, make clear that no direct first-hand evidence had been found to support the charges. Authority staff had no doubts that the film gave a one-sided picture and was not impartial, but in view of the quality of the programme in terms of honesty and concern, and the degree of detachment which marked the commentary, transmission was approved after a preview.

A less generous view was taken of a 1971 *World in Action* investigation into Idi Amin's military takeover in Uganda, entitled *The Man Who Stole Uganda*. It portrayed the general as corrupt, despotic and crudely brutal: much as subsequent history revealed him to have been. But a number of prominent persons with first-hand knowledge of the country – including Lady Cohen, the widow of a former governor – protested to Granada that Amin was a friend of Britain and a fine man. They wrote to the Authority complaining of an incomplete, *parti pris* and largely misleading impression of the situation in Uganda. Although there was no reason to doubt the accuracy of the facts reported, the programme erred, in the opinion of the Authority, through sins of omission. It seemed to be one more example of an apparently persistent tendency in *World in Action* accounts of overseas countries to pick out only what was demonstrably wicked: a viewer relying on these programmes for his knowlege and understanding of the world would receive a very distorted picture. Granada argued in reply that the public had access to many other sources of information, and in the case of evils such as Amin impartiality was out of place.

A *World in Action* campaign on behalf of low-paid workers on British-owned tea plantations in Sri Lanka and southern India was featured in two programmes in March 1975. The Authority agreed that this was a wholly justified undertaking but, in response to a complaint from the Co-operative Wholesale Society (one of the owners whose estates were filmed), felt obliged to concede that the programmes fell short of expected standards of fairness and impartiality. The problem was one of context: 'It is perhaps an inherent drawback of a pictorial medium such as television that it communicates in terms of the concrete and particular and is ill adapted to the presentation of generalisations. For that reason the Authority believes that it is essential for programmes to set in context the particular details selected.'[28] The CWS was invited to take part in the second programme, but was distrustful and preferred to make its case in a long and detailed statement to the press, expressing disappointment that the Authority's judgment was not made known by an 'on-air' announcement.

On the thorny subject of Vietnam the members of the Authority were worried particularly by two programmes, one in late September 1970 from Granada and one in early October from Thames. The Granada programme, *The Quiet Mutiny*, took the form of a personal report by John Pilger of the *Daily Mirror* on the low and declining morale of US troops in Vietnam. It was based on a large number of conversations between Pilger and army conscripts, and it reflected an almost total disillusionment with Pentagon policy. Nowhere was there any mention or evidence of the crusading ideals with which the American government had embarked on the campaign.

The Thames programme took the form of a film report from communist North Vietnam by Clive Jenkins, a trade union leader who had recently visited that country as a member of a British trade union delegation. At the request of Authority staff it was prefaced by a statement that the film had been shot entirely in North Vietnam with North Vietnamese government permission and approval, and a tailpiece was appended emphasising that only one point of view had been presented. Further reports on Vietnam which would put the US and South Vietnamese viewpoints were promised.

Both these broadcasts raised in an acute form the problem of so-called 'one pair of eyes' programmes. Even when clearly so labelled, how could they possibly be squared with the law's requirement of due impartiality? But if they were disallowed would this not constitute an intolerable restriction on freedom of speech? On every occasion when an extreme view was broadcast and not challenged, must a complementary programme be devised so that an opposite opinion could be aired?

Radical Lawyer was an ATV documentary, broadcast in June 1972, in which Benedict Birnberg, a solicitor practising in south London, gave his personal, highly critical view of the way in which, in his experience, the law and police operated in relation to coloured people and members of the working class. Authority staff, invited to preview, were not persuaded that some balancing comments from the Chairman of the Police Federation and from a distinguished magistrate did enough to ensure due impartiality. But transmission was approved subject to the inclusion of an opening announcement that the opinions were personal to Mr Birnberg, 'although they do reflect attitudes which other people share'. Some reassurance was also found in ATV's intention to screen a documentary later that year which would feature a lengthy interview with the Metropolitan Police Commissioner.

On 8 June 1972 *Radical Lawyer* was made the occasion of an informal meeting of senior programme staff from the Authority and the five central companies to discuss the incipient problems of the 'personal view' programme. It seemed more than likely that this would be a developing strand in factual programming; and by no means an unwelcome one, despite the problems. During the following ten years there were to be many such meetings, not all informal. On four separate occasions during 1975 alone – twice at Authority meetings, twice at meetings with company managements – there were serious and prolonged exchanges of conflicting views about the claims of such programmes to a regular place in the schedules.

The most generally favoured solution was to encourage fuller use of the increased freedom implicit in the 'balance in series' concept. Theoretically at least, the idea of exposing a wide range of differing opinions over a period seemed attractive. Producers in the companies were convinced that a programme limited to a single viewpoint, persuasively argued and presented, would be more appealing to viewers than attempts to achieve a meticulous quasi-mathematical balance between opposing views on a controversial topic within a single broadcast.

Early in these discussions it was agreed that the viewer should always be made aware of what he was being offered: an invitation to view through 'one pair of eyes'; to hear one opinion among others on a topic of public interest and concern. Preferably therefore the programme should be clearly identified with a named person. For a similar reason it should not appear in the guise of a regular current affairs or documentary programme; or even, if it could be avoided, in one of the slots normally occupied by *World In Action, This Week* or *Weekend World*. Because they had been so scheduled was one reason why the Authority had found the John Pilger (*World in Action*) and the Clive Jenkins (*This Week*) programmes about Vietnam unsatisfactory.

Throughout the 1970s and into the 1980s it was with programmes from ATV featuring Pilger that the debates and heart-searching over personal view programmes were most concerned. Here was something patently different from the airing of opinion by someone personally involved in a particular issue. Here were no opinionated but relaxing traveller's tales like Alan Whicker's. Here was a highly talented crusading journalist deliberately setting out to convert the viewer to his way of thinking on some aspect of home or foreign affairs where he believed that those in authority were behaving in a wicked or criminally incompetent manner. It was apparent that he considered it to be one of his missions in life as a journalist to seek out such scandals and bring them to public attention with all the considerable hyperbole at his command.

Pilger dealt in this way with topics as diverse as the workings of English conspiracy law; nuclear power; the failings of the National Health Service; the evils of the remand system; the failure of land reform in Mexico; child victims of thalidomide; oil companies and North Sea oil; and the USA's alleged use of food aid as a political weapon. All were presented as documentaries with supporting evidence the clear purpose of which was to persuade the viewer that what was being claimed was not so much opinion as irrefutable fact, not so much one aspect of the matter but the whole truth. The evidence was selected uncompromisingly, very often for its strong emotional impact in visual terms (victims of famine or nuclear explosions, inmates of hospital wards or thalidomide cripples) and accompanied by a colourful and heart-wrenching commentary.

It might have been wiser to treat the Pilger programmes as a class of their own and not allow them to become confused with run-of-the-mill personal view programmes. For they were essentially exposures, differing fundamentally from other personal view programmes. Yet they were seen by most viewers as documentary or current affairs programmes of the normal kind. The device of labelling them as personal expressions of opinion made little difference.

Discussion led to the introduction of a new category of programme airing a variety of divergent personal views, but this proved to be no way out of the problem. The crowded schedules permitted only a limited number of such programmes, so that the regular appearances of Pilger assumed a disproportionate weight, and the other personal view programmes, however idiosyncratic, were different in kind from his. They were for the most part not documentaries presenting detailed evidence in support of an accusatory case. They were individual expressions of attitude or critical judgment towards matters within personal experience: in the case of such as Auberon Waugh and Jack Trevor, taking the form of pasquinades. Moreover, it was hard to accept that the expression of a committed right-wing opinion on one controversial issue of current political, industrial or social policy was adequately balanced in terms of due impartiality by an equally committed left-wing statement on a totally different issue several weeks or months later.

Programmes associated with the name of Kenneth Griffith were also in a class apart. When an accomplished Welsh film actor decides to become a creator of television documentary, when he writes his own scripts, produces and performs them himself with uninhibited *hwyl*, and when he impersonates historical persons according to his own highly subjective assessment of their characters and deeds, the result is a personal view dramatised documentary. With Napoleon Bonaparte, for instance, as the subject, such a programme could prove both entertaining and compelling. But once the same methods were applied to a character or events with a direct bearing on contemporary current affairs and policies, alarms sounded.

In March 1972 they sounded in the ears of Authority staff told of an ATV proposal to commission Griffith to write and present a documentary about the Irish rebel-patriot Michael Collins. Because of the obvious conections which would be perceived between such a programme and the current situation in Northern Ireland, the company was warned that great care would be needed if breaches of the Act's due impartiality provisions were to be avoided. A viewing of the finished product, together with a detailed study of the script, realised their fears and convinced Authority staff that it was unacceptable under the terms of the Act. They found it to be interpreting modern Irish history in a way which suggested that the Republican movement was the one and only Irish tradition: its effect was to idealise and romanticise the gunman or 'freedom fighter'. Their objection went beyond concern about its lack of impartiality. So far as the audience in Northern Ireland was concerned, it would be an incitement to violence and an offence to public feeling. But in this instance no Authority action was required. ATV's management reached much the same conclusion and the programme was not offered for transmission.

Kenneth Griffith and his friends were outraged by what they saw as a blatant act of censorship. A steady stream of complaint reached the Authority, the company and Members of Parliament. Sir Lew Grade responded with a viewing of the programme for interested MPs. This did not help Griffith's campaign at Westminster, but the controversy was kept alive under the slogan 'the public's right to know', which became the title of a Thames documentary in March 1973. This dealt not only with the suppression of the Michael Collins programme but also with a Griffith programme about Baden-Powell which could not be completed because of trade union policies against working in South Africa. In August 1974 Griffith issued a writ against ATV claiming breach of contract, and not until the end of 1976 was the incident closed in an out-of-court settlement.

A different kind of programme dependent on the role of a single dominant individual was the live, unscripted interview-cum-audience-participation show in which David Frost specialised. Here again the underlying problem was not only the possible breach of a statutory restraint but, more generally and widely, the use (or abuse) of media power: the use made of that power to influence the minds of his fellow citizens by a celebrity figure playing a central starring role before the cameras and having at the same time total control of the handling, shaping and presentation of the programme.

Described in the *Sunday Times* as 'ITV's most feared and most loved, most criticised and most publicised programme personality', Frost had made his reputation on the BBC's *That Was The Week That Was* and earned celebrity status with his performances in unscripted interview programmes for Rediffusion.[29] There had been, for example, an interview with Dr Peto (a general practitioner suspected, and later charged and found guilty, of the indiscriminate prescribing of dangerously addictive drugs) and another with Emil Savundra (found to be involved in shady insurance deals and later imprisoned). Both these programmes had stirred the media world, not least that part of it inhabiting 70 Brompton Road, by raising the bogey of 'trial by television' – not to mention possible proceedings for contempt of court.

In September 1968 the Authority issued to the companies a note of guidance on television 'trials'. It ruled *inter alia* that accusations must be disclosed in detail to the accused well in advance of the programme; that he must be told who was making the accusation and be allowed to have witnesses on his behalf present; that he must know whether there was to be an audience and, if so, its composition and role; that whoever was in charge of the programme must be separate from the accusers and conduct proceedings fairly like a judge or independent chairman; and, above all, that there must be no verdict.

The controversial Frost programmes made betwen 1968 and 1973 for London Weekend Television, the company which he had formed with Aidan Crawley, were not quite trials. Indeed it could be argued that they provided one solution to the due impartiality/personal view dilemma. If committed and arguable opinion was to be identified with a named, well-known person, then the very fact that Frost was so evidently in control of the studio debates, at the same time not hesitating to participate argumentatively himself on one side or the other, made it plausible to claim that it was personal views which were being propagated, not an illegally one-sided, considered editorial judgment on a sensitive contemporary political or social issue.

The best of the Frost series created all the excitement and drama implicit in a spontaneous clash between strongly held, personalised opinions about topics in which the viewers themselves felt involved. In letters ,and telephone calls to Authority and companies, viewer response was strikingly more voluminous than that provoked by any comparable broadcasts at the time. It was, almost without exception, highly critical of one or other of the speakers and reflected the desire of many viewers to join in the studio debates.

Among the controversial personalities interviewed in the first two series of programmes were: Enoch Powell; Bernadette Devlin; John Braine, the novelist; the Nazi Baldur von Schirach; Ian Smith of Rhodesia; Cardinal Heenan; Mohammed Ali; Tariq Ali; US presidential candidates Nixon and Humphreys; Christian Barnard; Spiro Agnew; and Norman Mailer. Topics under discussion included the contraceptive pill, the Northern Ireland conflict, labour relations in Britain, and capital punishment. Interventions by the Authority were few, and in February 1972 it even took the unusual step of passing a resolution congratulating the company and the Frost team on the Northern Ireland programme.

There was, however, understandable caution about a proposed interview with the United States anarchist Jerry Rubin, leader of the Youth International Party (Yippies), who had been involved in a recent conspiracy trial resulting from violent riots in Chicago during the Democrats' national convention in August 1968. When the programme was screened in November 1970, a group of British Yippies managed to infiltrate the studio audience. On a given signal they jostled Frost out of the way, took charge of the microphone, indulged in obscene language and offered round a 'joint'. For a brief period near-riot conditions prevailed.

This led to questions in the House of Commons and a full report to the minister, but generally a relaxed view was taken. Order had been restored and the programme was not faded out. Companies due to transmit a recording later in the week did so. If the intention of the programme had been to make viewers better acquainted with the nature of the character and doctrines of Rubin and his disciples, its purpose had been achieved. In May 1973, on the other hand, Authority and company staffs agreed that it would be undesirable to proceed with an invitation to another United States citizen. He was a doctor of medicine, and his subject was the female orgasm.

The network's third current affairs programme, LWT's *Weekend World*, caused fewer problems than *World in Action*, *This Week* and the Pilger and Frost programmes because of its policy of early consultation with the Authority and the particular care with which it handled controversial subjects. But the commotion over its interview with the IRA spokesman, David O'Connell, a few days before IRA bombings in Birmingham in 1974 caused the Home Secretary, Roy Jenkins, to remind the House of Commons that he had powers to dismiss all the governors and members of the broadcasting authorities, however much he would deplore so doing.[30]

To what extent Authority interventions and non-interventions were well judged is an open question. The anti-establishment bias of much of ITV's current affairs coverage was evident, and the Authority was a liberal but establishment-minded body. Some in the companies therefore believed that it was itself lacking the due impartiality which it demanded of them. After a long investigation the Annan Committee came to the conclusion that the Authority's statutory obligations to supervise programme content should stand, but urged it to intervene less often and in less detail.[31] Yet those obligations, as interpreted by the courts in the McWhirter case,[32] required 'precisely the kind of intermeddling which the committee went on to deplore'.[33]

8

WARHOL AND POULSON

In January 1973 the Authority made two decisions which precipitated *causes célèbres*. One was to transmit; the other not to transmit. Assent to the showing of a documentary on Andy Warhol was its first decision ever to be challenged in the courts. Rejection of a programme about John Poulson less than a fortnight later called in question television's investigative function.

On 14 January two popular Sunday papers carried alarmist reports about a documentary programme scheduled for network transmission on the following Tuesday in the regular ITV documentary slot at 10.30 p.m. It was a production made for ATV by the photographer David Bailey about Warhol, the American pop artist and film-maker who had become a cult figure after establishing his reputation with a painting of a Campbell's Soup tin. The programme had been shown to the press in a preview the previous week. 'Andy Warhol shocker for ITV,' reported the *Sunday Mirror*. 'This TV Shocker is the worst ever,' announced the *News of the World*: 'Millions of viewers will find it offensive.' The next day the London *Evening News* echoed with the question, 'Should TV millions see this shocker?', and the *Daily Express* hastened to join the chorus of condemnation, although its writer had not seen the film.

On that day, fortuitously, members of the Authority were in formal session discussing an unrelated matter: the award of Independent Local Radio contracts. During a break in proceedings they enquired what the fuss was about, and the Director General reminded them of intervention reports in which they had been notified of discussions between Authority staff and ATV about the programme, about changes which had been agreed, and about a viewing of the amended version by himself and his most senior programme colleagues. He saw no good reason why the Authority should now go back on its earlier acceptance of its staff's recommendation that the documentary as amended was acceptable for transmission outside family viewing time. He thought that the two or three details singled out in newspaper reports, such as the use of four-letter words and the generously proportioned lady who undressed and used one of her naked nipples as a paint brush, were of relatively minor significance and quite defensible in the overall context of a fifty-minute programme which aimed at reporting unsensationally on the work and ideas of the egregious Warhol.

Members were satisfied with this explanation. Not so the 'clean up TV' campaigners stimulated by the press. Lord Longford, Mrs Whitehouse and other crusaders of the Festival of Light were interviewed and reported in the newspapers the next day, when the broadcast was due. All expressed shock, horror and dismay. More practically, Ross McWhirter, a broadcaster, journalist, editor of the Guinness Book of Records and member of the National Viewers' and Listeners' Association (NVALA), immediately applied to the Court of Chancery for an injunction restraining the Independent Broadcasting Authority from transmitting the programme. His claim was that, on the evidence of the newspaper reports, the Authority was about to act unlawfully. The Television Act required the members of the Authority to satisfy themselves so far as possible that nothing was included in their programmes which was offensive to public feeling, decency and good taste. The newspaper reports seemed to him to offer evidence that this statutory duty was about to be ignored. As a member of the viewing public who had paid his licence fee, he believed that he had a right and an interest in preventing the intended unlawful offence to his feelings of decency and good taste.

McWhirter's application was refused. Nothing daunted, he went to the Court of Appeal. There the Master of the Rolls, Lord Denning, was well known for his championship of the rights of the private citizen against the sometimes arbitrary uses of power by public institutions. When the Authority's counsel argued that, because the plaintiff had no specific personal interest in the matter other than one which he shared with the viewing public at large, he could have no legal status for his action without the consent and co-operation of the proper defender of the public interest in such matters, namely the Attorney-General, Lord Denning's response was therefore not unexpected. He agreed that the question of the appellant's *locus standi* raised important matters of principle. But on the evidence before him material offensive to public feeling was about to be broadcast later that same day. By not seeing the programme for themselves the members of the Authority had not 'so far as possible satisfied themselves'.

Lord Justice Lawton agreed. He did not think that the Authority had behaved reasonably. As for McWhirter's status as a private citizen, he thought that, even with long-established and acknowledged legal procedures, account had to be taken of the changing conditions of modern life. Only Lord Justice Cairns, while agreeing that on the available evidence transmission seemed likely to be against the public interest, insisted that a private individual could not be allowed to seek enforcement of the public interest without the concurrence of the Attorney-General. So McWhirter was granted an interim injunction by a majority of two to one and the programme was not transmitted.

Clearly the matter could not be allowed to rest. Wider issues than the fate of one fifty-minute documentary were at stake. It was no longer merely a question of the nature of this programme (which neither McWhirter nor the Appeal Court had seen), nor even of the conduct of the Authority in approving its transmission. Not only were an individual's rights against a statutory body and the proper means of defence of the public interest against mistakes and misbehaviour by such bodies under review; the injunction had raised issues affecting the rights and limits of free speech, freedom of information, and freedom of artistic expression. The anti-censorship lobby reacted strongly, and the hue and cry now came from a different quarter and ran in the opposite direction. The television critic of the *Financial Times*, Chris Dunkley, voiced a general apprehension in describing the injunction as 'one small step for Ross McWhirter and one massive retrograde stride for British television in particular and the concept of free speech in general'.

Virtually all the national and regional press now entered the fray, some on the side of free speech, some as stalwart defenders of the British viewer against the moral poisons of bad language and sexual licence. Others were critical of the Authority for allowing the situation to arise. Thousands of members of the public made their views known on one side or the other. The stream of letters and telephone calls to the Authority and companies was on an unprecedented scale. Many of the letters were from organised groups, writing either individually or collectively, and, to judge by the repetitious wording, part of the Festival of Light or NVALA campaigns. Since it had not been publicly transmitted, none among this host of protesters and few among the pontificating commentators had actually seen the film. One telephone caller protested to the Authority with the words: 'I've been urged to complain to you about something, but I'm not sure exactly what I'm supposed to be complaining about'.[1] 'That a film should be banned without first being seen would be a cause for mirth if the principle were not so important,' lamented ACTT in a public statement: 'Broadcasters and film-makers are very conscious of the unceasing attempts of unrepresentative and eccentric minorities to impose their own standards on society as a whole.'

Lord Shawcross, Chairman of Thames Television, incautiously allowed himself to be interviewed. He placed on public record his personal distaste for all that he had heard about the programme and expressed delight at a decision by the court which had prevented the damage to Thames's public reputation which transmission would have entailed. Since serving as Attorney-General in the Labour administrations of 1945 and 1950 as Sir Hartley Shawcross he had earned his nickname of Sir Shortly Floorcross by moving politically to the right to become the scourge of the trade unions and the permissive society.

The immediate response to his statement was a round robin signed by 287 Thames programme-makers publicly dissociating themselves from it: 'We wish to make it clear that the present high standing of Thames's reputation is based in large part upon the creative freedom we have hitherto enjoyed. This decision. . . undermines that freedom and can do nothing but harm to us who work for Thames Televison.' Later Shawcross was to receive from Lord Aylestone a letter of instruction about the respective responsibilities of the Authority and programme company managements, together with a gentle reproach about the unwisdom of making such public statements.

The injunction was interim only, and on 19 January the Appeal Court agreed to further hearings. The first of three took place on the 26th, by which time the Authority members, their General Advisory Council, the three Lords Justice and the Attorney-General, Sir Peter Rawlinson, had all seen the programme. After their viewing the members had unanimously reaffirmed the decision previously reached on the strength of reports from their staff that the programme was suitable for transmission in the 10.30 p.m. documentary slot, and that it complied with the requirements of Section 3(1) (a) of the Television Act 1964 (not offending against good taste or decency). The GAC, representative of the viewing public, resolved (with two dissentients out of eighteen) that the Authority had adopted proper procedures in authorising transmission; that IBA staff had been right in recommending transmission (one dissentient); and that there was no reason why the decision to transmit should be revoked (one dissentient).

By the last of the hearings, on 5 February, all three judges were agreed that McWhirter was wrong. He should, and even in the short time available could, have obtained the consent and co-operation of the Attorney-General. Because he did not do so he did not have *locus standi* and the injunction could not stand. The law was what Lord Cairns and everyone else had believed it to be when Lords Denning and Lawton had nonetheless granted the interim injunction. But since the Attorney-General had now said that he was prepared to be associated with the action, in so-called relator proceedings, the court felt free to examine once again the behaviour of the Authority on the assumption that such proceedings had in fact already started (which, strictly speaking, without the issue of an amended writ they had not).

Their lordships felt themselves better able to do this because they had received two affidavits from the Chairman of the Authority explaining the structure of ITV and the relationship between the Authority and companies and describing the established patterns of procedure whereby Authority approval for company productions offered for transmission was obtained.[2] They had heard of the conclusions reached by Authority

members and their advisers on the GAC. Above all they had seen the
programme themselves. As a result they were able to show themselves,
with minor variations, as of one mind.

As individuals they all had reservations about the degree of offence
likely to be caused by certain incidents in the film. But, in the words of
Lord Denning, 'my views do not matter, unless they go to show that the
IBA misdirected themselves or came to a conclusion to which they could
not reasonably come. I am certainly not prepared to say that.'

Lord Justice Cairns felt obliged to reiterate that in insisting that the
Attorney-General must be a party to the action 'I do not consider that I am
upholding a mere archaic piece of red tape... The requirement... is a
useful safeguard against merely cranky proceedings and a multiplicity of
proceedings'. Since McWhirter's application for an injunction and his
appeal were misconceived, his lordship felt disinclined to answer the
hypothetical question whether on the evidence available it would have
been right to grant the injunction. But because further evidence was now
available, and because the parties were anxious to have an answer, he was
prepared to consider whether an injunction should be granted in a relator
action. He did think that the film contained indecent material and was
surprised that neither the eleven members of the Authority nor seventeen
out of eighteen members of the GAC thought it offended against good
taste and decency. 'But they did ... The Authority are the censors, I am
not. I would therefore not grant an injunction in relator proceedings.' Lord
Denning added that the General Advisory Council was 'as representative
and responsible a body as one would find anywhere'.[3]

Lord Justice Lawton felt that statutory bodies should be allowed to get
on with the job which parliament had given them without the distraction of
'fighting off interfering busybodies', but he believed that, notwithstanding
the earlier recommendations of senior staff, the Authority would have
been wiser to view the film for themselves once the press reports which
had caused such widespread alarm had appeared. In his view too the
programme did contain indecent material and, although he was not
prepared to say that the Authority had 'crossed from the permissible to the
unlawful', they had got perilously near to doing so. They would be 'unwise
to assume that the frontiers will always be pushed nearer licentiousness'.
Nevertheless, albeit with difficulty, he recognised that a reasonable
Authority *could* conclude that the programme complied with Section 3(1)
(a) of the Television Act 1964.

Lord Denning then ruled that the courts had 'no right whatever' to
interfere with the Authority's decisions 'so long as they reach them in
accordance with the law'. McWhirter's appeal against the Chancery
Court's original refusal to grant him an injunction was dismissed, the ban on
showing the programme was lifted, and he was to pay half the Authority's
costs. But he was given leave to amend the writ and join the Attorney-

General in relator proceedings against the Authority. This he chose not to do. He left for the United States on other business, and Mrs Whitehouse's NVALA set about soliciting public subscriptions to meet his costs.

The Warhol film was now free from legal restraint, but Lord Shawcross was threatening to raise the matter in the House of Lords, and he was not the only company chairman to have serious reservations. There had been an exchange of letters between Lord Aylestone and Sir John Burgess, Chairman of Border, who felt that his company's local image would be gravely damaged if four-letter words were broadcast. Lord Townshend, Chairman of Anglia, had viewed the film with four of his boardroom colleagues. Even before the grant of the interim injunction, they had issued a public statement declaring their refusal to have the programme shown in the Anglia region because it would be a breach of the law. On 24 January a letter was sent from Anglia's lawyers to the Attorney-General's office explaining the company's attitude.

Every ITV company was under a regular obligation to include 'mandated' programmes in its schedules and to screen them, sight unseen, on trust. In so doing they had to rely on the integrity of their fellow contractors and, in the last resort, on the degree and quality of the supervision exercised, whenever deemed appropriate, by the Authority and its officers. The Warhol documentary was a mandated programme, and some companies clearly believed that in this instance supervision had been deficient.

The Authority's Chairman was roused to respond to the dissident company chairmen regretting the misplacing of his confidence that 'chairmen of ITV companies who have worked with us over the years, who have never accused us of breaking the law in the past and who have their own programmes. . . scrutinised by the Authority, to be sealed with approval or very occasionally rejected or altered, would. . . prefer to trust the IBA's judgment rather than that of two Sunday newspapers.'[4]

It was against this background that Lord Aylestone and his Director General faced the assembled senior company executives at a meeting of the Programme Policy Committee on 14 February. Here the issue was seen as a national one and not an occasion for pursuing regional disagreements. In the face of a public attack the Authority was keen to close ranks and avoid a fragmented network, but not by the use of *force majeure*.

Agreement was facilitated by the Director General's reminder of the accepted practice that any company could, with Authority consent, opt out of showing a mandated network programme provided it offered in substitution a programme of equivalent quality, meeting some specific regional interest or need. This option was available in the present instance. No company would be compelled to show the Warhol documentary if its directors and management were, on grounds of conscience, unwilling. None took advantage of the offer. By a unanimous decision in favour of solidarity the programme was screened in all regions, but only after a cooling-off period of six weeks.

Shown at last on Tuesday, 27 March, it attracted an estimated 14.5 million viewers – substantially more than might have been expected in normal circumstances for a serious documentary at 10.30 p.m. on a Tuesday night. According to research based on interviews conducted on the morning after, most viewed out of curiosity and the vast majority disliked the programme, only 1 per cent finding it 'extremely interesting', 2 per cent 'very interesting' and 7 per cent 'fairly interesting'. Just under half gave up before the end, 70 per cent of these finding the programme boring, stupid, silly or rubbish and only 6 per cent claiming to be offended. But 77 per cent of the whole sample disagreed with the statement 'I thought it was in good taste', 60 per cent of them strongly. On the other hand the statement 'I found it to be offensive' was accepted by 24 per cent and rejected by 56 per cent, while 84 per cent agreed with the statement 'I thought all the fuss was about nothing'.

This episode was an expensive and somewhat bizarre illustration of the impossibility of defining such shadowy concepts as good taste and offence to public feeling to the satisfaction of all. Its outcome may be judged a vindication of the Authority's attempt to move with the times and a victory over the prurience of the popular press. The programme itself was earnest, worthy and unexciting. It used Warhol's own work to explain his and its purpose, in the belief that his art and films should not be ignored however distasteful they, or his life style, might be. The Director General replied to one protesting correspondent:

It is the right and the duty of the broadcaster to inform its audience about such phenomena so that people may judge for themselves as to their importance and their value. We do not believe that any useful purpose would be served by pretending that they did not exist because they were alien to many people. The programme did not set out to offend, to corrupt or to persuade. As should be the case with all documentaries, it set out to inform.[5]

Also exposed by the Warhol incident was the dependence on mutual trust in a system of national broadcasting provided by fifteen resolutely independent companies legally and contractually subordinate to a supervisory body. The strain on relationships manifested itself in some lingering bitterness. In May Maurice Edelman, MP for Coventry North, told the House of Commons: 'When Anglia Television was disinclined, and indeed refused, to show the Warhol film it was obliged by a *diktat* inspired by the major programme companies, with the support of the IBA, to do so. . . the IBA took upon itself the task of obliging the company to present a programme which it had condemned because it felt that it would undermine the quality of life.'[6] These privileged assertions, at variance with the truth, were made nearly three months after the Authority's unequivocal assurance given to all companies that they were free in conscience to take their own decisions about showing the programme.

Meanwhile the Authority had become the target for protests of a different kind. Granada intended to mark the tenth anniversary of its current affairs series, *World in Action*, with a programme focusing on the bankruptcy hearings which had been probing into the affairs of John Poulson in the county court at Wakefield for the past seven months. The *World in Action* team had spent many months in painstaking research investigating the background, and the programme was to occupy an exceptionally long, sixty-minute slot on Monday, 29 January. When the Authority refused to approve it, press comment was heavy, continuous and predominantly unfavourable to the Authority. But, in contrast to the Warhol experience, this critical barrage raised no corresponding clamour among the viewing public.

Poulson was an architect who, from a relatively modest start in the West Yorkshire town of Pontefract, had built a large and lucrative business empire in architecture and engineering through a number of separate companies, operating in Britain and overseas, concerned with public works, property dealing, building design and management consultancy. As a precaution against losing membership of his professional body, the RIBA, he had had these companies registered in his wife's name.

He was a prominent member of the Methodist Church, a freemason (Provincial Grand Deacon of the West Riding), an active Conservative party member (at one time Chairman of the South Yorkshire branch) and an associate of more than one leading Tory politician, including a former Chancellor of the Exchequer, Reginald Maudling, who had been chairman of two of his overseas companies. He had also established useful relationships with senior civil servants, trade union leaders and Labour politicians in local government, including Dan Smith, the Labour council leader in Newcastle and a director of Tyne Tees Television, who was at one time on his pay-roll. Despite all these connections the empire had collapsed, some £250,000 in the red.

Proceedings at the Wakefield court were concerned with the disappearance of very large sums of money and the prospects of recovering some of it for the benefit of creditors. The apparent scandals brought to light had been widely covered by the press. Acting for the creditors was Muir Hunter QC, whose ruthlessly uninhibited questioning of Poulson and some of his associates revealed the whole network of personal and business relationships, implying more than a hint of undue influence over the award of contracts, if not of actual bribery and corruption.

A bankruptcy hearing is not subject to the otherwise customary restraints of a court of law. It is not a trial; there is no judge, jury or verdict, and persons named are not necessarily called to testify. So when the proposed *World in Action* programme reproduced the courtroom ex-changes with the use of actors, many of the statements and innuendos reported seemed to Authority members like a series of acts of character assassination. Indeed the Bar Council had felt moved to question Muir Hunter in a private hearing; only to conclude that, in the particular circumstances, no charge of professional misconduct could be sustained.

There were disturbing similarities between a programme so closely identified with courtroom proceedings and the familiar bugbear, trial by television. Yet the programme was doing little more than reporting accurately and verbatim what had been said in court and already reported in the press. This knowledge did not, however, dispel Authority members' misgivings; nor did the programme-makers' justifiable claim that their purpose was the defence of standards of probity and integrity in public life. If the full charge could not stick, this still looked like trial by television at second hand.

Lawyers consulted included eminent counsel, and all were agreed that indisputably defamatory statements were being published about named individuals. Advice that a plea of justification and fair comment would most probably succeed against an action for libel did little to allay Authority worries. The bankruptcy hearings were still unfinished. Once they were concluded some of the relevations might well lead to criminal charges against some of those named. This led some members to suppose, erroneously, that the Authority might find itself in contempt of court.

Members also had moral doubts. Whatever might be the eventual outcome, was it proper for ITV channels to be used for the wider dissemination of defamatory statements about people in public life, however well justified and legally defensible those statements were be-lieved to be? Their unease derived from a sense of unfairness, which was not a valid reason for rejection under any specific requirement of the Television Act. They therefore relied on a seldom, if ever, used general provision, namely Section 4(2). This gave the Authority the broadest possible editorial freedom to reject any programme of which it happened to disapprove.

The members' decision was reached at a meeting on Thursday 25 January and made public the following day. That weekend the *Sunday Telegraph* reported what it described as growing unrest among the ITV companies at constant interference by the Authority: unrest which it claimed was felt in all companies from top management to studio floor. The insistence of the Authority on the inclusion of Verdi's *Macbeth* in an otherwise predominantly popular Christmas schedule was cited as a recent example of gratuitous meddling. The *Sunday Times* referred to a fear among television journalists that they were witnessing the start of a new trend towards bans on investigative reporting. The *Bradford Telegraph*, the newspaper for which the researcher on the Granada programme had worked, denounced the Authority's decision as outrageous. Alan Sapper of ACTT made a public declaration of his union's opposition to censorship in whatever guise or on whatever pretext.

Granada's request for a meeting with members to view and discuss the film together on the Monday, in the hope that they would be sufficiently reassured to sanction transmission that evening as scheduled, was firmly rejected. Authority staff thereupon circulated a paper informing members of the company's distress at this brusque refusal and warning of the likelihood of industrial action,[7] even though Granada had issued a statement emphasising its undesirability ('the implications of using industrial action to affect decisions... concerned only with broadcasting and editorial policy are very serious').

Lord Aylestone then agreed that Denis Forman, Granada's Managing Director, could come to see him on the following Thursday, 1 February, in order to make the case for a revised version of the programme which embodied material changes arising from discussion with Authority programme staff. The threat of a blackout was withdrawn pending the outcome of that meeting, but on the Monday pickets appeared outside the Authority offices in Brompton Road and statements of ACTT's position were handed in on behalf of the Manchester and London Granada shops. The NUJ declared its support.

On the same day enquiries were made on behalf of James Prior, Leader of the House of Commons, about the possibility of a viewing of the programme by MPs. Authority staff replied that the matter was still under consideration by members, whose responsibility it was by statute, and that strong exception would be taken to any action by MPs wishing to pre-empt its decision.

After the meeting with Granada on 1 February the Authority issued a press statement to the effect that members had read the script and viewed the revised film and were still concerned whether, as a matter of broadcasting policy, it would be right to give wider publicity to accusations against named individuals who had had no proper opportunity to defend themselves. However, the statement added, both the Authority and Granada were going to give further thought to the points made at their meeting and the Authority would be considering at its next meeting whether the programme might be shown at a later stage.

The next day Granada held a viewing for the press and issued a factual background paper. This expressed disappointment that the Authority had not been persuaded to change its mind and allow transmission on the following Monday, as the company had hoped. But, it continued:

> there can be no question about this: in banning the Poulson film the
> Authority acted within their legal powers ... We, as a management, did

everything in our power to persuade the IBA that they were wrong and we were right. We failed to do this and we have to accept their verdict . . . Within the last fortnight there have been suggestions that the judiciary and the House of Commons should take over certain powers excercised by the IBA. Fortunately, as we think, this is most unlikely to come to anything and the IBA will contine to be completely independent . . . Granada believes that the right way to run ITV is to have a broadcasting authority that is independent of parliament and independent of the companies. In their evidence to the SCNI the ACTT took the same view and expressed the opinion that the IBA should be tougher and take more initiative.

The company therefore deplored the union's intention to take action which challenged the IBA and the company management and laid itself open to the charge of itself acting as censor.

But according to the *Sunday Times* it was the Authority which was practising political censorship and this was a cause for general alarm: imputations of bad, even illegal, conduct were apparently permitted, but not if they touched the nerve of British legal and central government.[8] In this the *Sunday Times* found itself at one with the *Socialist Worker*, which declared: 'It did not take long for the assorted Labour hacks, Tory businessmen and academics who make up the IBA to agree that the programme was politically dangerous. . . They are direct instruments of a corrupt class which does not want its dirty linen washed in public.'[9]

The *Sunday Telegraph* went further. It drew attention to two Authority members in particular. Sir Frederick Hayday had been a senior official of the General and Municipal Workers trade union, and the Chairman of the Durham Police Authority, Alderman Andrew Cunningham, whose name had been mentioned during the Wakefield proceedings in connection with the award of several Poulson contracts, was a senior official of the same union. Baroness Sharp had given evidence at Dan Smith's Old Bailey trial for bribery in July 1971, when she had been his only character witness and testified to her complete faith in his integrity.

On Monday 5 February Eddie Milne, the Labour MP for Blyth, tabled a question to the minister. He wanted to know whether members of the Authority were required to make a declaration of interest before appointment.

That same evening came the threatened industrial action by the television technicians' union acting as censor in protest against censorship. During the half hour between 8 and 8.30 p.m. when the revised Poulson programme had been expected to be shown, screens in most ITV regions remained blank. Among the major companies only Yorkshire Television managed to stay on the air with a substitute programme in a 'shock night for TV millions' (*Daily Mirror*). While expressing some doubts about the blackout, the *Mirror* suggested that the whole affair offered clear evidence that the IBA needed replacing by a 'National TV Council'.

The *New Statesman* spoke of the thorough disenchantment of ITV workers and their conviction that the sooner the Authority was recognised as incompetent to do the job for which it was set up the better. There followed a brief analysis of Authority membership, pointing out that of the forty-six members since 1954 twenty-eight had been to public schools and twenty-six had titles, and that until the recent appointment of Christopher Bland at the age of 34 the average age of Authority members had been 59. Television professionals were reported to be asking: What qualifies these people to act as judges?

Meanwhile Granada and Authority programme staff had been working towards a solution. On 13 February Brian Young received a letter from Denis Forman detailing further changes which Granada were making and emphasising that for the time being no firm transmission date was envisaged. If timing had been a major factor in Authority objections, that obstacle was removed.

A copy of the new script embodying all the proposed changes was received at Brompton Road on 2 March, distributed to members and discussed at a specially convened meeting four days later. Those unable to attend at short notice sent comments. It was noted that the changes all tended to emphasise the importance, in the public interest, of the general issues raised by the case. Attention was concentrated less on the allegedly dubious behaviour of named public officials and politicians. The verbatim reproduction of certain relevant passages from the Wakefield hearing performed by actors had been replaced by less dramatic third-person reports in indirect speech.

In view of the sparse attendance no decision was reached. Instead it was agreed that the matter should be fully discussed and resolved at the next regular Authority meeting, due to be held on 22 March. In time for that meeting came another letter from Forman. This gave a new version of the programme's opening announcement, informing viewers that the programme about to be seen would be examining the issues raised by the Poulson bankruptcy hearings and their relevance to the rules governing relationships between public servants and outside business interests; and that it would look at the case for an obligatory register of the business interests of Members of Parliament. The programme's title was to be changed from *The Friends and Influence of John Poulson* to *The Rise and Fall of John Poulson*.

After its meeting on 22 March the Authority announced that the revised programme had been approved for transmission, subject to a check on certain points of detail. No hint was given of the argument which had taken place round the table. Evelyn Sharp was not to be deflected from her conviction that the programme would damage the reputation of local government and ruin Dan Smith, whom she admired as a builder of good, cheap houses. Indeed the conflict had developed, to some extent, into a duel between her and Forman; the one determined that the programme should not be shown, the other equally determined that it should. Another member of the Authority held to the view that, as a magistrate, she could not in conscience sanction a broadcast containing defamatory statements. That was a point of principle, and it made no difference that those defamed had already been widely discredited and the risk of a libel action was negligible. When the matter was put to the vote there was a tie at five-all. The missing member was absent on a business trip overseas. He was telephoned by the Director General and the programme was saved by his casting vote from South Africa.

Broadcast on 30 April, it was a model of its kind. 'No longer can any accusation be made of innuendo and unfairness,' reported the *Guardian*, although its television critic, Peter Fiddick, regretted that the long delay had blunted its impact. *The Times* praised the programme for attempting to give both sides of the story and for having 'cleverly pieced the Poulson story – or some of it – together and told an extremely complex tale in a simple and easily understood form'.[10]

The tale of the programme itself, like that of *Andy Warhol*, had been scarcely less complex.

9

VIOLENCE, SEX AND BAD LANGUAGE

Were the growth in television viewing and the growth of violence in society during the 1960s and 1970s parallel developments or cause and effect? Was this a prime example of the evil influence of the intrusive box or of psychological need for a scapegoat? On both sides of the Atlantic research studies multiplied, but differences of opinion on their interpretation tended to exacerbate the controversy rather than clarify the issues. In a long-running and unresolved argument the broadcasters were forced on the defensive and required to prove their innocence.

ITV and its critics were both motivated by a strong belief in the power of television for good or evil. Few would have agreed with Sir Ivone Kirkpatrick, the ITA Chairman who shocked the Pilkington Committee by doubting the influence of television on the values and moral attitudes of society. Society, he had coolly observed, would be largely what it was with or without television.[1] In support of this heresy he might have cited the example of Japan, where the incidence of violence on television was exceptionally high and in society exceptionally low. But common-sense suggested to most people that at least some television programmes exercised some harmful influence on some viewers. The real questions were, how often, how much and how many.

'Conflict is of the essence of drama, and conflict often leads to violence in many forms, and when television seeks to reflect the world – in fact or fiction – it would be unrealistic and untrue to ignore its violent aspects.' This justification of the portrayal of violence, whether physical, verbal or psychological, was published in *Television Programme Guidelines*, a collation of its rules and regulations issued by the Authority in September 1977 for the guidance of ITV's programme-makers.[2] Their attention was there drawn to the ITV Code on Violence, whose production and periodical review were among the responsibilities imposed on the Authority by Act of Parliament, where it was specified that special regard must be paid to programmes broadcast 'when large numbers of children and young persons may be expected to be watching'.[3] As the 1971 Code observed: 'A civilised society pays special attention to its weaker members.'

Towards the end of the 1960s a wave of anti-television agitation had washed across the Atlantic from the United States where city riots, campus battles and the political assassinations of the Kennedy brothers and Martin Luther King had roused alarm about the growing resort to violence. Consumer bodies protested vehemently about the high level of screened violence, which the competing networks, so far from denying, defended as a popular and profitable ingredient in their schedules.

On 23 September 1969 a US National Commission on the Causes and Prevention of Violence (chaired by Dr Milton S. Eisenhower) published its *Statement on Violence in Television Entertainment Programs*. This attracted wide attention, as did the US Surgeon General's Scientific Advisory Committee on Television and Social Behavior which was set up in the same year and produced five volumes of inconclusive evidence at a cost of $1 million.

The Eisenhower Commission noted that there had been plenty of crime and violence in American society before the coming of television and uncovered no single explanation for its various causes. But it found research evidence which strongly suggested that violence in television programmes did have adverse effects on viewers, particularly children, and employed some harsh words about those responsible:

Each year advertisers spend $2.5 billion in the belief that TV can influence human behavior. The TV industry enthusiastically agrees with them, but nonetheless contends that its programs of violence do not have any such influence ... TV entertainment based on violence may be effective merchandising, but it is an appalling way to serve a civilisation.[4]

This argument ignored the distinction between advertisements, which are designed to persuade and sell, and programmes, which are not. But whether or not valid in the context of either American or British broadcasting, such sentiments fell on receptive ears in Britain, and broadcasters were obliged to take account of acceptance by an influential minority of the still unproven assumption of a causal connection between violence on television and violence in society. ITV's Programme Policy Committee dwelt on the subject at meeting after meeting, especially in relation to the young. Was the accepted demarcation line between family and adult viewing times strictly enough observed? If 9 p.m. was right on weekdays, was 7.45 p.m. too early on Sundays? Should the watershed be standardised at 8 p.m. seven days a week?[5]

In May 1969 the first episode of Granada's *Big Breadwinner Hog*, a fictional series portraying gang warfare in the East End of London, created a public hullabaloo. It showed in use, in convincing detail, the weapons actually employed in this kind of warfare, including an all-too-realistic acid-throwing scene, and this provoked a volume of spontaneous protest greater than ITV had ever experienced. The Authority felt it appropriate to make a public apology. It previewed all subsequent episodes and, after the fourth, decided that the series would be best shown late at night. It was sad, but salutary, to discover that this series designed to expose the brutality of organised crime was assumed by many correspondents to be motivated by a desire to glorify violence. A subsequent research survey indicated that few people thought that *Big Breadwinner Hog* would attract anyone to violence, but the question remained: how far was it possible for a production of this kind to convey to mass audiences that it was not only telling a story, but also commenting responsibly on the society in which they were living?

Early in 1970 the government took an initiative. On 23 March the Chairmen and Directors General of the ITA and BBC answered a summons to the Home Office where the Home Secretary, James Callaghan, told them bluntly that he thought their codes admirable but not effectively implemented. He was concerned about violence in entertainment programmes, not news, and threatened to appoint a controlling body if they failed to put their houses in order.

The Authority responded on 1 May with a memorandum which stressed the amount of ITA-financed research being undertaken and declared its policy to be one of vigilance.[6] A Standing Working Party on the Portrayal of Violence was established, composed of three senior programme executives from the companies, three senior members of the Authority's programme staff and three members of the Authority's General Advisory Council, under the chairmanship of Bernard Sendall, Deputy Director General (Programme Services). Regular meetings were held between October 1970 and July 1971, when a revised Code was submitted to the Authority to replace the one which had been operative since 1964. This was adopted and issued in October 1971.[7]

The Working Party continued to meet and in January 1973 presented the Authority with an Interim Report, which was published in June.[8] ITV, it was convinced, had no alternative but to work on the assumption that the portrayal of violence on television did have harmful effects; and no evidence had been uncovered to support the assumption, fostered by the Hays Office in Hollywood's golden days, that portrayal of the use of violence for good ends was less harmful than portrayal of its use for evil ends. There was even a strong possibility that the reverse could be true, since the example of acts of violence by the good guys might weaken viewers' inhibitions against its use by themselves. The report also rejected the assumption that 'sanitised' violence was relatively harmless because of its stylised character. It stressed that horror in costume was still horror.

In identifying the immature and the emotionally insecure as the most vulnerable of viewers this report found that these tended to be among the heaviest viewers of violent programmes. It did not recommend any amendment of the 1971 Code, but it did recommend the trial use of a warning symbol (an outline rectangle – black or white – at the bottom left-hand corner of the screen) on programmes which might disturb such viewers. This experiment, conducted by ATV in the Midlands from August 1973 and later in the Southern area, was commended by the Annan Committee but never generally adopted. The 'White Dot' symbol warned of explicit sex as well as violence, and Mrs Whitehouse attacked it as a 'kind of licence to try offensive material'.[9] 'Will symbols turn you on?' the *Birmingham Evening Mail* asked its readers.

In an article on violence in 1975 Sendall argued that television should consistently explore positive ways of helping to improve the quality of life and so prevent crime. Mirroring life, warts and all, was not enough:

> Let us remember that what helps people – particularly the young – to eschew violence is growth in imagination, awareness of others, sensitivity towards pain in others. Violence on television ought not, as we all agree, to be there for its own sake. But programmes about such horrors as famine in Ethiopia and Bangladesh, documentaries about the unrich, the unhoused and the unloved, plays which arouse our compassion for people and enlarge our understanding of them, all of these are television which can achieve a true catharsis provided they have the quality.[10]

The Working Party's Second Interim Report, published in May 1975, concentrated on violence in news, current affairs and documentary programmes. It examined the policy and hours of family viewing time. Emphasising the joint responsibility of broadcasters and parents for children's viewing, it came down against a watershed later than 9 p.m. even though children were tending to go to bed later and view longer.[11] Once again no revision of the 1971 Code was thought necessary, and in the following year (1976) the Working Party was deemed to have exhausted its original terms of reference and, after six years' work, was disbanded.

The Annan Committee, reporting in March 1977, expressed some scepticism about the validity of research findings on violence but judged, like the Home Secretary seven years earlier, that the broadcasters had a case to answer: their codes were admirable but not always heeded. THIS CODE, said ITV's in capital letters, CANNOT PROVIDE UNIVERSAL RULES. THE PROGRAMME MAKER MUST CARRY RESPONSIBILITY FOR HIS OWN DECISIONS. Although producers were told that they must justify, first, the portrayal of violence and, secondly, the way in which it was portrayed, the committee found it exceedingly hard to believe that they had got the message: 'Occasionally nine o'clock resembles not a watershed but a waterfall.' *The Sweeney* was named as one of the programmes which had made the hour between 9 p.m. and 10 p.m. on occasions one of the most violent on British television, and it was an hour when many children would still be watching.[12]

Some news and current affairs topics were perpetually and inherently contentious. How, for example, could one report the news from Northern Ireland without publicising wanton violence and giving a terrorist movement the publicity which it sought and on which it thrived? The 1974 *Weekend World* interview with an IRA spokesman, followed within a few days by the IRA pub bombings in Birmingham in which twenty-one innocent people were killed and 160 injured, raised the dust of much public controversy. Because it was strongly hostile, the showing of the interview had been sanctioned by Joseph Weltman, the IBA's Head of Programme Services, at 1 a.m. on a Sunday despite protests from the Northern Ireland Regional Officer and, in view of the hour, without reference upwards.[13] The members of the Authority subsequently concluded that his decision had been the right one, that such decisions could be taken only on an *ad hoc* basis, but that in future any plans for a programme which explored and exposed the views of people who, within the British Isles, advocated and used criminal acts for the achievement of political ends should be referred to the Authority before any arrangements were made for filming or videotaping.[14]

Some members of the public wanted tranquillised television. They wrote to the Annan Committee to complain that the broadcasters were carrying terrorism and violent political protest all over the world 'like a plague'.[15] The committee found no solution, but thought that showing battered babies and tortured animals was unnecessary: broadcasters could reflect the real world by simply reporting harrowing events. It recommended that warnings of the portrayal of violence should be made more explicit and that 'the broadcasting authorities should monitor the amount of violence in their programmes and should publish regularly their findings and report thereon'.[16]

Thus admonished and prodded, ITV's policy-makers resumed their discussions. What more could be done? PPC meetings were enlivened with a diversity of reactions, mostly indignant: the elimination of violence would result in the portrayal of a censored view of society and an antiseptic programme policy; violence was admittedly included in programmes sometimes for ratings, but more often for realism; broadcasters were being bullied because of the dubious findings and woolly generalisations of sociologists asking the public loaded questions. If account had to be taken of minority opinion and ITV must be seen to be doing something about it, precisely what kind of violence did these viewers find offensive, and in what kind of programme? Why should broadcasters be defensive about violence which they knew that the vast majority of viewers found, not offensive, but entertaining? *The Sweeney* and the BBC's *Starsky and Hutch*, denounced for violence by Annan, not only attracted enormous audiences; they were highly rated in the Appreciation Indices.[17] Would the public really prefer anodyne, emasculated entertainment, or was a campaign by small pressure groups, inflated out of all proportion by the press, going to deprive viewers of what they liked best?

In September 1977 the Authority held a seminar in Newcastle at which the producer of *The Sweeney*, Ted Childs, mounted a strong defence. Why pick on my programme, he wanted to know: he had seen more savagery in an episode of *Poldark*, transmitted at 7.25 p.m. on a Sunday, than one often saw in *The Sweeney*. *The Gangsters*, *The XYY Man* and even the BBC's much-praised *I, Claudius* were, he claimed, all dramatically geared to a higher level of violent and emotionally disturbed human behaviour.

The Authority then called for a paper and devoted a meeting to discussing it. *Television Violence – The Last Ten Years* was written by Stephen Murphy,[18] one of the IBA's senior programme officers, who had been Secretary of the British Board of Film Censors. It set public protest against media violence in an historic context. The cinema had attracted waves of protest in 1936, 1950–51 and 1971–73. During the second wave the BBC had been forced to drop a popular radio series: *Dick Barton – Special Agent*. The recent waves of protest against television had both originated in the United States: in 1967 and 1976. Murphy told members that research 'does seem to have established that there is a correlation between violent crime and heavy television viewing. What has not been established is a causal relationship.' In other words, aggressive people liked watching programmes portraying aggression and this might or might not reinforce their aggressive tendencies to such a degree that they committed violent or more violent acts as a result. 'It is fair to say that the researchers have not yet demonstrated to general satisfaction that television violence is a major factor in violent crime.'

Murphy drew attention to the 'information explosion' – aptly named in this context. More and more news media were reporting more and more violence around the world. In fictional programmes, on the other hand, there had been no increase in the portrayal of violence, but technical advances had heightened the impact. Action shots were more convincing when filmed on location, and this had been facilitated by the development of the light-weight camera. Zoom lenses made close-ups of violent actions easier and cheaper. Colour made blood red. A new generation of directors and scriptwriters was employing these new techniques with great skill:

A close examination of the scripts of *The Sweeney* reveals a number of episodes without any violence, and few episodes containing a great deal of violence. But the scripts are so taut, the performances and the direction so good, that what violence there is has considerable impact. In this sense, British television is the victim of its own insistence on the pursuit of excellence.

At its special meeting in November 1977 the Authority agreed that it remained concerned and that its present activities of inquiry and vigilance should be fully maintained.[19] A reconvened Working Party held its first meeting the same month. This time it included two outside members: the writers John Bowen and Lord Willis. Evidence was heard from Dr William Belson, a leading researcher in this field, and from senior officers of the Metropolitan Police worried about the excessive number of fictional killings by policemen. Violence was also discussed at an IBA Research Consultation held early in 1978, and the Working Party reported in May, proposing some minor amendments to the Code.

After Annan the publication of three major research works kept the controversy on the boil. A comprehensive Home Office review of research into the behavioural effects of filmed violence offered some comfort to broadcasters by pointing out that none of the results published by a leading television research organisation, the Centre for Mass Communication Research at Leicester University, 'can be unequivocally interpreted to show a direct link between television violence and delinquency', despite the influence which television was likely to exercise generally over values and behaviour.[20] 'Social research has not been able unambiguously to offer any firm assurances that the mass media in general, and films and television in particular, either exercise a socially harmful effect, or that they do not.'[21]

Dr Belson's findings in his *Television Violence and the Adolescent Boy* (1978) were more positive: 'The evidence gathered through this investigation is very strongly supportive of the hypothesis that high exposure to television violence increases the degree to which boys engage in serious violence.'[22] Here a call was made for a substantial reduction in the amount of television violence available to the adolescent viewer.

The authors of *Sex, Violence and the Media* (H. J. Eysenck and D. K. B. Nias), published in the same year, were also satisfied by 'ample evidence that media violence increases viewer aggression'; but 'to say that TV has such effects is not to say that these are the only, or even the most important, influences which lead to violence and aggression; TV is just one of a number of influences, but withal a powerful and omnipresent one'.[23] The authors called on those who made and produced films, plays and television programmes to show more social responsibility.

On 26 October 1978 the IBA devoted itself to the subject in a 'Day on Violence' which began with a viewing session of examples of realistic and sanitised violence, proceeded to a discussion with programme-makers, script-writers and the Editor of ITN, and ended with a long internal debate on by now familiar issues.

In February 1980 an approach to the BBC for common action bore fruit in the publication of the two Codes between one set of covers, with a joint foreword by the two Chairmen: *The Portrayal of Violence on Television: BBC and IBA Guidelines*. Later in the year Dr Barrie Gunter was appointed an IBA Research Fellow for two years: his work, *Dimensions of Television Violence*, was published in 1985.

Thus anxiety and argument about this insoluble problem persisted throughout the 1970s and into the 1980s, with governments and broadcasting authorities, programme-makers and pressure groups, academic sociologists and commercial researchers, press and public all drawn into a never-ending debate. Researchers diligently counted the number of violent incidents screened, assessed their impact on viewers, and recorded viewers' opinions. By the mid-1980s more than 700 pieces of research had been published suggestive of a link between watching violence on television and enhanced aggressiveness among viewers, but a link of what kind?[24]

To the programme-makers some of the principles of content analysis applied to programmes by media sociologists seemed laughable. In identifying televised acts of violence how could one put the same undiscriminating label on a saloon-bar shoot-out in a Western film and the cold-blooded execution of a civilian terrorist in Vietnam, on a hard rugger tackle and the wrestling scene in *As You Like It*, on rioting on the streets of Belfast and a Tom and Jerry cartoon?

It became apparent too that viewers did not perceive programmes as researchers analysed them and that even the same incident would be perceived very differently in different settings. Familiarity of surroundings was a powerful factor: a bout of fisticuffs in a James Bond film, however rough, would make relatively little impact by comparison with even a modicum of violence during a domestic brawl in, for example, an episode of *Coronation Street*. Viewers were found to be equivocal, torn between abhorrence and relish depending on act and setting.[25]

The nub of the problem may be simply stated: whatever the degree of violence in society, it would be magnified by television, and this was inherent in the medium. The average viewer was soon inured, possibly desensitised, and it was noticeable that the general viewing public were not parties to any widespread concern or demand for a cutback.

What angered and distressed the viewer far more were questions of bad language and sex. An opinion poll conducted in 1973 found 20 per cent of respondents in favour of a total ban on nudity, sex scenes, swearing and blasphemy in television programmes, while only 14 per cent wanted scenes of violence between adults banned.[26] Later, sex dropped to third place in perceived offensiveness, but bad language stayed well in the lead, as an avalanche of protests to the Annan Committee testified.

In the cause of family viewing the Authority fought round after ineffectual round against these ugly sisters. In April 1969 members of the Programme Policy Committee were reminded of the need to exercise restraint in regard to bad language and the exploitation of sex themes and sex jokes. In November 1970 the Authority assured them of its awareness that public standards in sexual behaviour and language had become less certain and uniform, but emphasised that offence was still being given to substantial sections of the audience by the unnecessary use of bad language and over-explicit sex,[27] and it circulated a paper, *Good Taste and Bad Language*.[28] Company representatives excused their programme-makers on the grounds that writers of comedy and drama were not amenable to the dictates of managements.[29] Representations to the Guild of Screen Writers were thought counter-productive.

The Consultation on family viewing held in October 1970 provided a good illustration of the chasm which, despite the Authority's best endeavours, yawned unbridgeably between the creative worker in television and the strait-laced viewer. According to Lady Twiss, Chairman of the Central Television and Cinema Group of the Mothers' Union's Social Problems Committee: 'TV came into our sitting rooms and we somehow expected it to behave as we expected our families to behave. When it did not, which was very often the case, we were shocked, embarrassed – extremely embarrassed at times – and cross.' To which an unrepentant drama producer responded: 'Good! It does people a great deal of good to be shocked occasionally', and Lady Twiss replied: 'I represent the customer, and the customer is always right.' But to the producer she represented no more than an unrepresentative pressure group.

On 25 September 1971, in an organised backlash against the permissive society, the Nationwide Festival of Light, whose council members included Lord Longford, Malcolm Muggeridge, Cliff Richard, Mary Whitehouse and four bishops, attracted 35,000 people to a rally in Trafalgar Square. In December it sent a delegation to the BBC and on 20 January 1972 another, under its Chairman, Colonel Orde Dobbie, to the ITA. At 70 Brompton Road the delegation was received 'with interest and sympathy' by the Chairman, Deputy Chairman, Director General and Deputy Director General (Programme Services).

The Festival of Light reminded these eminences of their 'clear duty to exercise leadership in certain fundamental moral issues' because of the massive influence exerted by television. While a large proportion of ITV's output was quite unexceptionable, it argued in a subsequent memorandum that:

the case for moral purity and Christian values has tended to have less than its fair share of time and emphasis. The snide 'knocking' attitude has been too common. Undue prominence has been given to the abnormal, the deviant, the obscene, the permissive and the destructive in both drama and humour. This in spite of the fact that 'good, clean entertainment' such as *Coronation Street* and *David Nixon's Magic Box* do very well in the ratings ... The changes we have noted on our screens have almost all been in the same direction – towards a more permissive, more violent, more coarse and more superficial picture of life.[30]

This memorandum was a lengthy and detailed indictment which demanded a more considered response than the four-letter words (such as 'bosh')
irately pencilled in the margins of a surviving copy. It took the Authority
from April to October to muster and refine *Some Comments on the
Memorandum Submitted by the Nationwide Festival of Light Delegation*.
While courteously phrased, this contained some sharp rebuttals and
equally high-principled counter-assertions. To the argument that television
was not just a mirror but an agent of change the ITA replied that television
was not so much a principal creator and inspirer of change in social
attitudes, behaviour and ideas as an instrument for spreading knowledge
and experience more widely. A more open attitude to the portrayal of
sexual activity and spoken vulgarisms was signalled by events such as the
publication of *Lady Chatterley's Lover* in paperback, following the court
judgment in *Regina v. Penguin Books Ltd* in 1960, and whatever might be
thought about the rights and wrongs of that judgment it would be absurd to
attribute any influence in the affair to television.

Some Comments pointed out that television was part of a total environ-
ment and argued that it was not the duty of broadcasters to indoctrinate.
Stung by the charge that ITV was turning viewers into voyeurs, the
Authority reiterated its stated policy of not permitting the gratuitous
display of overt sex and nudity on the screen. It saw no useful purpose,
however, in censoring all representations of physical love and the naked
human body and went on to defend, item by item, the specific instances of
nudity to which the Festival of Light had objected.

In light entertainment the Authority was less sure of its ground and
confessed to some individual errors of taste. *The Benny Hill Show* had
been the target of one of the Festival of Light's most scathing onslaughts:

> Cheap laughs from obscene puns, suggestive *double-entendre*, perver-
> sion or promiscuity have usually been the province of the schoolboy
> lavatory-wall humour. We see no reason why Mr Hill, for all his wit,
> should be allowed to project this kind of nauseating stuff into the
> nation's living-rooms. Many viewers switch off, agreed. But why should
> they be denied entertainment when the Authority is committed to avoid
> what is offensive? And for those millions who do not switch off, the
> coarsening, de-sensitising process goes on by slow degrees sapping the
> will, eroding the sense of shame, eating away at the protective walls of
> reticence and the vision of high ideals.

Against this broadside the Authority defended itself by pleading the
tradition of British music hall and end-of-the-pier shows (while having to
concede that living-rooms were not, in fact, music halls or situated at the
end of piers).

The Festival of Light asserted its belief in the existence of a national way of life, but this, if it had ever existed, was no longer discernible in the 1970s. As an enlightened regulator, the Authority was navigating a midstream course between Mrs Grundy and Lady Chatterley under a hail of missiles from both sides. In 1972 complaints from the public were greater than ever, and monitoring confirmed an increase in the incidence of swearing, blasphemy and other verbal crudities, some in family viewing time, some even within children's programmes. The Authority told the companies that there were too many occasions 'where it is difficult to discern any grounds in artistic integrity or dramatic necessity for a degree of verbal self-indulgence that contributes little or nothing to the success or popularity of the programmes in which they occur'.[31] At a PPC meeting Sir Lew Grade gave an assurance that ATV took every precaution to avoid offence and admitted to only occasional slips, while Denis Forman of Granada felt, less reassuringly, that recent years had seen great progress in liberalising the language and reviving healthy Anglo-Saxon words.[32]

Restraint by the writers of television drama was less evident than the Authority's own restraint in refraining from tampering with the scripts of series like *Man at the Top, The Main Chance* and *Budgie*. The bad language in *The Substitute*, a play transmitted on 27 June 1972, was passed as acceptable by Authority staff, but Lord Aylestone tuned in as an ordinary viewer and listened to some of the dialogue with deep disgust. So did Bernard Sendall, who instructed his staff that the amount of bad language in drama was no longer to be tolerated: 'Doubtless in company with many thousands of others I switched on at 10 p.m. for the news, not knowing programmes were running late because of the boxing. The first words that came out of the screen were a stream of foul language.'[33]

In the spring of 1973 the volume of complaints was still growing, and the companies were again urged to take positive action: 'The Authority is in no sense seeking to apply unrealistic and outmoded constraints on programme creators. Our attitude over the Warhol documentary and *A Point in Time* should be sufficient evidence of that. Nevertheless. . .'[34]

The Warhol programme had brought Mary Whitehouse out in full cry, somewhat debasing her currency by lobbying viewers to complain.[35] Whether judged an angel of light or an enemy of freedom, she attracted admiration for sheer staying power. The 1970s and 1980s were the years of her maturity as a campaigner. Undeterred by a hard-fought and unrewarding duel with Sir Hugh Greene's BBC, she battled on against what she and her supporters found to be flagrant examples of godlessness and permissiveness in society in general and the television services in particular.

Against ITV, as against the BBC, she maintained an unremitting barrage of complaint. Alerted by members of her National Viewers' and Listeners' Association (NVALA) – a section of the community easily shocked – she adopted the sound policy of holding those at the top responsible for the sins of their subordinates. Throughout the decade she pestered the Chairman and Director General of the Authority and company chairmen and managing directors by letter, telephone and, in extreme cases, telegram. In even more extreme cases she invoked the intervention of the Home Secretary or the Attorney-General.

On the taste-and-decency shelves of the IBA's archives are files bearing such labels as Doubtful Plays, Infidelity Plays, Swearing in Programmes, Drinking in Programmes, Alleged Cruelty to Animals, X Certificate Films, Horror Films and P. J. Proby. These monuments to the Authority's attention to its social responsibilities are supported by numerous files of correspondence with NVALA. The subjects of those relating to this period run from protests about, for example, Mick Jagger in *Frost on Saturday* (1968) and *Banglestein's Boys* (1969) – 'the most obscene play' one NVALA member had ever seen – through a peculiarly reprehensible Christmas in 1973 (*Carry on Christmas, Billy Liar* and a *Benny Hill Show* 'preoccupied with breasts and bottoms') and *There'll Almost Always Be an England* in 1974 ('14 bloodys, 2 buggers, 1 sod it, 1 bastard and 1 sodding thing') to the climax of 'fucking Christians' in Dennis Potter's *Blade on the Feather* in 1980. Lady Plowden's long and painstaking defence of the use of this offensive term in the context of the play left Mrs Whitehouse almost as shocked by the 'attempted justification' as by the phrase itself.

Mostly the Authority was not her real enemy: rather, she was invoking Brompton Road as a bastion of virtue against the sallies of what she saw as the trendy media shockers employed by the companies. This occurred in January 1973 over a programme on contraception in a Granada series for schools, *The Facts are These*. In the *Daily Telegraph* the producer was reported as saying: 'I think we must now accept that many young people are going to sleep with each other between the age of 16 and the time they marry. The programme does not pass moral judgments on them. It does not say sex is wrong.'[36] In response to her appeal after reading this, Brian Young assured Mrs Whitehouse that the programme was 'an utterly responsible contribution to the moral education of secondary-school children'. He went on:

> I am grateful for your kind remarks about your confidence in me, but I am bound to observe that you are less than fair to those responsible, staff and advisers, at Granada. Neither I personally nor the IBA need to provide a barrier to protect the young of this country from corruption by the educational broadcasters working in that company. They made this programme with our consent and approval, with the help of the Health Education Council, and with the enthusiastic support of the mature and moderate men and women who serve our Schools Committee.[37]

In October 1976 she was pressing the Attorney-General to issue a statement to the effect that the Authority was failing in its duty to observe the good taste and decency provisions of Section 4 of the Independent

Broadcasting Authority Act 1973, citing in evidence *Ladies Night* (an ATV regional programme about which no one complained except Mrs Whitehouse), *Hitting Town* (a play about incest: ten complaints received), *Violent Summer* (a critically acclaimed film: seven complaints received) and *Bill Brand* (an eleven-part serial which attracted thirty complaints in all).

The Authority felt no need of third-party statements to reinforce its wholly adequate powers of control under the Act. In a letter to the Home Office it made the point that if, as Mrs Whitehouse seemed to suppose, the Authority's duty was to suppress anything which could give offence to anyone, then it would have had to suppress both ITV's winning entries at that year's Prix Italia, the most prestigious of international television festivals. These programmes, both made by Thames, were *The Naked Civil Servant*, a play about a homosexual, and *Beauty, Bonny, Daisy, Violet, Grace and Geoffrey Morton*, a documentary about shire horses which contained an explicit sex scene of a mare being served.[38]

As with violence, the Authority's policies on decency and taste were acceptable to the majority and gave occasional grave offence to minorities. This was known from attitude surveys conducted at least once a year, which regularly included questions about offensive material heard or seen on television. Additionally, during the first three months of 1976, a special survey was held at the request of the Authority's Complaints Review Board. This required every ITV company to log and report all viewers' complaints received during the quarter. When analysed, these revealed that the preponderance of concern was not about offensive material at all. What really annoyed viewers were matters affecting schedules: changes in the time or day of the screening of favourite programmes; programmes starting earlier or later than announced; and programme substitutions.

In June 1976, following this survey and more discussion by the Programme Policy Committee, the Authority made a formal Review of Family Viewing Policy in Relation to the Requirements of the Act. Nothing much had changed, it transpired. In situation comedy and light entertainment many battles were still being fought over what was or was not acceptable, mostly within the companies. Authority staff, who were in daily communication with producers, lamented the incurable belief of some writers that the attention of audiences was best held by the regular insertion of innuendo into comedy scripts. The main problem, as the officers of the Authority saw it, was to persuade writers that their own best interests were served by taking into account viewers' opinions about offensive material. They pointed to the shining example of *Morecambe and Wise*, but others could point to the BBC's equally popular *Till Death Do Us Part*. In drama expletives were not easily expunged from the dialogue of modern, realistic plays, and scenes of couples naked in bed were becoming a routine feature. Nevertheless, as company spokesmen argued, television was far behind the cinema in permissiveness.

Caught in the cross-fire between accusations of boldness and timidity, the Authority, after due consideration, decided not to tighten its family viewing policy, which required that programmes shown before 9 p.m. must be suitable for an audience which included children. This did not mean the same degree of control at all times before 9 p.m. and none afterwards. Control was applied progressively throughout the evening and related to promotions and 'trailers' and advertisements as well as programmes. It was not assumed that no children would normally be viewing after 9 p.m., but parents were expected to assume responsibility for what programmes their children watched after that time. Members were in full agreement that a schedule of programmes so innocuous that it would at all times contain nothing unsuitable for an audience of children would be indefensible.[39]

Sex In Our Time was certainly not for family viewing. It was a series of eight one-hour programmes made by Thames for transmission at 11 p.m. (London only) during the autumn and winter of 1976. Should the Authority approve transmission and face an outcry from the guardians of public morality, or should it face derision by suppressing discussion of a topic being openly aired in other media? Thames had knowingly undertaken production without prior consultation. It then submitted Authority staff to shock treatment by inviting them to preview some frank programmes already completed and read some franker scripts for programmes as yet unmade. The Authority's staff responded by insisting on the withdrawal of the series from the company's autumn schedule pending further consideration. Thames had taken the precaution of securing advance press publicity for the series, and this ensured that its suppression would become a public issue.

A liberal stance had been adopted in approving other Thames programmes dealing with sex. *The Naked Civil Servant* had been one example; *Problems*, an educative series about sexual adjustment transmitted at midnight, was another. *Sex In Our Time* did not pretend to be educative and it took no moral line: its standpoint was one of dispassionate observation. One programme – about the commercialisation of sex – dwelt uncritically on the activities of pornographers, and it seemed that this might unintentionally serve as an advertisement for their business.[40]

Seven out of the eight programmes had been made when the Authority referred the matter back to the Thames board for 'clarification' of the company's own judgment on what its programme department was intent on screening. Cuts and re-editing were considered, but the company then issued a press statement: 'On reviewing all the material, it has been decided that the programmes could not be transmitted without offending some viewers, even at a late hour. Thames will not, therefore, include these programmes in its forthcoming schedules.' Predictably, this caused a rumpus within the company. ACTT members threatened industrial action and the press reported 'Threat to black out TV over sex show' (*Daily*

Mail), 'Censorship in Our Time' (*Financial Times*), and 'Virgin Screens' (*Guardian*).[41] But if the assessment of a senior programme officer at the IBA was sound, parliament had left the Authority and the company's directors no choice. 'In my view,' he reported, 'to transmit these programmes would be to drive a coach and horses through Section 4(1) (a).'[42]

The brief television appearance of an uncouth pop group gave the Director General an opportunity to defend the Authority's policy of cautious liberalism in a speech in February 1977. Frankness about sex on television could, he believed, break the ice in families and lead to less hypocrisy. As for bad language:

> I dislike as much as anyone else the casual blasphemy, the cheap coarseness; but, when watching television, I know that I am eavesdropping on many different situations. If I ask that every conversation in a play or a documentary shall be carried on with full awareness that millions are listening in, I have a prescription that will totally banish all realism. It is a question of finding the right balance between what actually happens and what is tolerable for the viewing group. Inevitably some viewing groups will claim that the broadcasters allow too much, and some who are concerned with realism will claim that they are too fussy. In this context, I wonder now whether the appearance of the Sex Pistols on television was quite the total disaster that some imagined. Ninety seconds of a most unpleasant life-style and most unpleasant language came on screen, thanks to a mistake or two undoubtedly; and as a result several million people saw the kind of thing which was being offered as 'entertainment' for teenagers. The result appears to have been that, far from being attracted by all this, the public asserted that it didn't want more of it – and bang went the Sex Pistols. No bad thing you or I might say. Yet there were those who wrote to tell us that the work of years, by many devoted teachers, inculcating good habits of speech and behaviour, had been undone by this brief moment of bad behaviour seen on television. This is to assume (as too many moralists do) that the result of seeing disagreeable elements in life is that people like and imitate these. Not so. People are sturdier and more sensible than their would-be protectors allow.[43]

The Champions (Granada) was a play about football fans judged fit for screening only by a majority vote of members of the Authority after a preview. Rich in offensive but authentic language and transmitted therefore at the late hour of 11 p.m. (on 12 March 1978), it offered an opportunity to test public opinion, which, as so often, proved ambivalent. A thousand adults were interviewed the next day. Of those who had seen the play 79 per cent thought ITV was right to screen it, while 45 per cent condemned the bad language.[44]

Another charge levelled against the broadcasters was blasphemy. In March 1977 a letter from the Archbishop of York notified the Chairman of the IBA of a motion passed by the General Synod of the Church of England by 205 votes to 9. This criticised ITV and the BBC for repeatedly allowing the name of Jesus Christ to be dishonoured in their broadcasts. In the eyes of the Church the use of the name of God or Our Lord as a crude expletive was too prevalent on television.

But Lady Plowden in reply refused to accept the validity of this criticism, arguing that close monitoring over the years provided no evidence to support use of the word 'repeatedly'. Occasional errors of judgment did not justify such a sweeping condemnation. 'We use our influence,' she wrote firmly, 'to create a climate of opinion within our organisation which recognises, and is responsive to, the viewpoints expressed in the Synod's debate.'[45] In drama, she claimed, angry invocations to 'Jesus Christ!' and 'God Almighty!' were uttered naturally but not excessively by characters usually portrayed as having no religious beliefs.

It was the use of secular expletives which was less defensible. The pendulum was swinging away from the tame 'damns' and 'bloodys' of yesteryear towards racier colloquialisms like 'piss off' and 'sod off' and vulgarisms like 'bugger' and 'shit' which were emerging from bars and clubs into more open use. This new generation of common swear-words aroused protests and complaints not provoked by the old, and in drama and documentaries no satisfactory criteria could be devised which would separate swearing as an authentic element from swearing as a self-indulgence.[46] What was called in question was not so much broadcasters' morals as their manners.

The number of offended viewers picked up in the IBA's regular surveys dropped after the mid-1970s. But in June 1979 members were once more expressing concern about a decline in standards of taste in family viewing time,[47] and at the beginning of the 1980s, while complaints generally were fewer, those categorised as 'taste and decency' were rising again.[48]

In an age of changing moral attitudes and shifting boundaries of taste, some lapses by broadcasters may perhaps be excused. Television would be unimaginative and unambitious if it never sought to cross cultural frontiers, including the frontier of taste. On the other hand, violations of the taboos of language and behaviour are more deeply upsetting in the intimacy of the family circle than in the privacy of a book or in public performances in theatres and cinemas.

Those points were made by the Annan Committee, which delivered its verdict in these words: 'The true offence occurs when broadcasters do not learn the lesson about the current state of taste, and repeatedly violate it because they do not care or want to shock for the sake of shocking. Have the broadcasters been guilty of this offence? Certainly a sizeable part of the public think they have.'[49]

10

POLITICAL PROPAGANDA

Politically, the 1970s were enlivened by four general elections, an extensive reorganisation of local government and three potentially momentous referenda.

Broadcasters were faced with new legislation which complicated rather than simplified the law relating to electoral broadcasts. The provisions of Section 9 of the Representation of the People Act 1969 contained two important prohibitions. Before the latest time for the delivery of nomination papers no programme could be broadcast 'about a constituency' if any candidate took part in it. After closing time on nomination day broadcasts in which one or more candidates took part might proceed only with the consent of all candidates, whether or not they themselves took part. These rules raised nice questions of interpretation. What did 'taking part' mean? How should 'about a constituency' be defined? Who qualified as a candidate before nomination day? The Authority issued to the companies a series of notes of guidance on all such matters affecting the appearance of candidates in election broadcasts.[1]

Then there were political broadcasts over whose presentation and content the broadcasting authorities had virtually no control despite their responsibilities under the law. These had long been one of the more interesting anomalies of the broadcasting scene in the United Kingdom. They existed before the creation of Independent Television and were assimilated into the system as a reluctant inheritance.

An agreed number of minutes in each calendar year was made available to the main parties represented in the House of Commons for Party Political Broadcasts (PPBs), each party's share being proportionately related to the number of votes received at the previous general election. Additional time was afforded for a similarly agreed number of party statements during a general election campaign (PEBs), when time was also made available for broadcasts by any minority party fielding at least fifty candidates, even if not represented in parliament.

Further airtime was granted on request for government statements by the Prime Minister or other responsible ministers. These 'ministerials' were purportedly uncontentious, 'purely factual' expositions of some new item of legislation or development in national policy. In this context an unstated constitutional principle of parliamentary democracy differentiated between a politician addressing the nation as spokesman for the government in the exercise of its duties and the same politician speaking as the representative of a political party in power. If therefore he not only stated and explained but also sought to defend, justify or persuade, another kind of ministerial followed as soon as was practicable: a broadcast reply by the Leader of the Opposition or a shadow minister. Under the agreed formula this then triggered a ministerial of the third kind: a programme in which official spokesmen for the government and the opposition parties appeared together to discuss and argue the issue. The annual budget broadcast by the Chancellor of the Exchequer was normally followed instead by solo broadcasts from the two main opposition parties.

Most members of the broadcasting community regarded these party 'commercials' and the arrangements whereby they were brought about as an unnecessary encumbrance. They believed that left to their own devices, given their commitment to due impartiality and fair dealing in the treatment of all matters of current public policy, they could provide alternative presentations of party doctrine and policies which would serve the democratic process more effectively; particularly in the confrontational situation of an election campaign when party spokesmen sometimes seemed more concerned to use their allocation of privileged airtime for discrediting opponents instead of illuminating issues and elucidating policies.

Support for this belief was to be found in the coverage of two events unique in British political history: the referendum campaigns on continued membership of the European Economic Community (June 1975) and on Scottish and Welsh devolution (February 1979). In both cases arguments for and against cut across party lines, and there were no PPBs. As a consequence the two broadcasting bodies had a much more influential part to play in ensuring adequate and impartial informative coverage, all of which they devised and originated themselves.

Independent Television was under no legal or formal obligation to transmit any of the PPBs, PEBs and ministerials broadcast by the BBC by arrangement with the parties. But, to avoid the possibility of a requirement under the 1964 Television Act that PPBs and PEBs shown by the BBC should be simultaneously relayed by the ITV network, the Authority had bowed to pressure and given an undertaking to screen these broadcasts 'based', as the Postmaster-General of the day put it, 'on an offer made by British Broadcasting Corporation after consultation with the Authority'.

This annual 'offer' had originated in an *aide-mémoire* described in its preamble as 'the official record of the terms of the agreement reached in February 1947 between the government, the opposition and the BBC'. It was the product of discussions held after the first post-war general election between the BBC's Chairman and Director-General and senior representatives of the outgoing Conservative government (Churchill and Eden) and the newly elected Labour government (Herbert Morrison and Arthur Greenwood). These took place in the Prime Minister's room at the House of Commons, and both parties' Chief Whips were present. The tone was set by Greenwood, the Lord Privy Seal, when he said:

> The government do not think it desirable to reduce to written rules the principles which should govern the BBC in regard to political broadcasting... The principles adopted must depend on good sense and goodwill. It is as impossible to formulate exhaustive principles on paper as it is ... to define what conduct is unbefitting an officer and a gentleman. The few short principles indicated should therefore by regarded as a mere *aide-mémoire* of the conversation. [2]

The agreement was thus without statutory or other force, and it was concerned with the use of 'the wireless'. In 1955 doubt was expressed about ITV's eligibility to become a party to it because 'with commercial television we are, by definition, not dealing with gentlemen'. [3] A revised version was produced in 1969, but this confined itself to ministerials: most importantly, it removed the right of reply from the BBC's discretion and made it automatic on request in cases where there was no general consensus of public opinion. Otherwise the 1947 agreement improbably survived unchanged as an instrument of television policy. Both 'mere' *aide-mémoires* are reprinted here: as Appendices D and E. Neither, it should be noted, regulated coverage of speeches at the annual party conferences. There, by omission, politicians trusted broadcasters to be even-handed in treatment of what were assumed to be newsworthy events. Controversial statements in those settings were not thought to necessitate the requirement of a right of reply.

There was, in theory at least, nothing automatic about the resultant arrangements for the annual series of party political broadcasts. Although ignored in the 1969 *aide-mémoire*, they were the most substantial element of propaganda on British television. Their number, their timing and the total airtime which they occupied were the subject of often heated debate but always eventual agreement at regular meetings, held at least once a year, with the Lord President of the Council in the chair.

Present were senior spokesmen from the parties, normally the Chief Whips and sometimes the Prime Minister and Leader of the Opposition. The broadcasting authorities were represented by their Directors General and senior programme officials, and the ITV delegation included one chief executive to speak on behalf of all the companies. What began in 1946 as a small *ad hoc* gathering grew into a committee of between fifteen and twenty, with the three main parties even introducing professional media advisers. There was reluctance to enlarge it further. Only in 1974 did the Scottish National Party and Plaid Cymru win the right to be represented. In 1981 the newly formed Social Democratic Party was allowed only to address the committee.

The constitutional status of this Committee on Party Political Broadcasting is not easy to define. Despite the attendance of a Lord President and sometimes a Prime Minister, the conspicuous absence of the minister with specific responsibility for the broadcasting services was an indication that the committee should not be regarded as an integral part of the state machine for the supervision and ultimate control of television. PPB committee meetings were periodical encounters of broadcasting top brass with contending political parties rather than with the government of the day and the official parliamentary opposition. The pattern of broadcasts which they arranged for the year ahead was the product of two sets of compromises: on the one hand between the parties and their separate demands; on the other between the politicians and the broadcasting bodies. Agreement was seldom reached without asperity.

Usually there was a previous discussion between the Authority and the BBC in order to present a common front. Where the interests of the two services were at variance a workable compromise was most often, but not always, reached. In October 1971 Brian Young, on behalf of ITV, felt unable to go along with a suggestion from Charles Curran, the BBC's Director-General, that the parties should be told to accept a reduced annual allocation under threat of having none at all if they refused.[4]

Party broadcasts for the 1970 general election campaign (PEBs) were agreed on the basis of five each for Labour and the Conservatives and three for the Liberals, all of ten minutes. The Scottish and Welsh nationalist parties (SNP and Plaid Cymru) were allowed one five-minute broadcast each, to be shown only in their own countries. A similar allowance of time was allocated to the Communist Party, which fielded more than fifty candidates.

The Authority, the companies and, without doubt, the viewers were pleased that the total of 145 minutes represented a reduction from the 165 minutes of the 1966 general election when some of the broadcasts for the three main parties lasted fifteen minutes. ITV's own election campaign coverage (with 1966 comparisons in brackets) totalled: ITN 3 hours (5½), networked programmes 2½ hours (as in 1966), regional programmes 20 hours (33½). Granada was off the air throughout the campaign because of an industrial dispute, and Harold Wilson gave this loss of coverage in the north-west as a reason for Labour's narrow defeat.

The Labour and Conservative PEB rations of fifty minutes each were unchanged during the remainder of the decade. The Liberals' thirty minutes were increased to forty for the second general election in 1974 but reduced again to thirty in 1979. The SNP's allocation rose steadily from five minutes in 1970, to ten in February 1974, twenty in October 1974 and thirty in 1979, but Plaid Cymru were restricted to ten minutes in both the 1974 elections and in 1979. In 1979 the National Front, Ecology Party and Workers Revolutionary Party were all able to claim five minutes under the fifty-candidate qualification.

These arrangements conformed to no clear-cut formula. They were, remarked Edward Short, Lord President, in September 1974, 'a matter of arrangement without any specific basis'.[5] Voting strength at the previous general election was not necessarily a reliable criterion. There were complicating factors such as the increasing volatility of the electorate, the evidence of mid-term by-elections and even, in the case of the SDP from 1981, the emergence of a new party.

In 1956, at the beginning of ITV's involvement, Sir Robert Fraser had initiated a series of talks with party leaders in the hope of reaching agreement on a better way of providing access to television for party propaganda within the constraints of the law. They failed; as did more than one such initiative in later years. The evidence suggests that this was largely because any party which demonstrated constructive interest was invariably suspected by its opponents of seeking to gain some tactical advantage.

In 1978 the two broadcasting authorities put forward a proposal to allot ten minutes for every 10 per cent of the votes won at the previous election. This would have given Labour and the Conservatives four ten-minute broadcasts instead of five and the Liberals two only. The five-minute allocation for minority parties with fifty candidates would have been abolished: it was deeply unpopular with the Scottish and Welsh nationalists because it meant that the National Front and the Communists, with no MPs, had access to the whole electorate and were given as much time as the SNP or Plaid Cymru to address the Scottish and Welsh electorates irrespective of the number of their candidates in those countries. It was therefore proposed that such minority parties should have their broadcasts shown in Scotland or Wales only if at least 10 per cent of their nominated candidates were standing for Scottish or Welsh seats. The major parties were not persuaded, however, and all these proposals were stillborn.

The annual series of PPBs unrelated to general elections followed roughly the same pattern as the PEBs. In each of the years 1969 to 1972 140 minutes were divided in the ratio of five broadcasts each for Labour and Conservatives (2 × fifteen minutes and 3 × ten minutes) and two of ten minutes for Liberals plus one of five minutes each for the SNP and Plaid Cymru in their own countries. Towards the end of 1970 the broadcasting authorities, with Lord Hill in the lead, gave serious consideration to a substantial reduction, but the BBC wanted this linked to an agreement to abandon the established practice of simultaneous transmission on all channels. BBC insistence on this condition was based on the assumption that, with viewers switching away from PPBs by the hundred thousand, non-simultaneity would produce such irrefutable evidence of the unpopularity of these broadcasts that the political parties might themselves feel disposed to ask for fewer or even none at all.

Over this proposal the Authority adopted a high-minded stance, arguing that the decision to be taken was one of principle: either the broadcasts should be retained on a simultaneous basis as a service to the public or they should be abandoned because they were unpopular.[6] Mindful of the BBC's two-to-one advantage in channels, the Director General also struck an earthier note in describing agreement to the abandonment of simultaneity as unilateral disarmament.

Conservative Party spokesman were not averse to the notion of non-simultaneity, but Labour and the Liberals were against it,[7] and when the PPB Committee met on 17 February 1971 it was agreed that simultaneity should continue and the allocation of time remain at 140 minutes.

A year later (on 27 January 1972) the Director General reported to the Authority that another meeting of the committee had again agreed on maintenance of the *status quo*. It had rejected a suggestion by Edward Heath that the three main parties should have different broadcasts in each of the national regions of the United Kingdom in order to meet the challenges of the local nationalists. Also rejected was a proposal from the politicians that they should be allowed a larger number of broadcasts of shorter length within the same time allocation. In the following year, however, the Liberals were permitted to take their twenty minutes in the form of one fifteen-minute and one five-minute broadcast.

In May 1974 pressure from the politicians for an increase to 150 minutes met resolute and united opposition from the broadcasters, and a proposal by the broadcasters that more time should be given to the Liberals and the Welsh and Scottish Nationalists at the expense of Labour and the Conservatives broke the crust of restraint and brought latent hostility to the surface. An explosive outburst of anger from Heath was sufficiently violent for the strictly confidential proceedings of the PPB Committee to be leaked to the press. As *The Times* reported, the meeting had 'adjourned in an atmosphere of recrimination'.[8] 'Over the past twenty years I have been attending these meetings,' Heath is reported to have said, 'I find them the most useless and time-wasting I have ever attended.'[9] Bob Mellish, the newly elected Labour government's Chief Whip joined the fray to remind the two Directors General that the airtime under dispute was not theirs to dispose of as they thought fit. It belonged, he said, to 'the British electorate and the British people'. According to the *Guardian's* political correspondent, Heath's 'majestic' attack on the BBC delighted other party representatives.[10]

The outcome of this dispute was a diplomatically framed proposal from the committee secretary that each party's entitlement of PPB airtime should in future be related more specifically to votes polled at the previous general election: 'The rate might be a ten-minute broadcast for each two million votes... Any party with a remainder of votes exceeding one million votes would be entitled to an additional broadcast.'[11] He also suggested that 10 per cent of the local votes should give the national parties in Scotland and Wales the right to ten minutes of time. These proposals were approved at a meeting of the committee on 11 September 1974, and in 1975 the allocation to the three main parties, based on their polls at the previous general election, was 6:6:3, each broadcast lasting ten minutes. This annual apportionment continued throughout the remainder of the 1970s.

The broadcasters' lack of control over the content of PPBs was troublesome. In October 1969, for example, Jeremy Thorpe, leader of the Liberal Party, used one to ask viewers for their votes in three forthcoming by-elections and for contributions to party funds. The Authority was reassured by a ruling from the government's law officers that the reference to by-elections was not in breach of the Representation of the People Act. But the secretary of the PPB Committee had to be informed that appeals for funds were not permissible, and this point was duly minuted at its next meeting.

Yet on 1 September 1971 another direct appeal for contributions to party funds was made, this time by Ian Mikardo for Labour. It provoked a strongly worded note to the BBC from Francis Pym, the Conservative government Chief Whip, and the Corporation was obliged to eat humble pie. After the Thorpe broadcast it had undertaken to ensure that party spokesmen recording a PPB in BBC studios should be told of the decision then taken. On this occasion it had failed to do so.

To ITV Mikardo's broadcast was even more offensive. He named several large companies who were not only contributors to Conservative Party funds but also advertisers on ITV, urging viewers not to buy their goods. A stiff letter from Brian Young to Charles Curran was drafted enquiring what steps might be taken to ensure that the PPB feed which ITV was trustingly taking sight unseen from the BBC did not contain material in breach of 'our Act or your Charter' and did not go against accepted BBC or ITA practice. But it was then judged that less formal representations might be more effectual and the letter was not sent.

On Labour Day 1972 the Labour Party committed a gaffe when a PPB by Edward Short on the subject of unemployment showed the premises of Landis & Gyr Ltd, meter manufacturers of Acton, West London, which according to the commentary had gone out of business. The statement was untrue, and the ITA and BBC received letters from solicitors acting for the firm. A public apology and payment of legal costs were demanded. Discussions between lawyers representing the broadcasters, the party and the firm then agreed a form of words for an apology by the Labour Party, and this was broadcast in the break before *News at Ten* on 15 May. The Authority associated itself with this apology, but the BBC in its transmission did not. All costs were paid by the Labour Party.[12]

These incidents were doubly embarrassing for ITV. Not only was the network accepting material for retransmission without any foreknowledge of what the political parties were intending to say or show; it was also receiving this material from another broadcasting organisation, taking it on trust that care was being exercised to avoid the inclusion of indefensible matter. As long ago as 1961 advice about the Authority's legal responsibilities for the content of such broadcasts had been sought from Counsel, and his advice had been unequivocal. Noting that a party political broadcast was not unlikely to contain defamatory matter, he did not think that the defence of innocent dissemination was available either to the Authority or to the companies.

In 1973 substantial changes were made to the system and structure of local government. Partly as a consequence there was an unusual bunching of local government elections: English and Welsh counties and the Greater London Council on 12 April; Metropolitan districts and the Welsh boroughs on 16 May; and non-metropolitan districts on 7 June. The political parties manoeuvred accordingly to get their PPBs strategically placed in relation to one or other of these polls. They also pressed for an increased number, even if individually shorter, and a special concession was made on the understanding that it was not to be taken as a precedent.

Other special circumstances arose early in 1974 with the miners' strike and the curfew on television hours. In order to accommodate a Labour Party PPB within a restricted schedule it was decided that the *News at Ten* on 23 January would have to be reduced to a ten-minute bulletin. After a strongly worded protest this was accepted by the Editor, but not by ITN's journalists. On learning of the decision the day before, they voted to withdraw their labour from 3 p.m. on the following day unless the news was restored to twenty minutes. The Father of the NUJ Chapel wrote to the Director General accusing the Authority of failing in its public responsibility during a period of national emergency. It was left to Harold Wilson, who was to be the main speaker in the broadcast, to defuse the crisis. He withdrew unconditionally his right to have it shown on all channels. ITV's original schedule was restored and the BBC alone transmitted the PPB.

Ministerials too gave rise to problems and argument. In June 1969 ITV exercised an 'all or none' option by not taking ministerials concerning changes in the Industrial Relations Bill. Instead extracts from government and opposition ministerials were included in the main ITN news bulletin on the basis of their news value. A year later the newly elected Conservative government formally requested that ITV should take all such broadcasts and agreement to do so was endorsed by the Authority at its meetings in October 1970 and by the companies at the November meeting of the Programme Policy Committee. This agreement was, however, qualified by insistence that ITV transmissions would not have to be simultaneous with the BBC's.

On 23 October 1972 Edward Heath, then Prime Minister, gave an interesting reason for rebuffing a suggestion from Harold Wilson that he should make a ministerial broadcast about an EEC summit meeting. 'I was not proposing a ministerial broadcast on TV which would give the right honourable gentleman the Leader of the Opposition the right to reply, because I have my own audience and he can find his.'[13]

A year later Bob Mellish, the Labour opposition's Chief Whip, was writing angrily to *The Times*,[14] and more formally to the secretary of the PPB Committee,[15] about a different ploy by Heath to deny his political opponents a platform from which to criticise his policies. To inform the public about important emergency economic measures being taken by the government, the Prime Minister chose a press conference instead of a ministerial broadcast. It was held in London at Lancaster House. His ten-minute statement plus exchanges with journalists lasted for an hour, and the proceedings were televised in full by the BBC, while ITN screened the opening statement but no more. 'It was,' wrote the *Evening Standard*'s political editor, 'a government propaganda exercise of the first order with newspaper journalists playing bit parts for the benefit of the Prime Minister and the television outfits.'[16] BBC News and ITN broadcast extracts in their regular bulletins and each offered the Leader of the Opposition between one and two minutes for his comments.

Mellish's anger was understandable. It was under the previous Labour government and at Wilson's instigation that the *aide-mémoire* agreement had been amended to give the opposition an automatic right of reply to a ministerial broadcast held to be controversial. The change had been made, Mellish wrote, in 'recognition of the increasing, now vital, part which television and radio play in the functioning of democracy. But democracy requires that everyone is given a chance to play by the same rules. Mr Heath has now changed those rules.' This was, however, not quite the case. Heath was escaping the commitment to an opposition ministerial through a loophole opened by Section 3 of the 1947 *aide-mémoire*. This had not been amended, and it explicitly excluded outside broadcasts from the agreed arrangements for a right of reply (hence the exclusion of party conferences).

In September 1974 Wilson invoked the precedent of a practice adopted by Attlee and Eden to claim a ministerial in order to announce a coming general election. This was in effect the first step in the election campaign, and it was regarded by broadcasters as a trick for obtaining more than the agreed airtime for party election broadcasts. It was thought to abuse the system by blurring the all-important distinction between a politician speaking in his role as a minister and the same person appearing as spokesman for one of the contesting parties. But the objections of the broadcasters were unavailing and the practice was followed again by James Callaghan in 1979.

During 1974, the year of two general elections, PEBs occupied 155 minutes of airtime in February and 175 in October. In March, July and November there were ten-minute budget broadcasts by government and opposition totalling sixty minutes. No fewer than four series of second-category ministerials took a further 120 minutes. By mid-November the grand total amounted to between nine and ten hours, and the parties still had in hand their PPB allocation for the year: another fifty-five minutes (in Scotland sixty-five). This they decided to use in the run-up to Christmas, although schedules planned to give viewers a much-sought respite from a full year of political in-fighting had already been prepared by the companies and approved by the Authority.

'In the interest of the parties themselves I cannot help feeling that four successive Wednesdays of party broadcasting at this time would be thought to be excessive by viewers whatever their political persuasions,' wrote Brian Young to Mellish.[17] The response from the parties followed the customary pattern. Labour, being in government, was prepared to consider the request sympathetically. The Conservatives, being in opposition, were not. Willy-nilly therefore, the broadcasts had to be accommodated.

It had been agreed by all concerned that the 1974 PPBs would be shown simultaneously by both services, but the BBC now claimed to be unable to screen them at the agreed time of 10 p.m. It insisted on 9 p.m. instead. Difficulties in making any other adjustments to BBC schedules at that late stage were said to be insuperable, but the problem was no less awkward for ITV's schedulers. The companies protested strongly to the Authority, demanding that those appointed to command them should also protect them and not allow the Corporation to break an agreement freely entered into.

But the Authority was chary of inviting intervention by the political parties and, so far from helping the companies, chose to break the impasse by issuing a directive compelling them to accept a 9 p.m. transmission time for the Labour Party broadcast on 20 November, even though it was already billed in *TVTimes* for 10 p.m. The subsequent Conservative PPB went out as scheduled at 10 p.m., but the two remaining ones (Labour and Liberal on 11 and 18 December) were again broadcast at the unwelcome hour of 9 p.m. A press statement informed infuriated members of the public that the change in advertised transmission times had been the result of BBC recalcitrance.

This imbroglio engendered a more realistic attitude towards simultaneity, which had been a persistently contentious issue at PPB Committee meetings down the years.[18] The Labour Party was the most resistant to any change. Wilson went so far as to claim that abandonment of simultaneous transmission would be a denial of democracy.[19] It would rob the Labour Party of its right to speak to the nation, Tony Benn claimed at a meeting of the Labour Party National Executive in February 1980. These were implied admissions that, as audience research consistently suggested, most viewers would rather watch any other programme than a party political broadcast. No other programmes registered such a low count on the Appreciation Indices. PPBs and PEBs were pills thrust down the throats of the electorate.

However, in 1975 the Conservative Party under its new leader, Margaret Thatcher, decided to experiment with non-simultaneity. Its PPBs were transmitted at 9 p.m. on BBC1 and BBC2 and at 10 p.m. the same evening on ITV. Research carried out over the following two years gave the Labour and Liberal simultaneous transmissions higher total ratings, but the Conservatives won significantly higher Appreciaton Indices (by attracting mainly party supporters).

By 1980, with the exception of one broadcast for each party immediately preceding local elections, all PPBs were being transmitted non-simultaneously, although, so far as the parties were concerned, only on an experimental basis. Conservatives, Liberals and the broadcasters favoured continuance, but in 1981 the Labour Party insisted that at least half of its annual allotment should be simultaneous. Fresh research had shown a larger fall in total audience size as viewers came to realise that they were being allowed to escape to a different programme, and by this time Appreciation Indices were not noticeably higher.[20]

Two of the other issues in contention throughout this period – the fair treatment of minority parties and the timing of PEBs – came together in February 1974 to spark a dispute which shook the rickety system of oral understanding and gentlemen's agreements by inviting the intervention of the law.

In the late 1960s and early 1970s the Scottish and Welsh nationalists gained more votes than airtime. In the 1966 general election SNP's share of the votes was more than 5 per cent; in Wales Plaid Cymru's was 4.3 per cent. Yet in 1969 their five-minute allocation of PPB time out of a total of 145 minutes amounted in each case to 3.5 per cent. Bitterness at this disparity was aggravated when on 9 April 1969 a fifteen-minute Labour PPB was used for an attack on these two five-minute parties. On 17 April a notice appeared on the House of Commons order paper in the name of Winifred Ewing (SNP) and Gwynfor Evans (Plaid Cymru) deploring 'this dishonest procedure' and asking the two major parties to prevail upon the broadcasters to grant more time to enable them to exercise their 'democratic right of reply' to the attacks made on them.

The ITA's Director General was moved to write a diplomatically phrased note to the government Chief Whip, John Silkin, suggesting that the broadcast had not been helpful to the Authority in its efforts to prevent the nationalists from gaining more time.[21] The Authority's objective was less time for the larger parties, not more for the smaller. It was also unsympathetic to demands for additional regionalised transmissions. Because transmitters and their areas of coverage did not and could not take account of the national boundaries, this would mean subjecting populations on both sides of Scottish and Welsh border regions to excessive doses of PPBs.

When in the 1970 general election the votes of the Scottish nationalists in Scotland rose to 12 per cent and Plaid Cymru in Wales to 11.5 per cent, the Authority took the view that retention of the five-minute allocation could be defended provided that the major parties were persuaded to accept a reduction in their annual total from 140 minutes to 105.[22] This proposal was put to the parties at a meeting of the PPB Committee early in 1972, but it was not supported by the BBC and was not agreed.

Pressure from the nationalists by correspondence and in the press continued. A memorandum from Plaid Cymru to the PPB Committee drew attention to the party's 37.1 per cent of the poll in a parliamentary by-election at Merthyr Tydfil on 13 April 1972. It demanded an increase of annual television time to thirty minutes and representation at meetings of the committee. But the UK party leaders resisted any concession, and the BBC was bound by an undertaking given in 1965 by its Chairman to the Postmaster-General that it would make no allotment of time for regional PPBs without the agreement of the three parties represented on the PPB Committee. Although the Authority was not hampered by this commitment, it was clearly not feasible for ITV to act unilaterally.

By the end of 1973, after intense discussions involving their representative Broadcasting Committees in Wales and Scotland, both Directors General were able to inform the Labour and Conservative leadership that members and governors were at one in wishing to increase the Plaid Cymru and SNP annual PPB allotments from five to ten minutes, and that should the UK parties block the proposal they would have no alternative but to make the whole matter public.[23] The proposal was agreed on 4 February 1974 at the committee's next meeting, but before it could be implemented the first of the two 1974 general elections intervened to bring the claims of the nationalists into the full limelight of publicity.

The timing and successive order of election broadcasts had to be meticulously scheduled to meet the requirements of fairness and due impartiality. Transmission times were chosen to allow each of the parties access to audiences of approximately equivalent size. For the three large UK parties these were to be at 10 p.m. simultaneously on BBC1, BBC2 and ITV, the first three broadcasts to be made successively by the government party, the main opposition party and the Liberals; the last three successively in the reverse order.

In Wales and Scotland negotiations over suitable transmission times for the nationalist parties' local broadcasts were conducted between local BBC, IBA and ITV company officials, and it was agreed to offer Plaid Cymru 8.50 p.m. on 26 February (two days before election day) for simultaneous transmissions to Welsh viewers. Since the production of the broadcast was to be in the hands of the BBC it was a BBC programme officer in Cardiff who communicated the offer to the secretary of Plaid Cymru and, on its acceptance, proceeded to make arrangements for the studio recording. A billing for the broadcast was then sent to *Radio Times* and *TVTimes*. This was on 14 February.

One week later, through the offices of their Chief Whips, the three main parties lodged a joint objection to the timing of this broadcast and to one scheduled at the same time for the SNP in Scotland. When the complaint was examined by the Authority's senior staff it was judged to be justified on two grounds. First, a transmission time of 8.50 p.m. offered a potential audience significantly larger than the one at 10 p.m. Secondly, to place Plaid Cymru and SNP at that point in the sequence of election broadcasts, which was the last evening on which they were permitted, would be likely to give these parties an additional unfair advantage. The BBC concurring, they were therefore offered 5.05 p.m. on Saturday, 23 February as an alternative. Examination of rating charts had established that the number of viewers at that time would be approximately the same as at 10 p.m. on a weekday. The SNP accepted the new time under protest. Plaid Cymru went to law.

On 25 February, three days before polling day, a writ was served on the BBC and IBA jointly, claiming an 'order of specific performance' of the agreement to transmit the party's election broadcast at 8.50 p.m. the following day. The application was heard by Mr Justice Bridge on that day. The only evidence before him was the affidavit sworn by Plaid Cymru officials detailing the course of events, including two items of which the Authority had not been aware. These were the nationalists' willingness to accept a compromise time and the allegation of a BBC threat that a refusal of the alternative time offered would mean no broadcast at all.

The judge ruled in favour of the plaintiffs and made an order that the defendants permit the broadcast to be made as originally arranged. The defendants appealed, and the appeal was heard the same afternoon before Lord Denning, Lord Justice Megan and Lord Justice Scarman, who not only dismissed it but reinforced Mr Justice Bridge's ruling in the terms of a mandatory order which required both broadcasting services not merely to 'permit' but 'to do whatever necessary' to see that the Plaid Cymru broadcast was transmitted at 8.50 p.m. that evening 'on the usual Welsh channels as agreed and advertised in *Radio Times* and *TVTimes*'.

During the proceedings the BBC's Director-General informed the IBA that he was contemplating a refusal to obey a court order if the case went against them, because no court had the power to order the BBC to transmit a particular named programme at a particular named time. The IBA had been a reluctant co-appellant and would not go along with this. Sir Charles Curran was persuaded that what he was contemplating would be a serious contempt, and the programme was duly transmitted by both services at 8.50 p.m. that evening. The SNP asked for a repeat of its broadcast at the same time, but was told that in view of its acceptance of the changed time this could not be permitted.

The issue before the courts was a simple one: Was there or was there not a legally enforceable agreement? The judgment given created foreboding among broadcasters lest it be used as a precedent in the general areas of news and current affairs, where programme preparations were unavoidably subject to constant, and sometimes rapid and unpredictable, change. Arrangements were customarily made with participants and performers on the basis of oral agreement, often by telephone and at extremely short notice, and it had always been assumed that these were subject to an unspoken proviso that no binding commitment could be given to any specified time of transmission. The same proviso was held to apply to timings published in the programme journals. Changes from billed transmission times were not uncommon. Mr Justice Bridge had taken the point that the imperatives of television broadcasting imposed a measure of flexibility, but did not think it a consideration which should influence his ruling on the central issue.

The appeal judges drew attention to the absence of any evidence from the defendants. They conceded the possibility that had such evidence been provided (in the form of affidavits from the BBC and ITV officials concerned) it might have been sufficient to contradict the implication which they found in the plaintiff's statement that a binding and legally enforceable agreement did exist. But in the absence of such contrary evidence they felt obliged to exercise their discretion in the sense most likely to serve the cause of justice.

It was unfortunate, too, that the role and status of the PPB Committee was not fully explained or understood. The Court of Appeal was led to infer, erroneously, that a meeting of that committee had issued directives to the BBC and IBA about the timing of the broadcasts by the two nationalist parties. This error of fact was embodied in Lord Denning's judgment, bearing the implication that the broadcasters were acting, not on their own initiative and according to their own judgment, but at the behest of the 'big boys' ganging up to protect their vested interests against a tiresome minority group.

Another error by Lord Denning was his use of the analogy of a television advertisement – an analogy singularly ill-chosen to serve his argument. Just as the agreement to transmit a commercial was a legally binding contract, he pronounced, so also was the agreement entered into between the broadcasters and the Plaid Cymru. But it was common knowledge that advertisements shown on Independent Television on the basis of contractual agreements between advertisers and programme companies were always subject to Authority approval and, on occasion, rejection. In exercising a right to approve or reject advertisements it was doing its duty under the law, just as it was attempting to do in reaching its decision that the time first offered to Plaid Cymru had been mistaken and could not be approved.

The Authority was critical of the BBC's handling of the case. Although it should have been spotted by Authority staff and corrected, the original mistake had been the BBC's, and the incident unpleasantly rubbed home the secondary nature of the IBA's role in party political broadcasting. At their first meeting after the judgment members resolved that 'arrangements should be made to ensure that the Authority was established more recognisably as an independent party in any future negotiations with the political parties on the subject of party political broadcasts at election times'.[24]

It was during this general election campaign in February 1974 that the Labour Party committed itself to renegotiating the terms of the United Kingdom's membership of the European Economic Community. Its manifesto declared that 'the people should have the right to decide through a general election or consultative referendum'. In the October general election it promised to 'give the British people the final say' on whether to accept whatever terms could be renegotiated.

Not until the end of January 1975 was it made known for certain that this final say would be a national referendum. As late as 20 January Harold Wilson had stated publicly that 'no decision has yet been taken ... on the issue of a referendum versus a general election,' although adding that 'there does not seem to be any desire in the country for an early general election following those of last February and October'.[25] A few days after that transparent understatement the government published a White Paper outlining plans for a referendum in which voters would be invited to give a straight yes or no to continued EEC membership.[26] A Referendum Bill was introduced on 26 March and received the Royal Assent on 8 May.

The Labour Prime Minister's problem was evident. Although both had dissidents in their ranks, the Conservatives and the Liberals were in favour of continued membership as a matter of official party policy. Wilson, however, was faced with conflicting views and a strong anti-market swell within his own party. In the Cabinet Roy Jenkins and Shirley Williams were decidedly for and Tony Benn, Michael Foot and Peter Shore passionately against. The TUC as a body was opposed to membership, but some leading trade unionists and at least one large union (the General and

Municipal Workers) were not. A general election in which this was the main issue was obviously out of the question.

The broadcasters had long since reached this conclusion. From the previous summer there had been discussions between Authority and company programme staffs about coverage of a referendum campaign. The various groups in the country for and against membership were beginning to hold public meetings and in one or two cases making tentative approaches to the broadcasting bodies. Before long they were to coalesce into two umbrella groups, calling themselves respectively Britain in Europe (pro) and the National Referendum Campaign (anti).

Whether or not there were to be formal statements on the model of the party election broadcasts, it was clear that the issue must be thoroughly aired in nationally networked and regional current affairs programmes. On 4 February 1975 a Programme Policy Committee paper was issued laying down draft guidelines.[27] Time given to representatives of the opposing groups would have to be on a fifty-fifty basis. Impartiality achieved through balance over a series would be acceptable, except during the final week before the vote. Special attention would need to be given to impartiality in the regular programmes featuring local MPs which were part of several regional companies' local output. Continuing observance of balance between the parties was necessary but would no longer suffice.

It was recognised that television coverage might be the decisive factor in an issue of the greatest significance to the future of the country. The task for the broadcasters was twofold. They must aim at offering the public as complete and objective a supply of factual information as possible, presented in as broadly intelligible, readily accessible, palatable form as their skills would allow. Then they must create sufficient opportunities for the public to hear and assess the arguments put forward by each side.

An ITV inter-company General Election Working Party had been set up in 1974. It included representatives of ITN, the central and regional companies and IBA programme staff. With experience of the considerable collective planning accomplishments of that body, it was decided early in 1975 to set up a similar Referendum Working Party to co-ordinate network contributions from the companies so that, without inhibiting their differing approaches, a coherent, if varied, ITV approach to the broadcasters' dual task could be planned.

At the same time the Authority's Regional Officers were called upon to submit from January onwards regular monthly reports on all programmes and programme items shown locally which dealt, however non-controversially, with Common Market issues. From 23 January ITN kept a log of all expressions of 'pro' and 'anti' opinion included in its bulletins. Polling day was 5 June.

On 26 April expanded ITN news bulletins covered a Labour Party special conference on the referendum with time evenly balanced between the opposing factions. From 1 May the midday news, *First Report*, featured a daily Referendum Postbag to which viewers were invited to submit questions for discussion on air by speakers for and against membership. During the period of the campaign *News at Ten* carried a series of five-minute films examining aspects of the ongoing debate. On 2 and 3 June referendum phone-in programmes followed *First Report* from 1.30 to 2 p.m. On 6 June ITN's *Europe: The Nation Decides* ran from noon until 6.30 p.m., reporting and analysing results as they came in and making the final announcement that two-thirds of the electorate had voted 'yes'.

Granada provided three major network productions. The *State of the Nation* team negotiated exclusive access to the work of the EEC Commission in Brussels. They were able to film meetings of the Commissioners. The resulting programme, shown from 10.30 p.m. to midnight on 19 May, offered viewers the chance to be fly-on-the-wall observers of the Community's decision-making process. In *The Last Debate* on 2 June twelve prominent MPs – Edward Heath, Roy Jenkins, Reginald Maudling, Roy Hattersley, David Steel and John Davies (pro) and Peter Shore, Enoch Powell, Douglas Jay, Judith Hart, Neil Marten and Douglas Henderson (anti) – debated the issue within the format of a Parliamentary Standing Committee, chaired by Betty Harvie Anderson MP. *In Search of the Market* was the film of a coach tour through Belgium, Holland, West Germany, France and Italy by a group of UK passengers escorted by Lord George Brown (pro), Clive Jenkins (anti) and two economists. This was shown in two parts on the evening of 3 June.

Thames's *This Week* screened an interview with the Prime Minister on 15 May and a week later a general report on the anti-market campaign, featuring Michael Foot, Enoch Powell and anti organisations in Bournemouth and Liverpool. On 1 June LWT's *Weekend World* gave a detailed analysis of the political and economic implications of membership.

In the regions, in addition to coverage in regular local news bulletins, there were some thirty-six special referendum programmes. Tyne Tees, for example, mounted a series of five thirty-minute specials under the title, *Division on Europe*. Anglia's *Farming Diary* featured MPs with agricultural interests arguing the benefits of the EEC to farmers. In two editions of HTV Wales's *Yr Wythnos* a mixed panel of pros and antis answered questions from a studio audience.

The substantial range and variety of ITV's well-prepared coverage amply explored the relevant economic, political and social issues and interests in so far as the limitations of the medium would permit, and the BBC mounted a similarly full exercise. That they should be required, in addition, to surrender precious airtime for broadcasts on the models of the over-familiar PPBs and PEBs seemed to the broadcasters a barely tolerable manifestation of the arrogant self-importance of the party hierarchs.

The White Paper had been tentative in this respect. It merely suggested that the broadcasting authorities might decide to make an equal amount of time available to the two main campaigning organisations for a series of short 'referendum broadcasts'. More ominously, it added that 'the government would welcome such an initiative'.[28] 'Initiative' was clearly intended to embrace 'compliance', and the broadcasting authorities did not, as they might have done, use this unprecedented occasion to make a bid for freedom and loosen the politicians' privileged hold on portions of airtime.

On 27 March, the day after the Referendum Bill had been presented to parliament, Neil Marten, the Conservative MP at the head of the National Referendum Campaign (to restore to parliament the exclusive right to make laws and impose taxes and to re-establish the UK's power to trade freely with any country), wrote to the Lord President, Edward Short, proposing that the PPB Committee be enlarged to include representatives of the two umbrella groups. Brian Young, on behalf of the IBA, found it hard to accept that that committee was the appropriate body to have the last word on whether sufficient organisations accepted the umbrellamen as fairly balanced spokesmen for their interests and opinions. But he did accept that in the last resort it would be the Lord President (presumably after consultation with his ministerial colleagues) who would be called upon to decide.

At a meeting at 12 Downing Street on 8 April it was noted that the broadcasting authorities would be offering each of the two umbrella groups four broadcasts, each of ten minutes' duration. It was agreed that one of the four could be fragmented so that separate versions might be shown simultaneously in England, Scotland, Wales and Northern Ireland.

A fortnight later, on 22 April, a further meeting was called. To this Neil Marten and his principal antagonist, Sir Con O'Neill, the former ambassador to the EEC in Brussels and now the leader of Britain in Europe, were invited. Whether either of these two meetings could be accurately described as meetings of the Committee on Party Political Broadcasting is veiled in doubtless intentional obscurity. Despite its venue and the presence of the committee's secretarial staff neither the Lord President nor any of the senior politicians from the three main parties attended. The latter were represented by their press and broadcasting officers, while the SNP, which had been invited, chose not to send a representative.

The broadcasters' offer as noted at the first meeting was accepted without dispute at the second. The toss of a coin was won by the National Referendum Campaign, which chose to go last in the series and the sequence thus became: Britain in Europe 22, 27, 30 May and 2 June; National Referendum Campaign 23, 29, 31 May and 3 June. All eight broadcasts were scheduled for simultaneous transmission at 9 p.m. with the exception of those of 30 and 31 May which were to be shown at 9 p.m. by the BBC and 10.30 p.m. by ITV.

These plans were frustrated by an industrial dispute affecting the ITV companies. On 23 May only LWT, Westward and Tyne Tees were able to transmit. On 27 May strike action prevented any ITV transmissions at all. On 29 May only HTV and Southern were on the air. On 30 and 31 May, 2 and 3 June all companies were able to transmit, with the exception of STV and Grampian on 30 and 31 May and of STV alone on 2 and 3 June. These programmes, lost to ITV viewers, were shown by the BBC and not rescheduled on ITV because at the time of their transmission by the BBC they had an unchallenged monopoly of television channels on the air.

Apart from some minor tiffs during the campaign, arising for the most part from the anti-marketeers' belief that the media were predominantly pro-market, ITV's relations with both groups were good. There was little of the atmosphere of smouldering suspicion punctuated by occasional flare-ups which had long been a feature of relations with the political parties. After the results of the poll had been announced complaints, so often a feature of election post-mortems, were notably absent. This may have been because there was only a single issue before the electorate and because the result was unequivocal. Or it may have been due to the success of the broadcasters in being impartial and being seen to be impartial.

Jointly financed audience research into the impact of the overall ITV and BBC coverage recorded that at least half the viewers interviewed were either very or fairly satisfied that the programmes were unbiased. The numbers of those who were either rather or very dissatisfied on this point fell during the campaign from around 15 per cent to less that 5 per cent.[29] One critic, Sean Day-Lewis, writing in the *Daily Telegraph*, judged that 'though most of ITV was off the air for the final complete week of the campaign, its one channel managed a far more positive approach than the BBC's two'.[30]

The handling of this highly important and extremely controversial national issue by the broadcasting authorities offered a convincing demonstration in support of their oft-repeated contention that the political parties could safely abandon their claims to annual series of PPBs, in the knowledge that the presentation of diverse political issues on television would be fair and objective and that the provision of balanced and clear factual information would be in dependable hands. A second referendum experience was to offer an even more convincing demonstration of that conclusion.

With 1974 in mind, the Authority exchanged some gloomy correspondence with the BBC in the autumn of 1978 about prospects for the coming year. In 1979 there were to be not only the normal run of PPBs but also PEBs for general election broadcasts and direct elections to the EEC assembly, as well as referenda in Scotland and Wales on the proposal that those countries should be accorded a significant measure of devolved independence from the United Kingdom government in Westminster.

IBA and BBC officials were agreed on the undesirability of allowing the concurrence of these events to bring about a substantial increase in the use of their channels for party propaganda. They decided on joint resistance to any extra party broadcasts in respect of the referenda. The issues, they believed, could be more satisfactorily covered by the regular, properly balanced current affairs output; and more fairly too, since there were elements in both major parties in disagreement with their leaders on this matter.

It was recognised, however, that the parties could not be prevented from using a normal PPB to publicise official party views on devolution. Nor could the government be prevented from broadcasting a ministerial statement should it so choose. If that were merely an encouragement to use the vote, it would present no problem beyond its effect on one evening's schedules. But if it were advice on how to vote, there would have to be a second broadcast from the opposition and a third including both major parties and the Liberals and, in justice, the SNP and Plaid Cymru.

At a meeting at 12 Downing Street in the first week of January 1979 agreement was reached much in line with the broadcasters' proposals. There would be no extras for the referenda. Instead, PPBs from the three UK parties taken from their annual ration would be shown at 9 p.m. on all channels on the three days immediately preceding the poll, which would take place on 1 March. The two national parties would be accommodated at times between 6 and 6.45 p.m. during the same period. There was little or no prospect of a ministerial broadcast.

An SCC meeting on 9 January welcomed the agreement, but its relief was premature. The agreement had been reached with the parties' broadcasting officers. Another Downing Street meeting, on 17 January, was attended by the two Chief Whips, who overruled their underlings and insisted on extra broadcasts during the week of the polls despite statements to the contrary by the Secretaries of State for Scotland and Wales the previous November. A compromise offered by the broadcasters a week later was rejected and, since Labour and the Liberals were officially in favour of devolution and only the Conservatives against, members of the Scottish Labour Party opposed to the proposal threatened legal action in the Scottish courts, pleading a breach of ITV's statutory duty to observe due impartiality.

On 2 February David Glencross for the IBA and Hardiman Scott for the BBC met Murdo Maclean, secretary of the PPB Committee, and a new proposal emerged. This called for the creation of umbrella groups, as in the EEC referendum. There would be a yes group and a no group irrespective of party labels, and it would be up to the parties to decide among themselves who would be the spokesmen. Each group would be offered one ten-minute broadcast in Scotland or Wales only. There would be no additional UK PPBs, but the possibility of using one from the existing allocation would stay.

Time was passing and the planning of schedules falling into disarray. On 7 February Glencross presented a full account to an Authority meeting, outlining the various proposals made and emphasising the continuing uncertainty. At the suggestion of the Director General the members agreed that in the event of a further rejection by the parties the broadcasters should revert to their first proposal.[31] Within hours of that meeting it was learned that the parties had indeed rejected the 'umbrella group' proposal. They were therefore offered, once again, one broadcast each to be taken from their annual allocation, and this time the offer was accepted by the Whips.

The anti-devolutionists were at once up in arms. The arrangement meant that the Scottish electorate would be saturated with Yes propaganda, said Teddy Taylor, the Conservative shadow Secretary of State for Scotland. He forecast 'a lasting and deep feeling of resentment among the people of Scotland' at this 'shameful situation', making nothing of the fact that a PPB by the Conservatives, who favoured a No vote, had the eve of poll slot.[32]

Members of the Labour Vote No Campaign in Scotland went further. On 13 February they presented a petition to the Court of Session asking for an interdict against the IBA on the grounds that the proposed transmissions would be in breach of duties laid upon the Authority by the Independent Broadcasting Authority Act 1973: specifically, the provisions relating to proper balance in Section 2 and to 'due impartiality . . . as respects matters of political . . . controversy or relating to current public policy' in Section 4.

On 16 February an interim interdict was granted prohibiting the IBA from transmitting party political broadcasts in Scotland during the referendum campaign. The Authority responded with a press statement reaffirming what it believed to be the correct interpretation of its statutory duty, which was to allow 'each party to express a view about the referendum issues in a PPB taken from its normal allocation'. But since the Scottish court had taken a different view, it went on, 'the Authority must inform all parties that all political broadcasts between now and 1 March are withdrawn from Independent Television'.

Lord Thorneycroft, Chairman of the Conservative Party, wrote to Lady Plowden on 20 February pointing out that as long ago as November the party had booked a PPB for 28 February 1979. A broadcast was being prepared for that date 'to go out on all transmitters'. Should the Authority fail to honour the agreement, the party would hold it responsible for any costs incurred.

On the same day a telexed message from Broadcasting House notified all BBC newsrooms, political editors, correspondents, information officers and others that, after considering the order of interdict against the IBA, the Corporation had decided to withdraw its intended PPBs in Scotland and Wales. This decision had been reached even though the BBC was not subject to the same statutory obligations as the IBA: 'although Lord Ross's ruling does not apply in law to the BBC, our constitutional commitment to due impartiality requires us to conform'.

Transmissions in England and Northern Ireland were still possible and this was considered by Authority members at their meeting on 22 February. In the light of Lord Thorneycroft's letter it was resolved to accept the Conservative broadcast in those areas, on the understanding that it did not deal with the subject of devolution; but the Conservatives were not to be allowed to save up the time for later PPBs in Scotland and Wales. Lady Plowden's letter to Lord Thorneycroft to this effect crossed with one from him withdrawing: 'We have reluctantly come to the conclusion that the only course left to us is to cancel our party political broadcast for the 28th and solve our mutual difficulties by taking a broadcast at the next available opportunity'.[33] So no PPBs at all were broadcast on this occasion.

It can scarcely be argued that the democratic process was impaired by the absence of party broadcasts on an issue which cut across party lines. STV, Grampian and Border in Scotland and HTV in Wales (in English and Welsh) used their daily local news bulletins for campaign coverage and mounted special devolution and results programmes, conscientiously observing the strictest impartiality and ensuring equal allocations of time to the pro and anti campaigners. An important part of all companies' efforts was a clear explanation of precisely what was being proposed in the Scotland Act and the Wales Act, and the edition of *Weekend World* networked on the Sunday before the poll outlined the facts and implications of devolution for viewers in the United Kingdom as a whole. As with the EEC referendum, the outcome was a decisive vote in favour of the *status quo*.

The interdict obtained in this campaign encouraged similar action only two months later during the run-up to the general election held on 3 May 1979. On 12 April the SNP applied to the Scottish Court of Session for an interim interdict against the IBA to prevent the broadcasting of network news and current affairs programmes in Scotland which did not give the SNP parity of treatment with the Labour, Conservative and Liberal parties. The application rested on the 'proper balance' provision of Section 2 of the Act, which had succeeded in February, rather than Section 4 which enjoined 'due impartiality' and which the Authority had always believed to cover questions of political balance. By the Authority, and by the law south of the Border, the 'proper balance' stipulation was taken to refer, not to impartiality, but to a balance between information, education and entertainment programmes.

On this occasion the court endorsed the Authority's view and the application was dismissed. In delivering his judgment Lord Robertson said that granting an interdict would have 'ridiculous results' and 'farcical consequences', since it would lead the Authority to breach its statutory obligations elsewhere. Such was the culmination of a long struggle by the minority parties for more equitable treatment in the allocation of airtime.[34]

Although party political broadcasts during the 1960s and 1970s could hardly be described as effective vehicles of persuasion, the parties remained wedded to them as a right of editorially unfettered access to the screen. The broadcasting authorities justified them on the grounds that the parties ought to be allowed to address the electorate directly through the medium out of constitutional propriety, which it might well be thought authoritarian to challenge. 'It may be sensible to point out to politicians that their broadcasts are not greatly appreciated by the electorate', ran one memorandum to the members of the Authority from their staff, 'but it might be best if any initiative to reduce or abolish party political broadcasts came from the parties and not from the broadcasters.'[35]

With elections popularly supposed to be won and lost on television, this wishful thought was doomed to disappointment. Instead, in the course of the 1980s, the parties were at last moved, through a more imaginative use of the medium, to make their propaganda less unappealing to the viewing public.

11

ENGINEERING: THE SYSTEM TRANSFORMED

Progressive developments in broadcasting equipment and techniques during the 1960s and 1970s made it 'far less necessary for an engineer to stand as middleman between director and viewer', as the IBA told the Annan Committee. In ITV's infancy engineers had 'determined what could or could not be in the studios; who could or could not receive the programmes; how close to the camera the actors must stand and in what temperature; how much make-up they needed; how far outside the confines of the studios the camera could venture'. But in the years of maturity their daily operational role dwindled.

> The engineers have provided the programme production staff with equipment which they can, in effect, virtually operate themselves. Within the studio complexes, the technical areas have increasingly been concentrated away from the studio floor with versatile switching systems to allow the recording and playback machines to be assigned to different studios and to be operated remotely. Transmitters and camera channels are no longer attended by a clutch of technicians, caring for their complex racks of electronics, one technician to every ten knobs or so. Today it is in the planning of the systems and networks, the development of entirely new facilities and the maintenance of technical quality that is to be found the true role of the engineer; simplifying operations, making them more economical, more tolerant and to more consistent standards of quality.[1]

During this period viewers in the United Kingdom enjoyed the highest technical standards of colour television broadcasting in the world, despite a loss in picture quality as video-tape progressively replaced film in the interests of economy and immediacy. The portable hand-held camera, developed in the USA for electronic news gathering (ENG), was introduced and used not only for hard news gathering but also for lightweight video recording in location work on current affairs and documentary programmes. In 1970 Outside Broadcast Units were equipped with

176

heavyweight studio cameras requiring substantial mounting and restricting the distance from mobile control to camera position to some 200 metres. By 1980 this had become a Stone Age operation.

The new role of the engineer in a new world of broadcasting technology was assessed as follows:

> The engineer ... still has an important, if less spectacular, part to play in any organisation concerned with broadcasting. It is the engineer who recognises the limitations of existing equipment, the constraints of bandwidth, radio-wave propagation and physics; who is likely to see most clearly and quickly the opportunities for television in new developments in other branches of science; who conceives and provides the new facilities with which the programme-makers can exercise their imagination; who seeks to provide video-tape recording with the traditional editing flexibility of film; who recognises that television will increasingly be based on digital rather than analogue techniques; who can foresee the technical and operational advantages and problems in direct broadcasting from satellites; who, if necessary, still restrains the enthusiasm of the programme-maker with the cooler judgments of science; who, in brief, uses his flair and imagination and knowledge and grasp of engineering-economics in the service of broadcasting – just as the programme-makers use their flair and imagination and knowledge and grasp of production-economics.[2]

From the beginning of ITV engineering was an ITA responsibility. Despite the regional emphasis of Independent Television, all ITV transmitters were owned, equipped and maintained by the Authority. To plan the transmission system and design it to meet national needs and requirements, a single controlling body was essential on practical grounds, and parliament chose to assign control of both message and medium to the same Authority. In the words of the Television Act 1964 the function of the Independent Television Authority was to provide television services 'of high quality, both as to the transmission and as to the matter transmitted'.[3]

This formal and physical command of the channels of dissemination distinguished the Authority from other supervisory bodies, such as the US Federal Communications Commission and the Canadian Radio Television Commission, with their purely programme control and licensing functions, and served to underline the Authority's legal responsibility as the publisher of every ITV programme and advertisement. Thanks to the prosperity of the system the Authority was able to impose on the companies rental charges sufficiently high not only to meet its 'high quality' obligations under the Act but also to permit transmitter facilities and related research 'when necessary on social grounds' and 'to provide a television service in many parts of the country where strictly commercial grounds would have precluded it'.[4] This entailed building a large number of transmitters in the regions of some of the smallest companies: sixty-six in the south-west, for example, compared with thirty-six in Greater London.

During the 1960s some hard re-engineering choices had to be made between colour systems and about definition standards.[5] In 1966 the consensus of opinion within ITV inclined towards the German-developed PAL (phase alternation line) colour system rather than the American NTSC ('never twice the same colour') favoured by the BBC or the French SECAM supported by the Russians. PAL's quality of picture was the most reliable, and ITV's arguments succeeded. But the Authority was also lobbying for colour on the existing 405-line VHF system. This, it was argued, would reach and attract a larger number of viewers more quickly and more economically than a switch to the 625-line system on UHF introduced for BBC2 in 1964. Research had shown that for every viewer who received a better picture on UHF three were dissatisfied.

Television-set manufacturers, who had been reluctant enough to meet the demand for black-and-white dual-standard UHF and VHF receivers, jibbed decisively at the prospect of making and marketing dual-standard colour sets. This meant that, if the Authority had its way, only 405-line VHF colour receivers were likely to be on the market, and BBC2, together with the long awaited ITV2, would be condemned to monochrome in perpetuity. Members were therefore persuaded to accept the much more expensive and cumbersome option of colour on UHF only. This necessitated conversion of the entire network to 625 lines and a long period of duplication until the single-standard VHF receivers could be fairly assumed to be no longer in use in significant numbers.

Major conditions were attached by the Authority to its agreement to this change of policy. It insisted that BBC1 also be duplicated on UHF; that the growth of colour services on ITV and BBC1 be kept in step; that the pace of growth in BBC2 coverage be reduced to allow ITV and BBC1 to catch up on UHF; that coverage on UHF be permitted to grow faster than had been permitted on VHF; that colour transmissions start simultaneously in London, Midlands and the North; and that all ITV regions be reached in colour within three years. Above all, the Authority required an early government decision to implement these proposals.

To obtain such a settlement, bullying tactics had to be employed on a government even more prone than most to deferring decisions. Happily Lord Hill was Chairman and on 8 November 1966 he led an ITA delegation to the Postmaster-General to expatiate forcefully on the dire consequences of any further official shilly-shallying over the launch of an ITV colour service. Hill enjoyed a controversial tenure of the chairmanship of both broadcasting organisations. On this occasion he fully earned the gratitude of broadcasters and viewers.

Three months later, on 15 February 1967, the government authorised colour on all three channels. BBC2 was equipped to launch its colour service before the end of that year: ITV and BBC1 had an opening night two years later, on 15 November 1969, from shared main stations at Crystal Palace (for London), Sutton Coldfield (for the Midlands), Emley Moor (for Yorkshire) and Winter Hill (for Lancashire). These were followed a month later by Rowridge, on the Isle of Wight, and Dover (for the South) and Black Hill (for central Scotland). With a choice of three channels and coverage of nearly half the population, Britain at once had colour transmissions and programmes on a more extensive scale than anywhere else in Europe. By the end of 1970 775,000 colour sets had been delivered to the UK market since 1967; by the end of 1973 the total had reached 6 million.[6]

This achievement and its development in a countrywide re-engineering operation by ITV, BBC and GPO engineers during the 1970s was a triumph which passed almost unsung. Because no major hitches or interruptions to the service were experienced, the ease of the transformation of its pictures into colour was taken for granted by the public. Ten years after the government had given the go-ahead, more homes in Britain had a colour television set than had a telephone, invented a hundred years earlier. By 1977 the new 625-line UHF system was available to 39 out of 40 of the population: of the fifty-one main high-power UHF stations planned, fifty were in operation by the end of that year. Standards of reception were generally good and provision had been made for a fourth channel.[7]

The IBA and the BBC were jointly congratulated by the Annan Committee on the speed and excellence of the operation.[8] They had co-operated to good effect in the co-siting of transmitters and the sharing of masts, some sites and masts being owned by one authority, some by the other. But the IBA's task was the more arduous, for initially BBC colour origination could be concentrated at BBC Television Centre and colour programmes distributed from this one centre over a fixed network of inter-city links, whereas ITV colour had to be launched as a fully switched regional operation. Even networked programmes had to allow for the insertion of advertisements in colour.

Success was not achieved without setbacks. The first threat of a major disruption to the timetable had come early in 1968 when a memorandum from Sir Robert Fraser to the companies opened with the typically terse warning: 'There is trouble ahead.' Contractors employed by the Post Office to install new links equipment were experiencing such serious design and development difficulties that the starting date for colour might have to be delayed from autumn 1969 to April 1970 or even later.[9] After consideration of a number of unpalatable alternatives it was decided to press for no postponement, even at the cost of more limited opening coverage. The set-manufacturing industry's strongly stated preference was for half a cake rather than no cake at all in 1969. This proved to be the right decision, for under pressure the problems were overcome.

Another serious reverse was the collapse of one of the main transmitter masts. At the time of its erection in 1966 the 1265-foot tubular steel mast on Emley Moor, overlooking the M1 south of Leeds, was one of the two highest man-made structures in Europe (the other being the Authority's mast at Belmont in Lincolnshire). Its sudden fall on 19 March 1969 not only deprived nearly 6 million viewers in the Yorkshire Television area of their programmes; it threatened the schedule for the introduction of colour throughout the country. Yet, despite snowy and icy conditions, the VHF service was restored to 70 per cent of the viewers within four days and to virtually all within four weeks, and colour on UHF was introduced on the planned date. The temporary replacement was a 680-foot triangular section lattice steel mast brought from Sweden. The permanent replacement structure, commissioned and built in 1971, became Britain's highest concrete tower. Rising to 1080 feet at a height above sea level of 841 feet and weighing 14,000 tons including foundations, it was an impressive monument to a near disaster.

Icing was generally believed to have caused the collapse, but a committee of enquiry attributed it to a form of oscillation which occurred at a low but steady wind-speed. Modifications to the similar masts at Belmont and Winter Hill, including the hanging of fifty tons of steel chains within each structure, prevented any recurrence. Assessing where the blame lay took rather longer than replacing the fallen structure. The Authority sued the main contractor, EMI, and the mast designers, BICC. After hearings in the High Court, the Court of Appeal and the House of Lords, a final judgment was delivered on 15 March 1980. EMI was found to be in breach of contract and BICC in breach of warranty and negligent. This established liability, but the size of damages remained the subject of further litigation. Under an out-of-court settlement reached in the autumn of 1983 the Authority accepted £3.2 million to cover damages, costs and interest.

Transmissions from Emley Moor had been fully restored in April 1969. On 7 September a delicate network operation took place during what was called S-Night. Transmitting the ITV service both in colour on 625-line UHF transmitters and in black and white on the existing 405-line VHF transmitters entailed a change in the production standard. It was necessary to adopt a standard of 625 lines at each studio, to distribute the 625-line pictures round the network and provide at each 405-line transmitter a converter to transform the 625-line signal to 405 lines for the existing service. But first there had to be an intermediate step to maintain the 405-line service until all studios had been converted for 625-line production. Each studio was provided by the Authority with 625/405-line standard converters. This enabled all companies to continue sending 405-line signals to the network after they had changed their production to the 625-line system. When all companies had made the change the standard converters were moved in a dramatic overnight operation to the transmitting stations to feed the 405-line transmitters. A duplicate service on VHF and UHF then became possible.

Initially only a small percentage of the population had either UHF or colour receivers and the expensive duplication of single programmes on two transmission channels was to last for more than fifteen years. This became a sensitive issue with political undertones. Even when some 800 UHF stations had been built and all viewers receiving a VHF signal had a UHF signal available, those who could not afford a new set would be deprived of a service when VHF transmissions ended. There was an understandable reluctance to come to a decision which would affect poorer people adversely and it was not until May 1980 that a Home Secretary felt able to announce plans for a phased closure of the VHF 405-line network. This was to be over a period of four years from 1982, but in the event the closure was accelerated and the last of the VHF transmitters were removed from service during January 1985 with virtually no complaint from the public. Following a recommendation by the Merriman Committee, that was the end of the use of the VHF band by ITV and the BBC. It was then set aside to satisfy a growing number of mobile radio applications.

The timetable for phasing out the VHF network was accurately assessed by a Television Advisory Committee appointed in 1971 by the Minister of Posts and Telecommunications to advise him on technical developments. The membership, under the chairmanship of Sir Robert Cockburn (Chairman of the National Computing Centre and a former Chief Scientist of the Ministry of Aviation), was drawn partly from government, partly from the radio and television industry and partly from outsiders. The IBA was represented by the Director General and, on the committee's Technical Sub-Committee, by the Director of Engineering. TAC's report, published in 1972, concluded that 405-line transmissions on VHF would have become totally ineffective by the mid-1980s because sets capable of receiving them (whose manufacture was due to end in 1975) would have reached the end of their useful life.

Compared with VHF, the UHF bands were substantially more costly for the broadcaster. This arose from a difference in propagation characteristics. The higher frequencies of UHF did not spread downwards over hills. High ground or man-made obstructions blocked their path. To achieve near-national coverage UHF required nearly ten times as many transmitting stations as were needed by VHF. Running costs too were much higher because the klystrons used to provide high power at UHF were much less efficient than the valves used at VHF.

It was also a worrying factor that the composition of each ITV company's regional audience was being determined as much by technical decisions as by social and ethnic considerations. The effect of this comprehensive re-engineering on the various regions within the United

Kingdom was examined by a Committee on Broadcasting Coverage set up by the government in May 1973 under the chairmanship of Sir Stewart Crawford, a retired diplomat. Its brief was to study the broadcasters' plans for coverage in Scotland, Wales, Northern Ireland and rural England, taking into account the particular needs of the people in those areas. Priorities, allocation of resources, technical feasibility and the financial implications were all within its terms of reference.

The committee reported in November 1974.[10] It found little wrong with the plans. These envisaged three phases of local relays to small pockets of the population. Under Phase I coverage was to be extended to reach potential audiences of 1000 or more; under Phase II a minimum of 500; and under Phase III communities of fewer than 500. With Crawford's encouragement and endorsement the first two phases were completed by the end of the decade, and Phase III was authorised in May 1980.[11]

This was certainly social, not commercial, engineering. Whereas the early main stations had represented a capital investment of a few pence per viewer (5p in the case of Crystal Palace), by 1978 a local relay, at some £50,000, was costing almost £50 per household or £17 per potential viewer: by 1980, on remote island sites, a cost of £80 per head was sometimes incurred.[12]

According to Crawford, the first requirement of viewers was a good picture; if the technical quality was satisfactory, they were content with network programmes at first, but soon became more selective and more aware of local loyalties and interests. Demand for these to be reflected on television came not only from individual viewers but, more forcefully, from organised groups promoting regionalism – political parties and regional authorities, local societies and associations.

The federal structure of ITV was better suited to satisfy this demand than a unitary body like the BBC, and Crawford noted that 'in England and Scotland ITV devotes substantially more time each week to meeting the particular needs of the people in the different regions than does the BBC'.[13] But some of this benefit was lost because the coverage of the relevant transmitters failed to match the areas for which the regional programmes were devised, and this distortion was accentuated during the changes from VHF to UHF because the coverage of the UHF transmitters was much wider in penetration than the VHF where there were no topographical obstructions.

The primary objective of the IBA's UHF policy was maximum coverage throughout the UK. In planning for this it sought to copy the established VHF areas, which themselves were designed to follow, as far as was technically possible, the natural regions within the country. Mostly this proved possible, or at least the effect of UHF allocations was marginal. Conspicuous exceptions were the Bilsdale West Moor transmitter in North Yorkshire, covering Cleveland and much of Durham to the north and

1. Retirement of the first Director General, 1970. *Left to right* Howard Thomas (Thames), Sir Robert Fraser, Lord Aylestone, Cecil Bernstein (Granada), Sir Lew Grade (ATV)

(Photograph courtesy of Thames Television)

2. Sir Robert Fraser, Director General 1954–70

3. Sir Brian Young, Director General 1970–82

4. Lord Aylestone, Chairman 1967–75

5. Lady Plowden, Chairman 1975–80

6.   Bernard Sendall, Deputy Director General (Programme Services) 1955–77

7. Colin Shaw, Director of Television 1977—83

8. Anthony Pragnell, Deputy Director General (Administrative Services) 1961–77, Deputy Director General 1977–83

9a. Howard Steele,
     Chief Engineer and
     Director of Engineering
     1966–77

9b. Tom Robson,
     Director of Engineering
     1977–86

10. Crawley Court

11. IBA Members and Officers, December 1979. *Sitting (left to right)* Lady Anglesey, Rev. Dr W.J. Morris, Mrs Ann Coulson, Lady Plowden (Chairman), Christopher Bland (Deputy Chairman), Anthony M.G. Christopher, Mrs Mary Warnock and George Russell. *Standing (from left to right)* A.J.R. Purssell, Anthony Pragnell (Deputy Director General), Professor J. Ring, Sir Brian Young (Director General), Tom Robson (Director of Engineering), John Thompson (Director of Radio) and Colin Shaw (Director of Television). *Absent:* Professor Huw Morris-Jones

12. Lord Annan, Chairman of the Committee on the Future of Broadcasting 1974–77

much of North Yorkshire to the south, and the Belmont transmitter in Lincolnshire, which covered Lincolnshire and West Norfolk as well as large parts of Humberside and South Yorkshire.

Bilsdale presented the Authority with an unprecedented embarrassment. Uniquely, its coverage extended over essential parts of the coverage areas of two companies, and this contributed to their coming together under the ownership of a specially formed holding company (Trident). Allocated to Tyne Tees, the Bilsdale signal penetrated far into Yorkshire, taking in even Harrogate and York itself. The position and height of the transmitter made a solution difficult. In the mid-1970s consideration was given to redesigning the station so that it could operate with separate transmitters and aerial systems for north and south. At that time the necessary frequencies were available, but the expense would have been great and viewers would have had to re-equip themselves with new aerials. The frequencies were subsequently used to ease the problem of Belmont by supplying an Anglia service in the Kings Lynn area.

Over Belmont there was a prolonged and exceptionally acrimonious political row following the Authority's announcement of the reallocation of the transmitter from Anglia to Yorkshire Television for the contract period beginning in July 1974. The change was explained by the IBA to the Crawford Committee as resulting from an assessment that more people living in the transmitter's principal service area had an affinity with Yorkshire. The assessment was based on an analysis of the population figures in different parts of the area. It was supported by information about viewing habits, by the fact that Humberside was being grouped with Yorkshire in a single Economic Planning Region, and by the recommendation of the Royal Commission on Local Government in England which proposed that South as well as North Humberside should form part of a greater Yorkshire region. But Crawford was not impressed by these arguments and doubted whether the change would best meet the needs of the people in the area. The committee concluded, however, that it would be satisfactory if special attention were paid to the provision of opt-out local programmes for the area. The IBA then adhered to its decision in the face of intense resistance from Anglia and local Members of Parliament.

Was the Belmont station wrongly sited? Planned in 1963 (before Yorkshire Television existed) to fill the gap between Tacolneston, Emley Moor and Bilsdale, it was intended to provide a power and aerial height which would penetrate into towns such as Lincoln and Boston and along the coast on both sides of the Wash to Hull and Grimsby. In engineering terms it was economical both in cost and in conserving frequencies. A more expensive plan, with lower masts and more stations, would have produced fewer overlaps, but it would also have left pockets of population without reception, and this would have conflicted with the policy of bringing the best possible service to the maximum number of people in the most cost-effective manner. Except in areas of unavoidable overlap every viewer was intended to have one, and only one, ITV signal, with no choice of station.

Crawford was in general agreement with not duplicating a service in an area simply to provide a choice of programmes or to remove anomalies in editorial coverage, but it wanted exceptions made where anomalies related to language or national borders. Typical of a language problem was the position in North Somerset and North Devon, where topography dictated reception from the Wenvoe transmitter in Wales across the Bristol Channel, whose output included programmes with Welsh content and sometimes in the Welsh language. Only after vociferous protests were additional relay stations built to supply the area with an English service. The situation had arisen because a planning authority refused permission for the Mendip main transmitter on the English side of the Bristol Channel to be built on the site which would have given maximum coverage.

Low-power relay stations, too, sometimes decreed a misdirection of regional programming. In Northumberland the relay station at Haltwhistle picked up the Border signal instead of the Tyne Tees and in Cumbria the Alston relay retaliated by doing the opposite. Although Border Television was based in Carlisle, most of the southern part of the new county of Cumbria received Granada programmes from Winter Hill. In Scotland it was argued that Aberdeen would have been a more appropriate source than Glasgow for the programme feed from Fort William, but this was hotly disputed.

On rare occasions, where there was a choice and no obvious solution, the IBA conducted a poll to seek the views of the people living in the area. This happened at Swindon; and at Corfe Castle, where a public demonstration was arranged to show the differences in technical quality which could be expected from the alternative sources. In some parts of the country UHF transmitter coverage was less extensive than VHF, so that viewers in what had been overlap areas lost freedom of choice. The Dundee–Perth area was allotted to Grampian rather than Scottish; north-west Kent to London rather than Southern; Weymouth to Westward rather than Southern. Relay stations depended upon receiving a satisfactory off-air signal which they could reradiate and if this was available only from one source there could be no choice. Bringing signals from different sources by microwave link was considered prohibitively expensive until, at the start of the new contract period in January 1982, a number of relay stations in Cumbria and on the Lancashire/Yorkshire border were reallocated by this method.

The crucial question which all this raised for the Crawford Committee was whether ITV had too many or too few companies. In the end it reached the same conclusion as the IBA: that the number was about right. There was clearly a good argument for a single company to cover Scotland, for example, but there were also sound reasons for three. Generally, Crawford was struck by:

the strength of the ties which have developed between the smaller programme companies and the people living in the areas they serve. This appears to be reflected in the audiences obtained for their programmes of local interest and in the amount of local advertising. If the country as a whole were served not by 15 programme companies, some of them small and lacking financial strength, but by a smaller number of large and financially strong units, the result would certainly be no drop in the quality but a considerable reduction in the attention given to the particular interests of the people living in the remoter and less urbanised parts of the country. This would in our view be a serious loss.[14]

The committee attached the highest importance to the completion of UHF coverage in Northern Ireland, and it recommended that Grampian should make special provision for Gaelic speakers (70,000 according to the 1971 census, although not all in Grampian's area). But its major recommendation was for the fourth channel in Wales to be allotted to a separate service in which Welsh-language programmes would have priority, and that this should be introduced as soon as possible, without waiting for a decision on the fourth channel in the rest of the United Kingdom. Between the 1961 and 1971 censuses the number of Welsh-language speakers had fallen sharply: to 542,000. The survival of the language was seen as the key to the preservation of Welsh culture, and regional television was to be recruited to mount the rescue operation:

It was put to us forcibly that if the young watch mainly English-language programmes the decline of the Welsh language will continue; if they have the opportunity to see more programmes in Welsh of comparable quality and at the peak family-viewing times, there can be a real chance of survival. The need for more programmes in Welsh is seen as urgent if the present decline is not to go beyond the point of no return.[15]

This sense of urgency was not shared by the government, which responded with another committee: a working party to examine practicabilities. Chaired by J. W. M. Siberry, a former Under- Secretary at the Welsh Office, this recommended that the BBC should build and operate six main stations in Wales for a fourth-channel Welsh service, and that the IBA should build and operate some eighty relay stations for the service. It proposed that the production of the programmes be shared between the BBC and HTV, each providing 12½ hours a week, and that the service be financed by a grant-in-aid from public funds.

The S4C solution adopted by the government four years later was rather different from Siberry's proposals. The BBC and HTV were to share in the provision of Welsh-language programming, but the broadcasters were forced to bear most of the cost: in ITV's case, through a proportion of the

Channel Four subscription imposed on all the companies. In 1979 the whole of the task of engineering the transmitter network was assigned to the IBA as part of its responsibility for the new nation-wide fourth channel system. This recommendation by the Annan Committee had been endorsed by a White Paper in July 1978 as the only practical option. A bill was introduced in December and became law the following April as the Independent Broadcasting Authority Act 1979. Only then could a firm set of specifications for the new channel be issued, estimates invited and orders placed with manufacturers prepared to commit themselves to a rapid rate of production and installation. The major contracts, worth more than £16 million, were awarded in November to Pye, which had supplied the first UHF high-power tramitters, and to Marconi.

Building the major part of a network before the start of a service allows a degree of flexibility in completing stations and time for proving them before their performance comes under public scrutiny. It is particularly desirable when the equipment is of a new and untried design. Demands for a quick, modest start for the fourth channel were therefore resisted in favour of a 'big bang' policy of not launching this new service either in Wales or in the rest of the United Kingdom until 87 per cent coverage overall, including substantial coverage in each region, was available. During the build-up main stations were completed at the rate of two a month. At the launch on 2 November 1982 31 main stations and 107 relays were in operation. A hundred of the relays were in Wales, providing almost full coverage throughout the Principality when S4C was commissioned. Nowhere in the world had such a major transmitter network been built in such a short time.

Britain was a pioneer in other technical developments in broadcasting during the 1970s. In 1968 a research team of ITA engineers began a long-term study into what was to become the subject of much experimental and development work in television and radio: the application of digital techniques. During 1971–72 important advances were made in the field of digital standards conversion and this work was brought to fruition in February 1975 when an agreement was signed giving Marconi exclusive world-wide manufacturing and marketing rights in the Digital Intercontinental Conversion Equipment (DICE) developed by the Authority's engineers. In failing to agree on lines of standardisation television had committed an error similar to that made in the previous century in the matter of railway gauges. In the absence of a common international standard, conversion equipment became essential, and DICE met this need.

Originally designed for the conversion of 525-line pictures to 625-line, it was first used to enable Europe to see Richard Nixon winning the US presidential election in November 1972. In March 1973 it was installed in ITN's studios. By 1975, after further development, it had become a two-way system capable of converting colour pictures from the USA and Japan (on the 525-line NTSC system) to or from the 625-line PAL or SECAM systems used in most of the rest of the world. It was, the IBA claimed, 'by a comfortable margin, the world's fastest computer'. The digital technique which it employed eliminated the necessity for careful alignment and adjustment and effected conversion without any perceptible impairing of the picture. For this work the development team under John Baldwin received the Geoffrey Parr Award of the Royal Television Society and the Pye Colour Television Award in 1973.[16]

Other ITV successes in the development of digital techniques included digital automatic measuring equipment (DAME), micro-processor-based information display and retrieval systems for use in IBA Regional Operations Centres, and – most notably – a completely new service which made use of the television screen but differed from the normally accepted concept of television. This was teletext, known in the ITV system as Oracle.

Oracle was first demonstrated by the IBA on 9 April 1973, ahead of the Ceefax system developed independently by the BBC. IBA, BBC and industry agreed a joint technical specification, and experimental services were initiated by both organisations in September 1974. In the following year the ITV companies assumed responsibility for running an Oracle service. Editorial units were established at LWT and ITN with injection facilities from Thames. Oracle pages originated in London were broadcast in other regions, and regional editorial units were later installed at Scottish and Channel Television.

Using the television screen to display words and simple graphics, teletext offered something different from the sound of radio and the pictures of television. It exploited the unused time in a conventional television transmission – the blank periods between individual fields of picture information – to accommodate bursts of very high-speed data signals. It was a marriage of television and computer technologies.

Teletext required no additional transmitting stations, no new inter-city links or distribution facilities; it made no claim on the radio-frequency spectrum. Development and software costs, however, were both high. The collection, editing and processing of information proved an expensive operation. With no need to interrupt scheduled programmes, teletext enjoyed an advantage over radio and television in being first with the news, but only at the expense of teams of editors working in shifts to feed new pages into the system. These could be inserted in a matter of seconds, each teletext page displaying up to 200 words. During its early years the number of pages was limited to about 300.

Teletext could piggy-back information on a television broadcast, and the advantage of providing Oracle subtitles for the benefit of the deaf and hard of hearing was soon recognised. Following research commissioned

from the University of Southampton, increasing numbers of ITV programmes were subtitled, but visibly so only to those selecting the appropriate Oracle page.

Revenue for Oracle was sought from advertisers and sponsors, but a
financially self-supporting service seemed far away. The number of viewers
was small and the cost of decoders correspondingly high. For those who
bought receivers suitably equipped there was access to a continuously
available news and information service, including stock exchange prices,
television and radio programme times, listings of theatre, exhibition and
sporting events, and games and puzzles for children. Editors quickly
became knowledgeable about what viewers most wanted and skilful in
presenting it in the most attractive manner possible within the limitations
of the system. But for many viewers teletext was for some time a toy to be
played with occasionally rather than a regular source of information.[17]

Unkindly dubbed 'a medium without a message', it nevertheless
encountered the usual alarmist resistance to technological advance. The
Newspaper Society, representing the regional and local press, told the
Annan Committee that teletext would be 'crippling in its financial consequences for the newspaper industry'.[18] Its proposal of a five-year morator-
ium on commercial development to enable newspapers to adjust to the new
competition was not well received by the committee, which did, however,
recommend an enquiry, after several years, at which the press would have
the opportunity to make representations.

During 1980 some 90,000 television sets with teletext equipment were
manufactured in Britain, and in mid-1981 no more than 150,000 homes
were equipped to receive the new service, but as technology developed and
existing receivers came to the end of their natural life the cost disincentives
of teletext receivers declined and the number of potential viewers grew. In
May 1981 the IBA published a Code for Teletext Transmissions, as
required under Section 29 of the Broadcasting Act 1980. This applied rules
on taste and decency, accuracy and impartiality, religious propaganda and
charitable appeals, competition prizes, sponsorship, subliminal techniques
and advertisements similar to those applied to conventional ITV broadcasts.[19]

In Howard Steele, the Authority's Chief Engineer and Director of
Engineering from 1966 until he resigned at the end of 1977 to become
Managing Director of Sony Broadcast Ltd, Independent Television had
someone of uncommon ability foreseeing, initiating and presiding over
all these developments. Steele was a brilliant engineer who was articulate
in explaining complex technical matters to laymen and making them sound
simple. He was ambitious and a serious contender to succeed Sir Robert
Fraser as Director General. A graduate of Imperial College, London, he
had worked for the Marconi Company before joining ABC Television in
the early days of ITV. As ABC's Chief Engineer he headed a team

responsible for building that company's television studio at Teddington and studios for Ulster, Grampian and Channel Television. His research and demonstrations of the respective merits of the NTSC, SECAM and PAL systems contributed convincingly to some of the arguments about which colour system should be adopted in Europe. At the invitation of the Institution of Electrical Engineers he led an IBA engineering team in presenting the 1975–76 Faraday Lecture in a remarkable exhibition which he took on tour to sixteen cities in the British Isles. Entitled *The Entertaining Electron*, this attracted audiences of more than 56,000, most of them young people enthralled by his demonstration of broadcasting technology. Awarded the Gold Medal of the Royal Television Society in 1983, he died of cancer shortly afterwards at the age of 54.

The re-engineering of the system was Steele's responsibility. At the time of the government's announcement in February 1967 ITV had thirty-one VHF stations in service, representing a building rate of some two and a half stations a year since its start. The UHF transmitter network to carry colour involved the construction and maintenance of fifty-one main high-power transmitter stations and 750 small relay stations. By introducing a concept of remotely controlled stations ITA engineers designed and built the world's first major colour network of entirely unmanned transmitters. Thus no increase in engineering staff in the field was required, but the acquisition of so many new sites and the design, installation and commissioning of so many new transmitters, masts and aerials and so much power and remote control equipment precipitated a big expansion of head office staff, and Steele took the opportunity to reconstitute his Engineering Division.

A Planning and Propagation Department was formed with the task of locating the new stations, identifying the frequencies to be used and negotiating for sites on which the stations would be built. Unlike some statutory bodies, the Authority possessed no powers of compulsory purchase; nor was it immune from planning regulations. Many stations were delayed by prolonged searches for a site with an owner willing to sell which was also acceptable to the planning authorities. Local authorities, while welcoming better television coverage, often raised objections to sites which the engineers had selected as the most suitable: in Weston-Super-Mare, for instance, the beauty of the surrounding countryside took precedence and reception was unacceptably poor as a result. In urban areas relay transmitters had to be sited on rooftops where access was sometimes not easily obtainable. On average the time lapse between a decision to build a relay station and its completion was four and a half years. In the case of one main transmitter – the Wrekin in Shropshire – it took eight years merely to acquire the site.[20] Sometimes new tower blocks and gasometers interfered with reception after a transmitter had come into operation, and at Findon in Sussex a new relay station had to be erected when a wood which had reflected the UHF signal was cut down.[21]

Once a site became available and building could begin, responsibility passed to the Station Design and Construction Department, whose finished product was a complete and fully working station, ready for service. The Station Operations and Maintenance Department then took over to keep the station in service. Its staff operated from maintenance bases round the country, monitored and controlled from head office, and later from four new Regional Operations Centres, each of which was designed to control some 400 stations.

To control the large new network linking studios and transmitters and to monitor the technical quality of the programmes, a Network Operations and Maintenance Department was created. Like those servicing the stations, its staff, some trained at the Authority's Harman Training College, had to master the intricacies of technological change from 405 to 625 lines, from valves to transistors to chips, from VHF to UHF, from monochrome to colour, from analogue to digital. Their job was to maintain an acceptable picture quality despite the frequent changes of programme sources and routings resulting from a regional operation. They had also to ensure compliance with the Authority's code of technical standards in the studios, which were owned, equipped and operated by the companies. This Code of Practice for colour transmission came to be adopted by more than fifty overseas broadcasting organisations.

Operations within the companies were guided by these quality control rules and procedures. Yorkshire Television, as a new contractor, was able to begin its service in Leeds in 1968 with the first purpose-built colour studio centre. Other new studios designed specifically for colour were built by ITN, ATV, Thames, Southern and LWT, while the other companies made extensive modifications to existing studios. Within a year of the launch the companies had taken delivery of, or had on order, 187 colour cameras, 76 colour telecines and 60 colour video-tape recorders. Most areas were soon transmitting fifty hours of closely monitored colour programmes a week.

Informing dealers and the general public about the building programme was a necessary public relations exercise at a time when one or two new transmitters were coming into service each week. An Engineering Information Department was formed to undertake this task. It released a flow of technical information, briefed the technical press, handled display stands at exhibitions and responded to the many letters from the public posing technical questions.

Lastly, an Experimental and Development Department was charged with designing equipment which would improve the technical performance and reliability of the transmitter network and with researching major innovations at a time of explosive development in broadcasting technology. This department was seen by the Authority as an important investment in the future and by the companies as an extravagance. It enabled Steele to gain for Independent Television a world-wide reputation in engineering. He made the development of Oracle and the acceptance of digital systems possible by increasing the staff of his research and development team from six to fifty-three. Knightsbridge was not an ideal situation for engineering research and he argued strongly for a headquarters of his own. The outcome was the purchase by the Authority of a 32-acre estate five miles from Winchester, the demolition of its Victorian mansion and the erection of purpose-built offices, designed by the Ware MacGregor Partnership.

Along with the Engineering Division, the IBA's finance and staff administration departments moved to the silvan setting of Crawley Court in October 1973. Numbers totalled 450, 200 of whom had transferred from London. Later they increased to more than 600. In 1974 the Engineering Division was employing 300 engineers and ancillary staff at the new headquarters and a similar number at its colour control centres, mobile maintenance bases and regional engineering offices.

In November 1975 the importance of this new Engineering and Administrative Centre at Crawley Court was recognised by an official visit from the Queen and Prince Philip. In the following year Sir Brian Young told Lord Annan:

> The effects of VCRs and multi-set homes, the promise of cable and satellites, etc., all require multi-disciplinary study with engineering/economics/programme creation interacting in the debate. Because of its energy and readiness for innovation, I would claim unhesitatingly that Crawley Court (IBA Engineering) is a centre of excellence on which we should build. Joint research and development with the BBC (which we abandoned in 1971, except for continuing joint work on service area planning) was a great dampener of progress.[22]

Steele's deputy and successor was Tom Robson, a less public and more down-to-earth figure, but no less determined. The successful construction of the Channel Four and S4C transmitter networks was a testimonial to his experience and perfectionism. His work in achieving agreement on a World Digital Studio Standard and acceptance of the Multiplex Analogue Components (MAC) system for DBS transmissions in Europe was rewarded by the Eduard Rhein prize in 1985. He retired in the following year after nearly thirty years' service with the Authority.

Towards the end of the 1970s the widespread use of video-tape recorders began to free the viewer from the tyranny of time; programmes could be recorded in the home and watched at will. Other technical advances were poised to bring benefits to the viewer in the early 1980s: among them the extensive re-equipment of studios to introduce automatic controls through the use of microprocessors; television broadcasting by satellite; and multichannel cables with optical fibres for cable television. The pace of change, breathtaking in the 1970s, was to accelerate in the 1980s.

12

DOMESTICATED ADVERTISING

Television is said to be all about programmes, but politics and engineering are only two of the factors behind the screen which affect the nature and quality of what appears on it. Television advertising has a full record of its own,[1] but a general account of Independent Television would be incomplete without due acknowledgement of the contribution made by advertisers.

It is known but sometimes forgotten that anyone wishing to watch ITV programmes must first pay a licence fee the whole of which (minus the cost of collection) passes to the BBC. Except for a relatively small supplementary income from the profits of its programme journal, from the sale of programmes overseas and from investments, ITV is dependent for its existence on the sale of airtime to advertisers, without whose support there would be no programme service of any kind. 'If the system failed to attract advertising revenue because insufficient people were watching,' the IBA told the Annan Committee, 'then it would lack the resources to produce good programming and would eventually collapse. Its feasibility rests on public approval.'[2]

Some have argued that Independent Television was not in fact independent because of its total reliance on advertisers; others have maintained that it was the advertisers who made Independent Television independent and that the separate financing of BBC and ITV served to protect the independence of both sets of broadcasters. 'Just as newspapers have additional strength because they can charge not only their readers but also their advertisers, so broadcasting stands on two legs; and the fact that the first leg is so dependent on government decisions makes the second leg more important.'[3]

Much of the early opposition to the introduction of commercial television had been founded on fears that the content of programmes would be influenced by advertisers, but this proved illusory from the very beginning of the service,[4] and in the mid-1970s the Annan Committee 'received very little evidence opposing advertising as a means of financing broadcasting'.[5] Certainly there was insistent pressure from advertisers for high-rating programmes during peak viewing hours, but this demand was not incompatible with ITV's function and objectives as providers of people's television. Programme-makers themselves wished to maximise audiences, and a safeguard against any pressure deemed improper was always at hand in the requirement that all schedules be approved by the Authority.

The television and broadcasting Acts also laid upon the Authority the duty of ensuring that advertisements were 'clearly distinguishable as such and recognisably separate from the rest of the programme', and within ITV advertising became a thing apart. Not only the statutory body but also the programme-makers were detached from the business of salesmanship. In some companies the sales department was run as, or as though, a separate company.

This isolation of the advertiser was partly a reflection of advertising's lowly reputation in some influential quarters, where it was condemned as 'an undesirable type of propaganda which cons the public, creates anti-social wants, promotes materialism, feeds off fears of sickness and off social envy, degrades the use of the English language and presents a stiflingly banal view of life'.[6]

Whatever the degree of truth in this, the integrity of the service was emphasised in repeated assurances that advertising was being strictly controlled: 'Through the exercise of its statutory powers and with the co-operation of the television companies and the advertising business the Authority has shown that the savage beast of advertising which it was once feared would stalk through our living rooms can be domesticated.'[7]

In the beginning the advertisements had been much criticised as intrusive. Television in the USA had been born with advertising; in Britain, where it had not, the commercials seemed unnatural. Before older viewers became accustomed to them and a new generation grew up for whom they were a normal part of television's service of entertainment and information, there was political pressure for less advertising as well as for tighter controls. An unsuccessful attempt to make six minutes the maximum in any hour was mounted by Labour MPs in June 1959 when Herbert Morrison introduced a Television (Limitation of Advertising) Bill under the ten-minute rule.

The Authority's original decision that advertising should be restricted to *an average* of six minutes in the hour remained unchanged down the years. Some categories of programme – schools broadcasts, church services, royal occasions – were sanitised as ad-free zones, so that extra minutes became available at other times. Heavy demand from advertisers for peak-time slots persuaded the Authority at first to permit a normal maximum of eight minutes in any clock hour. This was reduced to seven-and-a-half in September 1960 and to seven from January 1961; although a six-week return to the seven-and-a-half minute maximum was permitted in March and April 1972 to clear a backlog of advertisements after labour disputes in the power industry had interrupted the service.

Allegations that the ITV companies were repeatedly flouting IBA rules which restricted the amount and distribution of advertisements, and were thereby coining hundreds of thousands of pounds of revenue a year, were made in the trade magazine *Campaign* in 1973.[8] In the Anglia, ATV, Granada and London areas 104 breaches of the normal maximum of seven minutes per clock hour were said to have occurred in one week, and other examples purported to reveal minutage exceeding the permitted average of six minutes an hour throughout the day. Although denounced by the IBA as 'entirely false', these allegations were taken up by the *Sunday Times*, which claimed to have discovered instances of ten minutes or more advertising during a clock hour. In the House of Commons the government was urged to investigate whether the IBA was discharging its statutory responsibilities.

The truth lay embedded in the fact that the normal limit of seven minutes could properly be exceeded where the interests of programme presentation required flexibility. If the most natural break in a programme fell just before or just after the striking of the clock it was possible for the advertising over a two-hour period to split 8:6, 9:5, or even very exceptionally, 10½:3½ instead of the usual 7:7. The maximum daily average of six minutes was not affected.

The Authority had little difficulty in concluding that there was no question of any company having evaded its rules,[9] or in satisfying the Minister of Posts and Telecommunications that the companies did not benefit financially from the degree of flexibility which the Authority allowed.[10]

The natural break was a chronic grey area. According to the Rules as to Advertisements in the television and broadcasting Acts: 'Advertisements shall not be inserted otherwise than at the beginning or the end of the programme or in natural breaks therein'. But from 1955 onwards it became all too apparent that some breaks were less natural than others. Plays are traditionally divided into acts and scenes; films are not. It was possible to make entertainment programmes and documentaries in such a way as to accommodate breaks, following the American practice, but however conscientiously the Authority attempted to formulate acceptable regulations there was plenty of room for argument. Commercial breaks within news bulletins dated from the lengthening of the main evening news to half an hour with the introduction of *News at Ten* in 1967. In 1977 the agreement of CRAC and the Home Secretary was obtained to their insertion in religious programmes lasting for more than half an hour which were not services or otherwise devotional.

Time and custom brought public acceptance of the most natural of the breaks, but objections continued to be voiced to the time-honoured dramatic device of the cliff-hanger. In evidence to the Select Committee on Nationalised Industries in 1972 Tony (Anthony Wedgwood) Benn, Postmaster-General from 1964 to 1966, declared: 'The rules that were made at the time are in my view virtually completely ignored. A car will be skidding round the corner in a thriller and one has cornflakes and deodorants, then the skid is completed after the natural break.' On the other hand, he added, 'if you ask me as a viewer whether opportunities to make a cup of tea or telephone to my mother are welcome as a result of these breaks, I do not find them too bad . . . I do not feel as worried about this as I would have done in my earlier passionate speeches in the House about this in 1953 and 1954'.[11]

Five years later the Annan Committee expressed misgivings: 'Present practice has moved a long way from the intentions of the Act, which are that the amount of time given to advertising in the programmes should not spoil the viewer's pleasure'.[12] The committee thought that advertising breaks in *World at War*, for example, were inappropriate, while recognising that without them ambitious programmes of that kind would not be made. They recommended, nevertheless, 'that the IBA should reduce the number of natural breaks in those programmes in which such breaks might be inappropriate'.

In its response to the Home Secretary the IBA conceded that it would always be disputable whether or not programmes were hampered or spoiled by natural breaks. But the charge of frustrating the intentions of the Act was denied, and the Authority did not agree with the committee about which programmes were inappropriate. One which was indisputably such had been ATV's lengthy epic, *Jesus of Nazareth*, whose screening without advertisements was accommodated by a redistribution of advertising time throughout the evenings to avoid loss of revenue. A count of the 180 programmes in a typical week's broadcasting in 1974 revealed that 100 contained no internal advertising, sixty included one break and twenty included two.[13]

As to content, the advertising industry itself stipulated that all advertisements should be legal, decent, honest and truthful, and irreproachability was the Authority's aim. Yet opinions were apt to differ when one or other of those epithets came to be defined and applied in specific cases. It would be hard to argue convincingly that Camay did indeed make you lovelier each day or that a cigar called Hamlet was a valid definition of happiness, but it was generally accepted that fantasy and hyperbole amused and did not deceive. In questions of decency and good taste public sensitivities were constantly changing and judgment had to be largely subjective. A degree of undress was thought acceptable in commercials for toiletries and underwear but not for typewriters. In 1959 it was ruled that 'real models wearing foundation garments should not show any movement if their flesh was visible'. By 1969 they could wear 'girdles beneath gauzy robes with the wind disturbing the edges of the drapery'. By 1979 'nubile girls in the briefest of bras and panties curvaceously poling a raft down a river' were passed as fit for family viewing and found to give no offence.[14]

False teeth and lavatory pans too won acceptability over the years, but genital deodorants, sanitary towels and tampons formed another grey area. In the early 1970s some discreet commercials for these products were permitted on an experimental basis. A dermatological specialist advised that vaginal deodorants were unlikely to be more harmful than under-arm deodorants and saw no good medical grounds for their exclusion, but in October 1972 the Authority ruled against all these categories on grounds of taste.[15] Contraceptives and family planning generally were also taboo subjects for advertisers until the 1980s, despite incessant pressure on the Authority by some members of its General Advisory Council.

Hidden persuasion was sometimes detected by those addicted to Freudian analysis. The most discussed example was the girl seen receiving a Cadbury's Flake bar lovingly between her lips. Was this or was this not sexual symbolism? Fellatio was in the eye of a number of outraged beholders, but Lord Aylestone replied that if some people cared to make a chocolate bar commercial into a blue movie that was their problem, not the Authority's.[16]

In 1971 the Townswomen's Guild was assured that all housewives seen in commercials were wearing wedding rings; which was more than could be said for all housewives in drama programmes. The Mothers' Union, on the other hand, complained to the Annan Committee that the TV mother in commercials was always well groomed in her expensive kitchen, while her children had grasping manners and spoke badly.

Generally more rigorous standards were applied to commercials than to programmes. One extreme example was found in the investigation of a complaint from a clergyman that he could not enjoy a family film with his children because it was interrupted by an exhibition of violence in a commercial. This led to the discovery that the commercial portrayed boxers gently sparring and the 'family' film was an Alistair Maclean adventure story featuring thirteen violent deaths.[17]

Advertisements 'directed towards any religious or political end' were forbidden under the Act,[18] and in 1969 commercials for *The Bible Today*, a part-work publication, had to be refused (at a loss of £1 million), while advertising for a part-work on witchcraft and magic was accepted. It was an unfortunate example of ITV turning its back on God and making money out of the devil and all his works, but here parliament allowed the Authority no discretion. Earlier, Counsel had advised that advertisements for the modern English version of the Bible would be in contravention of the law; and later, in October 1975, a reference to *The Bible for Children* in an advertisement was declared to be unacceptable – to the amazement and fury of the publishers, who obtained a contrary legal opinion in vain.

In 1972 an advertisement for a *News of the World* feature article on industrial trouble-makers was rejected as a breach of the same section of the Television Act, which specifically disallowed advertisements having 'any relation to any industrial dispute'. This too led to a heavy but misdirected attack on the Authority for censorship.[19]

Government itself enjoyed no immunity from its own restrictions. In 1975 the Central Office of Information sought without success to place a series of advertisements on behalf of the Labour administration to publicise its anti-inflation campaign. The Authority was advised by Counsel that an advertisement designed to attract public support for, or to excuse or explain, the government's legislative programme must be regarded as directed towards a political end. Similar advice had been received and accepted in 1971, when the Authority rejected advertisements from the Conservative administration which did no more than invite viewers to go to a Post Office and obtain leaflets about Britain's entry into the EEC. Advertisements for pamphlets giving the case for and against the United Kingdom's membership were refused on the same grounds during the referendum campaign in 1975, but others informing viewers of their right to vote on the issue and the date of the poll were passed as acceptable.

Thus the Authority, relying on Counsel's Opinion for its interpretation of the wording of the Act, accepted advertisements from the government of the day when the government was exercising its executive function in carrying out the law as enacted by parliament, or when exercising the prerogative of the Crown, but not when promoting the merits of government policy or attempting to attract public support. Governments might buy advertising time to explain to the public how family allowances or social security benefits might be claimed, for example, but not why they had or had not been increased. In maintaining this thin and, to some, perverse distinction, sometimes in the face of government indignation, the IBA was honouring an intention dating from the first Television Act (in 1954) by denying television advertising to those who would use it to influence opinion about matters of public controversy.[20]

By the mid-1970s there were more than fifty other Acts of Parliament restricting, controlling or otherwise affecting advertising in Britain. Eighteen related to finance; four specifically to children. Those most generally relevant were the Trade Descriptions Act and Medicines Act, both of which dated from 1968. But none was as wide in its powers as the Independent Broadcasting Act of 1973. This followed the earlier Television Acts, and the Sound Broadcasting Act, in effectively making the Authority a statutory instrument of consumer protection. The Authority had the power and duty to exclude any advertisement which could reasonably be considered misleading and to decide which categories of advertisement and methods of advertising should be excluded from television and radio. It was also required to consult with the Minister of Posts and Telecommunications and carry out any directives which he might care to issue within the scope of the Act.

The law required the Authority to appoint an Advertising Advisory

Committee and a Medical Advisory Panel. The membership of the committee had to be representative of organisations concerned with standards
of conduct in advertising, together with representatives of the general
public as consumers, under a chairman who was independent of any
prejudicial financial or business interests in advertising. To the. medical
panel, after consultation with the minister and twelve professional medical
bodies, the Authority appointed seven eminent consultants to cover general
medicine, pharmacology, chemistry, dentistry and veterinary science. These
were supplemented by four 'second opinion' specialists: in paediatrics,
gynaecology, dermatology and ear, nose and throat.

From its earliest days the Authority had published Principles for
Television Advertising relating to specific categories. These developed into
a more formal Code of Advertising Standards and Practice, whose
publication became mandatory under the Television Act 1964. Drawn up
by the Authority in consultation with its advisers and the minister, this
listed general rules and followed them with detailed appendices on the
three categories of major concern: advertising to children, financial
advertising and the advertising of medicines and medical treatment. The
general rules covered conditions for guarantees and mail order offers and
prohibited subliminal advertising and advertisements from such undesir-
ables as money-lenders, matrimonial agencies, bookmakers, private in-
vestigators and undertakers.

In 1956 the Authority had itself taken the initiative in asking the
Postmaster-General to issue a directive forbidding advertisements for any
form of betting, including football pools. In 1970 it changed its mind about
football pools, but the Postmaster-General declined to lift the ban. An
approach to the Home Secretary in 1983 met with a similar refusal.

To those who enjoyed the privilege of being allowed to advertise, the
weight of inhibitions and restraints often seemed excessive. *Caveat vendor*
had displaced *caveat emptor*, and one advertising man compared writing a
television commercial to writing a sonnet: 'It's a triumph of the creative
instinct over man-made rules.'[21]

Yet advertisers were very far from being discouraged. By 1973 new
advertisements were running at the rate of 20,000 a year. The majority
(some 15,000) came from local advertisers, using simple five-second or
seven-second slides to publicise shops, restaurants, services and entertain-
ments, but despite their number commercials of this kind occupied only 6
per cent of advertising time. The big national advertisers who took the
remaining 94 per cent were spending up to £6,500 for thirty seconds. For
this expenditure they could reach some 8 million homes. The apparently
extravagant cost of delivering their message was therefore no more than
one-twelfth of a penny per home. No sales force, postal service or other
medium could offer equivalent value for money.

Each year more than 9,000 advertisements required vetting for compliance with the Code. Advertisers and agencies submitted scripts in advance to avoid the risk of making costly films which might then be found unacceptable. Scripts were cleared through the IBA's Advertising Control office and an advertising copy clearance department set up by the companies under the umbrella of ITCA. At this stage all medical scripts were passed to a relevant member of the Medical Advisory Panel, and ITCA retained the services of other consultants to evaluate claims in specialist fields such as finance and engineering. Finished films were checked in daily closed-circuit viewings by Authority and company staff to guard against any departure from agreed scripts and any unacceptability in tone or style.

Wider issues raised in the course of these daily procedures were discussed by a Joint Advertisement Control Committee. Meeting under the chairmanship of the Authority's Head of Advertising Control, this committee determined the interpretation and application of the whole canon of restraints and prohibitions imposed by the various Acts and the IBA's Code – subject to a final ruling by the Authority itself as the statutory body. Doubtful cases were regularly previewed by the Authority's Chairman.

These supervisory arrangements had been established in 1959 by Archie Graham, who joined the ITA as Head of Advertising Control after twenty-five years in the Post Office. Humorous and occasionally indiscreet, he was not above telling an audience of seething advertisers that, although they and a minority of the ITV audience might be young libertarians, his job was to impose a censorship based on middle-class, middle-aged morality. When he retired in October 1974, weary of the arguments and pressures in serving a divided society, his achievement in reconciling public control of standards with successful creative advertising was recognised by the award of the Mackintosh Medal, the industry's highest honour for services to advertising, as well as the OBE. His successor, until 1981, was Peter Woodhouse, a barrister who had been ITCA's Head of Copy Clearance, who served the Authority with distinction, and whose conscientiousness extended even to eating dog food himself to check the chunks. With so many grounds for censorship and so many awkward adjudications to be made, controlling advertisements was a thankless job, and Woodhouse too retired early, in ill health.

Apart from the other regulations, advertisements might be disqualified for offending the rule against any blurring of the distinction between advertisements and programmes. In 1972, without submitting a pre-production script, the advertising agency for British European Airways produced a finished advertisement depicting Alan Whicker inspecting and commenting favourably on BEA services in a style indistinguishable from his performances in *Whicker's World*. Permission to screen the advertisement was refused, and a new version had to be made in which Whicker was allowed to be glimpsed briefly but not to speak. In Scotland in 1974 the voice of Gordon Jackson, the *Upstairs, Downstairs* butler, was heard extolling the virtues of Scotsmac wine during an interval in an episode of the programme. Any repetition was immediately forbidden.[22]

A running controversy surrounded the television advertising of alcoholic drinks, which was proscribed in some countries. As with the ban on cigarette advertisements imposed in 1965, this was a matter ultimately for decision by government and governments were not persuaded by the anti-alcohol lobby. One relevant factor was the continuance of a voluntary agreement made by the competing distillers companies in 1955 not to use the medium. Television therefore carried virtually no advertisements for hard liquor – liqueurs were an exception – and great care was taken to keep advertising for beers, wines and cider clear of children's programmes.

In 1978, in lieu of a ban, the rules were tightened on the advice of the Advertising Advisory Committee after consultations with and between the Home Office and the DHSS. To the rule that liquor advertising must not be addressed particularly to the young was added a new one that no liquor advertisements should feature any personality who commanded the loyalty of the young. The rule that advertisements must neither claim nor suggest that any drink can contribute towards sexual prowess was strengthened by another prohibiting treatments featuring special daring or toughness which were likely to associate drinking alcohol with masculinity. Any mention of immoderate or solitary drinking, or even of buying rounds of drinks, was also prohibited.[23]

Such elaborate precautions took the sting out of any widespread public hostility towards the intrusion of advertisements into the home. At the rate of about a thousand a year comments and complaints were relatively very few, and most were satisfied by an explanation of the policies adopted. When a critical article in the *Observer* in 1973 invited readers to write to the IBA with their complaints,[24] only four did so: one about the Stork versus butter test, one about a car passenger seen not using a safety belt, and two against advertising in principle as 'odious moral blackmail'.[25]

From the Annan Committee television advertising in general received an almost clean bill of health: 'If a mixed economy prevails, then advertising is inherent in its operation, and any propensity which advertising may have to mislead or offend must be curbed by law and regulation. The curbing of advertisers on television is a main function of the IBA and ... we think that for the most part it carries out this function efficiently.'[26] The committee judged that the system of spot advertising had produced good results and ought to be allowed to continue.[27] It did not endorse the Pilkington recommendation that the Authority, not the companies, should sell the airtime.[28]

But a majority of the committee expressed itself as less than satisfied where ITV's advertising to children was concerned. This caused surprise, for it was a subject regulated through detailed inhibitions and prohibitions in the IBA's Code of Advertising Practice and ITCA's Notes for Guidance. For example: 'Direct appeals or exhortations to buy should not be made to children unless the product advertised is of interest to children and one which they could reasonably be expected to afford for themselves' (in 1975 the maximum affordable price out of pocket money was assessed at £1); 'advertisements for medicines specially formulated for children must not be transmitted before 9 p.m.'; 'personalities and live characters featured on children's programmes must not be used to promote products, premiums or services'; 'for reasons of dental hygiene, advertisements shall not encourage persistent sweet-eating throughout the day nor the eating of sweet sticky foods at bed-time. Advertisements for confectionery or snack foods shall not suggest that such products may be substituted for proper meals.' With dental caries in mind, the catch-phrase 'Start them early with Cadbury's Buttons' had been disallowed. So too was 'A Mars a day', on the grounds that a Mars bar was a sweet and not a valuable and legitimate part of a child's daily diet like a 'pinta' of milk.

For so sensitive an area complaints from the public were remarkably few. The dangers of fireworks aroused alarm in some parents, and war toys came under intermittent fire from pacifist groups. There was a small, but fervent, annual protest over the advertising at Christmas of toys which many parents could not afford.

The Annan recommendations – that there should be no advertisements within children's programmes or between two programmes for children and that advertisements promoting products or services of particular interest to children should not be shown before 9 p.m. – were not supported by Lord Annan himself. He was among the minority who pointed out that most television watched by children on ITV would continue to carry advertisements and that the IBA commanded a battery of safeguards against excessive exploitation of the audience.

Nevertheless, most members of the committee felt overwhelmingly protective towards young viewers. They were worried about the interruption of children's enjoyment of programmes and about the encouragement of covetousness and extravagance. They believed that 'television advertising makes a far greater impact on children than other forms of advertising' and that 'children are inclined to believe that what they are told on a television programme is not only true, but the whole truth. How then are they to distinguish between what they are told in a children's programme and what they are told in an advertisement?'[29]

In opposition the Authority had pointed out in evidence that most advertising in and around children's programmes was directed at housewives, not children. It detected in these proposals 'perhaps one further manifestation of the feeling that advertising is *per se* undesirable'. The Authority's religious advisers had agreed that there should be advertising within religious programmes, 'since it was wrong to insulate them and to disown our source of funds. The same argument can be applied to children's programmes.'[30]

The companies estimated that such a ban would cause an annual loss of revenue of £22 million, or 10.75 per cent of their income from advertisements. This was more than the cost of the ban on cigarette advertising, which had represented 7 per cent of total ITV revenue; 10 per cent in the case of some companies. Their reaction was predictably hostile:

Advertisements are part of life in Britain today, and the companies question the validity of the view that children should be shielded from them. Whether they watch television or not, children are exposed to advertisements daily – in children's magazines, for instance, in children's cinema shows and on posters on the way to and from school. ITV's controls of television advertising for the protection of children are stringent. If advertisements were diverted from television to other media they would be subject to less rigorous standards of control. The most popular television programmes among children are not the children's programmes. If the purpose of removing advertisements from children's programmes is to shield children from them, it would be ineffective, as the following figures show: 4–9 year-olds watch approximately 1.8 hours of children's programmes out of a total of 12.7 viewing hours per week; 10–15 year-olds watch 1.5 hours out of a total of 13.4 hours.[31]

In deliberating whether the recommendations should be acted upon, the Authority noted that the majority in favour appeared to have been influenced by research undertaken in the USA which had little relevance to the situation in Britain, and that few seemed to share their fears. Of the thousand or so letters and telephone calls received on the subject of advertising during 1976, only two reflected the committee's majority view. In the lengthy parliamentary debates on the committee's report the subject was never even mentioned,[32] and the government decided not to legislate. The Authority then exercised its discretion with a compromise, banning advertising breaks from all children's programmes which were more than half-an-hour in length.

This, it was estimated, would cause the companies no financial loss. But at a bad-tempered PPC meeting in April 1978 the companies' representatives expressed their disgust with some vehemence. They argued that other centre-break prohibitions related to types of programmes, not types of audience. To concede that there was any audience to which it was inappropriate to advertise was seen as a betrayal and the thin end of a very thick wedge.[33]

Post-Annan, as pre-Annan, the Authority's involvement in the increasingly sophisticated business of selling television time was minimal. It had no responsibility for the tariffs and was sometimes embarrassed by increases in the rates charged.[34] Each company set its own price for its own airtime, but under the law its rate card had to be drawn up 'in such detail and published in such form and manner as the Authority may determine', and all charges for advertisements had to be in accordance with the published tariffs.[35]

Rate cards were issued quarterly after each company had assessed the strength of its market. If prices were pitched too low, time would be fully sold but revenue not maximised. On the other hand, over-optimism and prices set too high would lead to heavy discounting. In view of the uncertainty of market conditions, which could swing unpredictably with the fluctuating strength or weakness of the national economy, flexibility in pricing was all-important and it was the practice in rate cards to make provision for various 'regulators'. These often involved hard bargaining. There were volume discounts; there were low rates for local advertisers; a proportion of low-priced spots might be offered to advertisers making full-rate bookings. It was in the nature of the system that all rates were quite unrelated to the companies' costs. They were entirely geared to the level of demand.

According to Brian Henry, Marketing and Sales Director of Southern Television:

> By the end of the sixties three seemingly unrelated factors had combined to set the scene for the introduction of sales methods which were to dominate advertising on Independent Television for the next fifteen years. They were, first, the disclosure of monthly revenue figures from which each company's strengths and weaknesses, in sales terms, could be calculated by the buyer of airtime; secondly, the continuation of government-inspired price controls, whether statutory or voluntary, which prevented rates from being increased fully to reflect the level of demand; and thirdly, the advent of real-time computer systems which alone could cope with a constantly changing price structure. All three factors, when seen against the background of a fixed supply of airtime which was governed by limited hours of broadcasting, and therefore could not be expanded, led to the introduction of the controversial 'pre-empt' rate structures which were eventually to be adopted by all the ITV companies.[36]

The fine-tuning mechanism known as pre-empt rates was first introduced by Thames in 1968. Under this system an advertiser buying a particular spot at the normal rate faced the possibility of losing it to one prepared to pay at the higher, pre-empt rate. By this means supply and demand were balanced according to the current state of the market almost up to the day of transmission. Some companies incurred the wrath of advertisers and the disapproval of the Authority by going further and introducing a double or 'super fix' charge which enabled a company to accept bookings for fixed spots at a 15 per cent surcharge on standard rates, subject to cancellation should another advertiser be prepared to pay a 30 per cent premium. In 1970, following criticism from advertisers and the Prices and Incomes Board, pre-empt rates were temporarily withdrawn, to be replaced by 'early booking' discounts. But pre-emption was back on the rate cards within six months.

In the autumn of 1975 the Price Commission rejected an application for surcharges on the rates for advertisements in and around the first television showing of the James Bond film, *Dr No*. This was a block-buster which attracted an audience of 27 million and which, even without surcharges, materially assisted the achievement of what was then a record monthly revenue figure for the network, income in October of that year exceeding £20 million.

In the following year, to sidestep the Price Commission's objections, a Guaranteed Home Impression formula, whereby if the size of an audience fell short of expectations an advertisement would be repeated without additional charge, was developed by London Weekend into a highly profitable Gold Star package. This set unprecedentedly high rates for advertising attracted by the Bond films, Royal Variety Shows and other exceptionally popular programmes. Audience sizes were guaranteed and prices increased or reduced accordingly if they were exceeded or fell short.

During a prolonged period of price restraint, when the companies were obliged to meet the requirements of a Prices and Incomes Board or a Price Commission in setting rates, the intricacies of these devices became ever more tortuous and resented by buyers in what, owing to the limited availability of airtime, was invariably a sellers' market. Yet in the autumn of 1977, when Thames, the largest company, was not permitted to increase rates to meet demand, it was feared that this would mean a rationing of time which would prove even more unpopular among advertisers than rationing by price.[37] ITV's dependence on advertisers had become balanced by some advertisers' dependence on ITV. Manufacturers and retailers of consumer goods and services had come to rely on the direct access to millions of housewives in their own homes – by 1980 potentially almost every housewife in the United Kingdom – offered by the network.

Commercial airtime in the early years had been dominated by the great washing-powder war between Procter & Gamble (Daz, Dreft, Fairy Snow and Tide) and Lever Brothers (Omo, Persil, Radiant and Surf). In 1963 these two and Beecham were between them contributing 25 per cent of ITV's revenue. Their money was the salvation of Independent Television, but their advertisements did little for its reputation.

In 1970 washing-powder manufacturers were still among the biggest spenders. The top twelve products advertised in that year were Radiant,

Ariel, Weetabix, Kellogg's Corn Flakes, Persil, Daz, Maxwell House coffee, Electricity Council home heat, Blue Band margarine, Stork margarine, Oxo and Fairy Liquid. But during the 1970s ITV broadened its revenue base by attracting major retail stores and finance houses. The growth of multiples – Sainsbury, Tesco, Asda – at the expense of small retailers was accelerated by advertising on television. By 1980 the twelve largest advertisers had become: Dulux, Woolworth, Guinness, Midland Bank, GPO, Milk Marketing Board, Nescafé, *The Sun*, Asda, Whiskas, British Airways and the Co-op.[38]

Between 1963 and 1973, when no new product category filled the gap left by cigarettes, the cost of television advertising fell by 9 per cent in real terms, but between 1973 and 1983 it rose in real terms by 32 per cent, stimulated by the demands of new advertisers. Existing advertisers were critical of the rate increases, although not to the extent of withdrawing their custom. Bitterness among advertisers and advertising agents at the companies' monopolies was endemic, and this intensified at the prospect of the perpetuation and extension of the companies' monopolies with the coming of a second channel financed by advertising.

In evidence to the Select Committee on Nationalised Industries in 1972 the Incorporated Society of British Advertisers (ISBA) stated: 'Whilst most advertisers would agree that the companies have not, on the whole, abused their monopoly position and have behaved responsibly, advertisers would not want to see a long-term solution that continued – and even strengthened – the monopoly position in the future.'[39] Six years later its attitude had hardened when it told the next Select Committee that fixed supply and increasing demand had resulted in surcharges over basic rates rising from 10 per cent to 40 per cent.[40]

In the following year the Institute of Practitioners in Advertising (IPA), representing the agencies, issued a condemnatory document entitled *An Examination of Airtime Selling Practices*.[41] This listed examples of what advertising agents identified as restrictive practices and abuses of monopoly which enabled the companies to 'maximise their revenues with minimum hindrance from the IBA or government'. The advertising industry lobbied persistently for a competitive commercial fourth channel, with the sale of airtime in the hands of a separate sales organisation. In 1979 the CBI joined ISBA in making representations to the Home Secretary, but even with a Conservative government which believed strongly in the value and virtue of competition a different interpretation of the public interest prevailed. 'Competitive advertising on the two channels would inevitably result in a move towards single-minded concentration on maximising the audience for programmes,' declared the Home Secretary, William Whitelaw, in dismissing the advertisers' case.[42]

Although giving the salesman the whip hand, the companies' monopolies were not absolute. In London, the most important ITV area for advertisers, there were two competing companies, and elsewhere it was always open to them to omit from their schedules any company or companies whose terms seemed too stiff or to choose between regions when test-marketing a new product. Television time was never sold nationally, and the companies competed fiercely for their market share.

Television was certainly the preferred medium for mass marketing, but newspapers and magazines attracted far more of the advertisers' money, and radio, posters, direct mail and the cinema were available. In 1968 advertising expenditure in the press amounted to £347 million, compared with £129 million on television. By 1976, swollen by inflation, these figures had become: press £809 million, television £307 million.[43] Despite its growth ITV's share of total advertising expenditure remained almost constant: 25 per cent in 1961, 26 per cent in 1968, 23 per cent in 1974 and 27 per cent in 1980.[44]

With the arrival of Lord Thomson of Monifieth at 70 Brompton Road in 1980, first as Deputy Chairman and then as Chairman, the paying customers had for the first time the ear of someone at the top with experience in their world. Lord Thomson had been Chairman of the Advertising Standards Authority and was well qualified to make a professional appraisal of their complaints about high rates, the arbitrary imposition of unreasonable terms and a general take-it-or-leave-it attitude by the companies. If the advertisers had to swallow the pill of no competitive selling of airtime on a second ITV channel, it seemed both prudent and just to offer them the opportunity of more and better consultation, and a new body was formed for that purpose.

The first meeting of the IBA's Advertising Liaison Committee took place on 27 June 1980 under the chairmanship of Lord Thomson, who expressed the hope that the committee would create a climate of mutual understanding which would enable any problems to be rapidly and amicably resolved. Membership was at the highest level. It included the presidents and directors of ISBA and the IPA, the chairmen of the ITCA Council and its marketing committee and, from 1981, both the Chairman and the Director General of the IBA as well as the Head of Advertising Control. Meeting quarterly without any executive or judicial function, it dealt in principles and grievances. The IPA quickly tabled a paper on a special sales force for the fourth channel but was told that this was a subject which could not be reopened. ISBA circulated a draft Code of Practice for the Sale and Purchase of Television Airtime, which led to the formulation of a set of principles agreed between ISBA, the IPA and ITCA and formally blessed by the IBA.[45]

On some topics there had been no disagreement between the parties. The SCNI report published in 1973, for example, had contained four recommendations concerning advertisements, all unwelcome and all rejected: a stricter definition of natural breaks; consideration to be given to the adoption of the continental practice of transmitting advertisements all together in large blocks; regular discussion programmes in which the claims of advertisers could be publicly challenged; and a regular allocation of time for public service announcements about health and safety measures.

The second of these threatened the financial viability of the system. The bunching of commercials was practised in countries whose television services relied on advertisements for only a fraction of their revenue. Since the principle of a service wholly financed by advertising had been accepted, block advertising had become an unrealistic option. The third recommendation, that the practice of the American Federal Communications Commission (FCC) might be followed and consumer associations given airtime to challenge advertisers' claims, was also seen as misguided. The FCC exercised no control over advertisements before they were screened: the Authority ensured that no misleading claims were broadcast.

Nor was the committee's fourth proposal well received. ITV and the BBC already had a long-standing arrangement with the Central Office of Information to hold a large stock of filmed announcements on road safety, blood donors, foster parenthood and other public welfare subjects for use, free of charge, as fillers in unsold advertising time or other gaps in programme schedules not required for programme promotion. Throughout the calendar year 1972 transmissions of these fillers had totalled 22,000 in the various regions throughout the network: to the value, if they had been paid for, of £2 million. This was applauded by the committee but, looking the gift-horse closely in the mouth, it had noted that the transmissions were rarely at prime viewing times.

The companies had never conceded, nor the COI expected, free time for particular government campaigns – on seat-belts or drinking and driving, VAT or the Rent Act – which were paid for in the press and other media. Airtime for these was charged at the usual rates. But if precious peak-time was now to be made available without charge, why, it was asked, should broadcasting be treated differently from other media in providing regular free facilities for public service campaigns.[46] The companies pointed out too that ITV was already doing more than the BBC, which was strongly opposed to any compulsory extension of the facility.[47]

The Authority pressed the companies for at least a more even distribution of the free announcements – fewer than 400 of the transmissions had been in the London area. The companies replied that advertising time was in greatest demand in the central areas and suggested an additional minute a day of advertising time.[48] But the Authority was not disposed to make an extra allocation, even to accommodate public service announcements.

In notifying the COI of the outcome of these discussions, the Authority identified the broadcasting of information and advice on matters of public interest or concern as one of its functions in normal programming – in news bulletins, documentary and current affairs programmes, discussion programmes and local magazines, all of which reached large audiences in and out of peak time. *Coronation Street* and *Crossroads* were sometimes plotted to involve characters in welfare problems and solutions of which viewers could take heed. So prime-time needs were being met by the combination of public service programming on the one hand and planned advertising campaigns on the other.[49]

Paid-for advertisements occupied 10 per cent of ITV's screentime. In compensation for their interruption of the programme service they brought two benefits to the viewer: the programme service itself, which came free, and an additional free service of information and, sometimes, entertainment. It was often said that, by increasing the price of the goods and services advertised, Independent Television cost the public an annual sum roughly equivalent to the licence fee collected for the BBC, but there was little evidence to support the validity of this thesis. It gained credence only by ignoring economies of scale and the advantages of making the availability of products widely known.

For if mass advertising increases sales, production costs per unit will fall and this is likely to result in a lower rather than a higher price to the consumer; although less so with soap flakes than motor cars. If demand for a product falls and machines become idle, the stimulus of mass advertising may restore demand and production to an economic level and avert a price increase. It could be argued too that money not spent on television advertising would not necessarily be saved, being spent instead in other media or on point-of-sale promotion. In a competitive market, moreover, the cost of advertising is not always passed on to the consumer.

As the Annan Committee recognised, advertising on a large scale benefits the national economy. It encourages investment in research and development and in new plant and new products, because mass sales are thereby created and created quickly. Increasingly in the 1960s and 1970s television advertising became an important ingredient in the whole process of production and marketing and thus in employment and prosperity. A fair claim could be made that Britain's television advertising was the best and most effective advertising of any kind anywhere. Advertisements on television often worked as they had never worked in the press. ITV's regional structure, with its separate company sales forces, created marketing areas, and marketing in Britain was revolutionised.[50]

It was also a healthy economic indicator (confounding prophets of woe) that during twenty years of growth in advertising on ITV there was no corresponding increase in total advertising expenditure when measured as a percentage of consumer spending. In 1961 advertising represented 1.9 per cent of consumer expenditure, in 1968 1.85 per cent, in 1974 1.72 per cent and in 1980 1.88 per cent.[51]

The undesirable features of television advertising were social, not economic. They lay in the raising of expectations, the fostering of materialism and the narrowing of choice. These were not features peculiar to advertising on television, but they were heightened by its scale and effectiveness. The incentive to keep up with the Joneses, or feel socially deprived if one could not, was enhanced by the power of a persuasive message delivered to the whole family in its own home. The strength of the medium helped manufacturers who could afford a large advertising appropriation: retailers felt obliged to stock brands advertised on television, and other brands were likely to become relatively slow sellers, in danger of becoming discontinued lines. With diminishing sales they, rather than the advertised brands, were liable to become more expensive.

On the other hand, the impact of some advertisements transformed social habits for the good. When the first toothpaste advertisement appeared – the very first of all television commercials – more than a third of the adult population of Britain never cleaned their teeth, and the sales of toothpaste grew with ITV.[52] During the 1970s many viewers were persuaded to stop smoking by hard-hitting advertisements placed by the Health Education Council. The problem of what girls were to drink in pubs with their boy friends was solved innocuously through a successful television campaign for Babycham.[53]

Creatively, advertisements have sometimes been judged superior to programmes, much more money being lavished on much less screentime. The discipline of the thirty-second spot allowed no time for self-indulgence. Making commercials became a well-paid art form which attracted leading film directors. Lindsay Anderson went into the business of selling Ewbank carpet sweepers, Iron Jelloids, Guinness and Kellogg's cornflakes. Joseph Losey helped to market Ryvita and Horlicks. Karel Reisz worked for Persil, Mars bars, Chanel and Campari; Ken Russell for Black Magic chocolates; John Schlesinger for Stork margarine, Polo mints and Eno's fruit salts.[54] In opening up new vistas of creativity television transformed the advertising scene. London became the creative capital of television advertisement production, and the development of the art of the television commercial in Britain has even been seen as the British equivalent of the French cinema's *nouvelle vague*.

After some initial reluctance leading actors and others in the public eye took their cue from Sir Compton Mackenzie, the pioneer promoter of Horlicks, and appeared on the screen endorsing products. For most, any lingering scruples were buried under sometimes unimaginably handsome fees. Good money could be made from contributing no more than a recognisable voice. Television jingles – by Cliff Adams, Johnny Johnson and others – were among the favourite tunes of the day, more familiar to some children than nursery rhymes; and reiterated slogans became national catch-phrases: 'I'd like to teach the world to sing'; 'Nice one, Cyril'; 'I'm only here for the beer'. During the 1970s Cadbury's Smash became famous for its Martians, Homepride Flour for its cartoon flour-graders in bowler hats and Fry's Turkish Delight for its oriental fantasies. Orson Welles became inseparably associated in the public mind with sherry, robots with Fiat, David Bailey with Olympus cameras and, most famously until retired in 1974 (although seen in Wales and the West till 1976), Katie with Oxo.

Not all the stars were human or cartoon characters. A nation of animal-lovers was entertained by an Esso tiger, a Lloyd's black horse, a Dulux sheepdog and Arthur the white Kattomeat cat who, after an expensive lawsuit over his ownership, passed away much mourned in 1977 at the ripe old age of 14. Most loved and enduring of all were the PG Tips chimpanzees, who performed as tea-drinkers from 1956 to 1969, were rested from 1969 to 1971 and then brought back by public demand. They spoke with the voices of Peter Sellers, Stanley Baxter, Bruce Forsyth, Irene Handl, Arthur Lowe, Bob Monkhouse and Kenneth Williams.[55]

These animals promoting the sale of petrol, banking, paint, petfood and tea established themselves among the public's favourite viewing; for, irrespective of the product being promoted, these and some of the other advertisements during this period demonstrated a degree of technical skill and artistry in the practice of visual and verbal communication which it was possible to enjoy and admire in their own right.

13

ASPECTS OF RESEARCH

Audience research is, first and foremost, audience measurement: how many people were watching a particular programme or advertisement or at a particular time? That, in Independent Television, was the foundation for the scale of charges from which the income of the system was derived. Other questions and refinements then followed. What kind of people were they? To what extent did they assimilate what was communicated? How informed were they by that documentary, amused by that entertainment programme, moved by the story of that drama, motivated by that advertisement? What were their reactions afterwards? What was the effect, if any, on their subsequent behaviour?

In the 1950s and early 1960s in the United Kingdom the answers to all such questions relating to programmes, except the first, were little sought. But during the 1970s American practice was increasingly emulated. Television research projects proliferated, many taking a predominantly sociological orientation. The medium had spawned a new academic discipline: media studies. While broadcasters investigated what the customers thought of the service, social scientists were intent on discovering what the service was doing to the customers. Different approaches to research were pursued at this time by the Authority, by the companies collectively, by the companies individually, by advertisers and their agencies, and by academic sociologists.

The Pilkington Committee in 1962 had chided Independent Television for an undue dependence on ratings. It called for 'continuing and perceptive audience research' and recommended that the Authority should 'engage in, or commission, research and development work'.[1] Dr William Belson, a leading specialist in the field, in turn chided the committee itself for preferring 'argument and armchair wisdom' to research, arguing that an earnest belief that different kinds of programmes were good or bad for people was no substitute for discovering whether they were really beneficial or harmful. In 1967 Dr Belson was still deploring the neglect of research and attributing wrong decision-making in television to ignorance of the relevant facts.[2]

Meanwhile the Pilkington Committee's recommendation for more – and more systematic – study of the needs, interests, attitudes and reactions of the audience had been adopted in the Television Act 1964 in the following terms:

The functions of the Authority shall include the making of arrangements for bringing the programmes (including advertisements) broadcast by the Authority and the other activities of the Authority under constant and effective review, and in particular for ascertaining the state of public opinion concerning the programmes (including advertisements) broadcast by the Authority and for encouraging the making of useful comments and suggestions by members of the public; and the arrangements shall·include provision for full consideration by the Authority of the facts, comments and suggestions so obtained.[3]

To help to meet this requirement the Authority appointed a specialist Head of Research to operate within its Programme Division. He was Dr Ian Haldane, who set about developing qualitative research and assessment. After a prolonged period of experimentation a joint ITA/ITCA steering committee awarded the contract for an exploratory scheme to a consortium of three market research firms operating under the acronym TOP (Television Opinion Panel). Designed to highlight the strengths and weaknesses of the ITV output in the opinion of viewers, in obedience to the Ascertainment Rule, it was launched in the autumn of 1969 with initially disappointing results. But when reorganised and refined it became accepted despite some scepticism in the companies. From 1974 it ceased to be contracted out and was run by the Authority's Research Department itself under the name of AURA (*A*udience *R*eaction *A*ssessment).

Under this system a specially designed television diary was supplied to a panel of 1,000 adult viewers in the London ITV area who had been selected as representative in terms of age, sex and social class. They were asked to rate each programme which they watched according to a six-point scale ranging from 'extremely interesting and/or enjoyable' to 'not at all interesting and/or enjoyable'. Their opinions then formed an Appreciation Index (AI) within the range of 0 to 100. A weekly report was published covering all ITV and BBC programmes, and from this it could sometimes be seen that a programme which attracted a large number of viewers had not been greatly enjoyed or that a low audience figure was compensated by a high level of appreciation.

The first regional survey took place in July 1973; after which London and the other regions alternated weekly, each of the regions outside London being covered twice a year through representative postal samples of about 2,000 viewers. A Children's AURA and a Teenagers' AURA enabled separate studies to be made of the opinions of younger viewers, and the original plan was to establish not only AIs, but also a series of evaluation scores for every programme over a period of time. This proved too ambitious, however, and while appreciation matured – notably through the addition of special questionnaires – evaluation by this means was abandoned.

It flourished, though, in another service. A number of the companies subscribed to the Television Audience Programme Evaluation (TAPE) which predicted the success or failure of a programme by measuring its intrinsic and competitive appeal. The brain-child of Mike Firman of the advertising agency Masius Wynne-Williams, it broke programmes down into component parts and awarded Predicted Audience Rating (PAR) scores on the basis of his personal knowledge and assessment of the drawing power of stars, feature films and story formats. A comparison of total PAR scores indicated the audience share which a programme might be expected to achieve when scheduled against a known programme on another channel. An unexpected degree of accuracy was obtained by this systematic alignment of subjective value judgments, which its critics condemned as a malign influence – 'a method of anaesthetising viewers with their own prejudices'.[4]

Apart from AURA the Authority fulfilled its statutory responsibility by commissioning an annual survey of public opinion. *Attitudes to Broadcasting* collected information on such subjects as the public's perception of ITV's political impartiality, its views on the standards of taste and decency in programmes, and opinions about the hours at which programmes unsuitable for children were screened. Asking the same questions at regular intervals revealed shifts in attitude, and this 'dipstick' or 'barometer' method detected trends where data derived from a single survey might have misled. Between 1968 and 1974, for instance, it demonstrated a steady growth of preference among viewers for ITV rather than BBC1.[5]

Numerous *ad hoc* research projects and surveys were commissioned in connection with the Authority's formal Consultations and with specific programme questions. Consultations on religious broadcasting, on children's programmes, on news and current affairs, on sport, on drama all included a presentation of some relevant and up-to-date research findings. Comedy and light entertainment were found difficult to research meaningfully. Researchers were never allowed to forget that much time and money had been expended in making the scientific discovery that comedies become successful by making people laugh.

Violence was the subject most expensively and extensively researched.[6] The Authority's initial grant of £250,000 for the study of a relationship, if any, between television and delinquency was a sizeable figure in pre-inflation 1963. It led to a Television Research Committee and further funding; and this committee, under the chairmanship of Dr Noble, the Vice-Chancellor of Leicester University, led in turn to the establishment of a permanent research centre at that university. Devoted to the study of the media and mass communications, it was directed by Professor James Halloran, who had been the Noble Committee's secretary. A Research Fellowship was also established at Leeds University under the supervision of Dr Jay Blumler, funded at first by Granada and then by the Authority, and another was instituted within the Authority's own research department.

Other research funding was channelled through a committee composed of senior Authority staff under the chairmanship of a member, Professor James Ring. This allocated a total of £50,000 a year in grants for projects relevant to broadcasting, whether at universities or elsewhere. Under another scheme the Authority financed Schoolteacher Fellowships for work on the use of schools broadcasting: practising classroom teachers were seconded to a university or College of Education, and the Authority reimbursed the Local Education Authority for their salaries and the fellows for their out-of-pocket expenses.

ASKE Research, a consultant company which had played a part in establishing AURA, made an important contribution in drawing general conclusions from an accumulated mass of data and formulating what became known as 'laws of viewing behaviour'. These findings about viewing patterns made by Professors A. S. C. Ehrenberg and G. J. Goodhart and their colleagues contradicted some of the received wisdom of established tele-pundits. ASKE demonstrated, for example, that political balance would not be effectively maintained if opposing statements were made in separate programmes in the course of a documentary series, because little more than half of those who heard one side of the argument would be likely to hear the other. Audiences for successive programmes might be similar in number, but their composition changed. This proposition that real political balance could be achieved only within a single programme (as originally required of ITV) was strenuously contested by programme-makers, and in requiring due impartiality in 'matters of political or industrial controversy or relating to current public policy', the Broadcasting Act 1981 ignored the ASKE finding and followed the precedent of Acts from 1964 in stating specifically that a series of programmes might be considered as a whole.[7]

The ASKE studies for the Authority, published in 1975 as *The Television Audience*,[8] attracted the attention of the Annan Committee, which was without the funds to commission adequate studies of its own. During the financial year 1974–75 the Authority spent £33,000 on AURA, £30,000 on ASKE's and its own surveys and £13,000 on financing other research projects, and the committee judged these to be 'relatively small sums given the total cost of programmes'. It criticised ITV and BBC research operations alike as 'too piecemeal, too narrow and too superficial', but found the IBA's research more sharply focused than the BBC's: 'The IBA

were better able to stand back from day-to-day audience research work'
and analyse changes in the public's viewing, outlook and tastes.[9]

The abundance of these well focused studies may be illustrated from a
period of six months in 1975 while Annan was at work. Between April and
October the Authority's Research Department produced reports on television coverage of the October 1974 general election and the EEC
referendum campaign, on AURA diary completion, on the reactions of
coloured viewers to the comedy programme *Love Thy Neighbour*, on
audience reactions to John Pilger's personal view programmes, on public
opinion of radio coverage of proceedings from the House of Commons and
on several aspects of radio listening patterns. Between August and
October it published research summaries on the appreciation of ITV
drama, audience sizes on Saturdays, viewing patterns in the Channel
Islands, social class and programme appreciation, an upward trend in
ITV's weekend performance and what might be described as the situation
comedy situation.[10]

Towards the end of the decade, in preparation for the award of new
company contracts from 1 January 1982, the Authority commissioned a
national survey, whose findings were reported in June 1979.[11] This
examined the attitudes of viewers all over the country towards ITV in
general and its local services in particular. All fourteen ITV regions were
covered, including Northern Ireland and the Channel Islands. The objective was to provide a picture of viewers' opinions of ITV nationally and to
compare and contrast them within and between the different regions so
that the performance of each contractor might be judged.

Topics investigated included viewers' evaluations of different types of
programmes shown on ITV and BBC1; awareness of individual ITV
companies and associations with them; interest in local news and the
viewing of local news programmes; awareness and evaluation of programmes made by the local ITV company; and an overall appraisal of each
company.

An assessment of the respective images of ITV and BBC1 was obtained
by means of a series of statements to which the respondent was asked to
react by saying whether they applied more to ITV or BBC1 or to both
equally or to neither. By this test ITV was the clear winner in 'showing a
great deal of interest in the people of this area' (ITV 51 per cent, BBC1
7 per cent), in having a very friendly approach (ITV 45 per cent, BBC1
12 per cent) and in being go-ahead and lively (ITV 42 per cent, BBC1 10
per cent). BBC1 was judged the better in being 'extremely professional in
its presentation' (ITV 13 per cent, BBC1 54 per cent). It may be that
friendliness and professionalism were seen as mutually exclusive. By
comparison with the BBC in programme categories ITV obtained its best
scores for local programmes, late-night and weekday afternoon programmes, and serials; its worst for documentaries, music and the arts, and pop

music. At the end of the questionnaire respondents were asked: If you were stranded on a desert island with a television set which would receive only one channel, which channel would you choose? 48 per cent chose ITV, 39 per cent BBC1 and 8 per cent BBC2. Of those who chose ITV, women substantially outnumbered men.

While AURA, ASKE and its annual and *ad hoc* studies were supplying the Authority with about as much information on public attitudes and programme appeal as it could absorb (and rather more than the Annan Committee gave it credit for), the companies in association continued to concentrate on the basic task of audience measurement. On this, at the time of Annan, they and the BBC were between them spending more than a million pounds a year.[12] This apart, the Authority did not hold the companies' research efforts in high regard. 'The number of genuine programme research experts in the companies can in my experience easily be counted on the fingers of one hand,' wrote Bernard Sendall as Chairman of an SCC working party on research liaison.[13] He thought their research record dismal, and the Annan Committee too expressed itself as 'disturbed to find that hardly any of the ITV companies had their own research staff, but instead drew on market research firms for specific studies'.[14]

'This is not a good way to help producers,' the committee judged, but to the companies it seemed a sensible arrangement. The money which the IBA was spending on its research came from them and the findings were available to them. Why employ additional staff to duplicate the work? Company researchers were housed within the sales department, responsible to the sales director, not the director of programmes. They were in the business of wooing advertisers. Their job was to equip the company's salesmen with facts and figures to that end. Most producers and directors still preferred the guidance of experience and hunch, and managements fought shy of projects whose findings were apt to offer no practical guidance at not inconsiderable cost.

The companies' kind of research was not *Religion in Britain and Northern Ireland* (an Authority study in 1968–69), but *How Advertising Helped to Make Krona Brand Leader* (the case history of a new margarine from Van den Bergh launched in October 1978 into the yellow fats market in the HTV and Westward areas) and *The Effectiveness of Advertising in Reducing Pedal Cycle Casualties*, the story of a campaign on Anglia in 1980 which was estimated to have reduced casualties by 17 per cent and saved the tax-payer £450,000 at a cost of £129,000. In November 1978, upset by a recommendation of the Annan Committee, ITCA published *The Effects of Advertising on Children*, the report of a study which it had commissioned. How to reach the light viewer cost-effectively was also a problem on which much company research effort was expanded.

Another, which grumbled on vexatiously down the years, was the overlap between adjacent ITV areas. The blurring of boundaries between the companies' coverage areas, accentuated by the introduction of UHF transmissions, complicated the allocation of advertising budgets. By agreement with the advertisers' representatives, each company laid claim to fringe areas with a density of reception as low as 15 per cent of homes. The result was confusion, and rationalisation was demanded by leading manufacturers who had come to base their sales and distribution territories on the ITV regions and required them to be discrete. But not until the early 1980s was a new, and more sensible, way of defining and calculating overlaps introduced.

No small part of the commercial success of Independent Television rested on the heavy investment in research made by the companies strictly in their own interest. Attracting advertisers to their own areas was the major preoccupation and the tools of the trade were surveys such as Southern Television's series : The Southern Shopper, The Southern Motorist, The Southern Holiday-Maker, The Southern Investor. Here co-operation did not come easily. The British Bureau of Television Advertising (BBTA), a jointly funded co-operative venture formed in 1965, fell victim to inter-company competitiveness and was disbanded in 1974.

Subjects for research of interest to all companies were the presence and attention of viewers, test marketing and new business, campaign scheduling and advertising impact, the respective merits of 'bursts' and 'drips' and the growth and effectiveness of colour (for which no premiums were charged). A Television Consumer Audit was established to estimate the quantity of product sales generated by television advertising. This reached the heart of the matter: did advertising on television really work?

A strike which took ITV off the air in the autumn of 1979 fortuitously provided an opportunity to assess the effect of the withdrawal of television advertising on product sales and to rebut the political charge that advertising was parasitic, a waste of money, of no benefit to the manufacturer or the economy. The advertising agency D'Arcy McManus & Masius evaluated the performance of ninety-six brands which would have been advertised on television but for the strike, and the results of its research demonstrated not only that advertising worked but also that it was possible to determine to what extent it worked. The amount of money not spent on advertising the brands because of the strike was £9 million, and during the period of the strike alone their sales fell by an average of 4.5 per cent, representing a loss of £25 million in sales revenue or £2.80 per £1 not spent on television advertising. 'Apart from the saving of direct raw material and packaging costs,' the agency concluded, 'the difference of £1.80 between the value of lost sales and the saving on advertising must largely represent an immediate loss of profit to the advertiser ... In the longer term, these losses will have become significantly larger for those advertisers who failed to make good the shortfall in advertising expenditure.'[15]

Periodic endorsements of this kind were helpful, but audience measurement, day by day and week by week, remained the prime commitment and common interest. Since 1961 it had been the responsibility of the Joint Industry Committee for Television Advertising Research (JICTAR), a body on which ITCA (the ITV companies), ISBA (the advertisers) and IPA (the advertising agencies) were all represented, thus dispelling the suspicion that the companies might contrive to inflate the figures to their advantage. It was funded four-sevenths by ITCA, one-seventh by ISBA and two-sevenths by IPA; the chairmanship rotated; and all the parties accepted the independence and impartiality of its findings.

In November 1968 the contract for carrying out this head-counting operation was awarded to AGB (Audits of Great Britain Ltd), a company formed by former directors of TAM (Television Audience Measurement Ltd) which had held it for the previous thirteen years. The implementation of the specification and methodology was closely supervised by JICTAR's Technical Sub-committee. Audiences were measured by meters attached to television sets in 2655 homes selected as a representative sample of the households in each ITV area. The meters recorded the times when the sets were switched on and off and the channel to which they were tuned. Additionally a diary was completed on a quarter-hour basis detailing the age, sex and other characteristics of those viewing. The BBC's very different method was street interviewing of between 2,000 and 2,500 people about programmes which they had watched or listened to on the previous day. Like JICTAR's, the sample of these Daily Surveys was small but it constituted a representative cross-section of the population and was changed daily.

The ITV system gave information valuable to advertisers which was not available through the street-interview technique, and its meter-monitored diaries could record set usage by channel accurately. Its drawback lay in the not necessarily valid assumption that because a television set was switched on someone was viewing; but this might be checked from the diary. The flaws in the day-after recall method (DAR) employed by the BBC were its reliance on memory, the fact that its sample was confined to those who were out and about during the daytime, and an inherent bias because interviewers disclosed to respondents that they were working for the BBC.

As statistical exercises both systems were proper and valid, but continual discrepancies between the two sets of published figures undermined public confidence in either, and it was an interesting phenomenon, much remarked upon, that each consistently produced figures favourable to the organisation which chose to employ it. The Annan Committee found this 'unedifying' and recommended a single audience measurement operation. This had been proposed by Sidney Bernstein and others in the earliest days of ITV, and evidence submitted to Annan by ITCA, ISBA and IPA all supported the principle of a co-operative research service, provided that it met the requirements of the advertisers as well as those of the broadcasters.

In anticipation that their heads would eventually be knocked together, the companies and the BBC had in the mid-1970s underaken what became known as the Quantum experiment, to test the audience research techniques available. The object was to understand thoroughly the pros and cons of what the various methodologies could provide, although it was apparent from the outset that ITV could not abandon minute-by-minute data regarded as essential by advertisers for measuring commercial-break audiences. Because it was broadly coterminous with a BBC editorial region, the Yorkshire ITV area was selected as the test bed, and the existing meter sample was matched against information obtained from the BBC's aided-recall technique and from specially placed programme diaries. The biases which emerged were not unexpected. For example, the BBC's DAR technique registered lower daytime audiences because the street interviews were taking place at that time.

Putting the Annan recommendation into practice proved a painful and protracted process. Annan reported in March 1977 and it was not until April of the following year that the two parties, under political pressure, could bring themselves to notify the Home Secretary, in a letter signed jointly by Sir Michael Swann (Chairman of the BBC) and Lord Windlesham (Chairman of ITCA's Council), of their intention 'to share research data sources' from July 1979. The new arrangements, they promised, would meet the Annan requirement of shifting resources from head-counting to qualitative and audience appreciation studies as well as continuing to meet the advertiser's needs in audience measurement.[16] Annan had foreseen that a combination of ratings research into one system would not only end confusion; it would free a large sum of money for use on more and better qualitative research. But this was not to be.

Dismissing the notion that achievement of a common system was simply an ideal, the Annan report had described it as an 'essential' requirement. 'Desirable' would perhaps have been more precise. Co-operation between competitive organisations is rarely essential. Joint enterprises with its rival were anathema to the BBC and research had become an instrument of competition. A similar recommendation for a single programme journal was strongly resisted by both organisations. Negotiations between them on joint research were conducted through a minefield of policy disagreements, technical problems and ill will. One major obstacle was the political strength of the BBC radio department within the Corporation, where radio retained the mystique of senior status. This frustrated the expectation that the BBC Daily Survey of Listening and Viewing would be gradually wound up and replaced by a JICTAR-like operation. The continuation of the Daily Survey for the benefit of radio ensured that there were no savings to be reallocated to other forms of television audience research as Annan had hopefully visualised.

Following the letter to the Home Secretary, joint ITCA/BBC steering committees were set up to prepare the way for a new controlling body: the main committee found a neutral Chairman in Sir Stewart Crawford, former Chairman of the Committee on Broadcasting Coverage. But discussions had barely begun when ITCA made its bid to secure exclusive coverage of Football League matches and the BBC in a sulk broke off all official contacts. The technical experts were left to negotiate quietly on their own, with no committee to report to, until the two organisations were officially on speaking terms again. The target date of July 1979 came and went. It was then hoped that a start could be made by the end of the year, but two years were to pass before the specifications of the new measurement and audience reaction services were agreed and a further year before they came into operation.

The new controlling body, BARB (the Broadcasters' Audience Research Board), was born in July 1980, 50 per cent owned by ITCA, 50 per cent by the BBC, under Sir Stewart Crawford's chairmanship. Its first contract was awarded to AGB. The advertisers and agencies, who had been equal partners in JICTAR, suffered relegation to membership of the Audience Measurement Management Committee (AMMC), a subordinance body. Policy and financial control were reserved to the main board on which they were not represented. Complementing AMMC was the Audience Reaction Management Committee (ARMC), the audience appreciation subsidiary, of which the IBA became a member: the new arrangements had made AURA redundant. Its six-point scale was taken over.

The research system unified under BARB began its service in July 1981, more than four years after the Annan recommendation, which it fulfilled in the letter but scarcely in the spirit. It would be hard to claim that, by that date, harmony and rationalisation had been achieved.

14

ANNAN: PREAMBLE AND
EVIDENCE

On 3 December 1969 the Labour government's Minister of Posts and Telecommunications (John Stonehouse) informed the House of Commons that he had not reached any conclusion about a new Pilkington. But he thought it right to inform the House that he had been considering the prospects of some commission or committee of enquiry into the long-term future of broadcasting after 1976, because there would be 'mammoth implications' to be considered and it was important that those considerations should be borne in mind well before the expiry of the 1964 Television Act and the BBC Charter.

In a memorandum from his ministry to the Authority the following month it was argued that technical developments, both extant and impending, needed to be evaluated and consideration given to whether these would necessitate constitutional changes in the broadcasting organisations. In addition: 'on general constitutional grounds, enquiry by independent committee is necessary in order that the two broadcasting authorities, who dispose of the most influential and persuasive of the media of mass communication, should be accountable, and seen to be accountable, to the public'.[1]

The argument was unexceptionable; not so the time-scale. The formation of the Pilkington Committee had been announced in July 1960; it had reported two years later; and after two White Papers and the necessary legislative process the resulting Act had taken effect in July 1964. Now, less than six years later, a further investigation was being proposed which would take, not four years, but seven. The committee would be appointed in the summer of 1970; it would be allowed three years to report; government and parliament would take two years to consider, prepare and pass legislation; and the broadcasting authorities would be allowed two years to make the consequential arrangements. The system under which the health of broadcasting was diagnosed by having its roots pulled up and examined at frequent intervals was becoming hallowed by custom and seemingly inescapable; but were seven years of instability essential to the democratic process?

221

At a meeting with the minister on 21 January 1970 the Chairman of the
ITA argued that, since there was to be a preparatory Technical Committee
in any case, a full enquiry which would throw the whole broadcasting
system into the melting pot was unnecessary. Could not essential changes
be agreed between the Authority and the ministry? Or, if there must be an
enquiry, surely it need not be set up as early as 1970? The BBC's Chairman
(Lord Hill) supported the second point, but not the first. Nevertheless,
Lord Aylestone was not alone in believing a further major enquiry to be
unnecessary. The Chairman of the previous one, Lord Pilkington, had
stated in the Lords a few months earlier that he saw no need for it, and the
Prime Minister himself (Harold Wilson) was said to be unconvinced.

In February Lord Shackleton, a member of the government, made an
announcement in the House of Lords which was interpreted as a cautious
shelving of the issue: 'Speaking personally (I cannot be certain as to what
the government may do if the pressure grows), I should greatly regret any
form of a public enquiry now.'[2] In May, though, the government's decision
'to set up a wide-ranging review of the future of sound and television
broadcasting' was announced.

In the Commons a Conservative MP (Paul Bryan) drew attention to 'the
low morale among the creative and executive staff in broadcasting,
largely due to the fact that ITV and the BBC, during recent years, have
been over-investigated, over-reorganised and over-restructured'. What, he
wanted to know, was the hurry to appoint a committee of enquiry before a
general election when there were still six years before the BBC and ITV
mandates ran out.[3] In the Lords a former Conservative Home Secretary
(Lord Brooke of Cumnor) identified 'the premature setting up of this
committee' as 'the clearest indication that we have so far had that the
government, in their hearts, expect to lose the next election'.[4]

The Chairman of what was then being described as Pilkington Mark II
was to be Lord Annan, Provost of University College, London, and
formerly Provost of King's College, Cambridge. He was summoned by the
minister, who told him, over a postprandial brandy at the ministry one
afternoon, that within a few years there would be no more television
screens or newspapers. Pictures on the living-room wall would screen
programmes from dozens of television channels, show anyone telephoning
or being telephoned and display the news by teletext. Dazzled by this
prospect, Lord Annan confessed that he had much to learn about telecommunications but promised to do his best.[5]

The general election on 18 June was won by the Conservatives, and on
23 July Christopher Chataway, the new Minister of Posts and Telecommunications, told the House of Commons:

I am not persuaded of the value of launching another major enquiry
into broadcasting at this time. I propose to ask my Television Advisory

Committee to undertake a study designed to identify the main technical questions and to report to me early in the new year. This will provide the basis for a more informed public discussion of the issues. I will then consider whether an enquiry into the structure of broadcasting after 1976 is desirable and, if so, what form it should take.

Lord Annan, he made clear, was being stood down on political not personal grounds. At a subsequent press conference Chataway claimed that his decision would release broadcasters from inhibitions over planning for the future. He saw dangers in getting into a ten-year cycle of major investigations: everyone would be 'waiting for Annan', as for years they were 'waiting for Pilkington'.

The Conservative government under Edward Heath continued to show little enthusiasm for a general enquiry. Eighteen months later Chataway was still talking in terms of nearer to 1976 because 'a state of more or less permanent inquiry' was not sensible for broadcasting.[6] The opposition view was forcefully expressed by Phillip Whitehead, speaking in the House of Commons on 2 August 1972 (at 4.30 a.m.). He declared that putting matters to a Technical Advisory Committee was not good enough for those who had to make the right choices in broadcasting: 'Those choices are inextricably political and social as well as technical ... The background is of philosophical uncertainty about the nature of broadcasting and its role in society as well as of brash technical opportunity.' The consensus under which the media had long operated was, he argued, breaking down. Matters needed to be 'put to the proof by exhaustive analysis'. In responding, the new minister, Sir John Eden, expressed agreement with his critics on the need for 'the most informed public debate on what the future should be from 1976 onwards', but he was awaiting the report of the Technical Advisory Committee and the matter was too important to be rushed.

Impatience on the left manifested itself in the formation of several pressure groups. One was the Free Communications Group, which told the 1972 Select Committee that the ITA had, throughout its existence, failed to exercise the authority with which it was entrusted by parliament. Another was The 76 Group, composed of ITV and BBC producers and other creative staff lobbying for radical change. A third arrogated to itself the title of The Alternative Annan, later changed to the scarcely less presumptuous The Standing Conference on Broadcasting, and set out to do what Annan would have done. In 1973 a TV4 Campaign was organised specifically to lobby against the allocation of the fourth channel to ITV. This sprang from a study group formed under the auspices of the Labour Party's Home Policy Committee. Meetings of this group were held between May 1972 and May 1974, chaired by Tony Benn until March 1974 (when he became a minister).

These bodies contained libertarians objecting to a closed system with no right of access to outsiders, but all tended to be influenced by the same authoritarian marxist element, at that time stridently articulate within the Labour Party, the trade union movement and Polytechnic departments of sociology, and among some of 'the disgruntled, the disenchanted and the failed'[7] in broadcasting. Accountability was what they most vehemently demanded of the broadcasting authorities, and by accountability they did not mean accountability to the people via their elected representatives in government and parliament. They were concerned about 'closed decision-making by unrepresentative cliques', about profits, and about what they believed to be the failure of the media 'to relate to the needs of society'. Their objective was control in the name of freedom. 'With commendable and unusual candour,' wrote one member of the Authority, 'some such advocates admit they would use this accountability for attempts to control British broadcasting and use it as a means of social change and, the changes having been secured, go on controlling it to ensure that there is no further change.'[8]

These critics of the system harked back to Pilkington:

The gravamen of the Pilkington Committee's charge against commercial television ... was that its main objective, which is to make money from advertising revenue, is incompatible with the true objectives of good broadcasting. It is not good enough to say that ITV is justified by its large audience figures. The search for a programme formula which pulls in the largest number of people militates against real freedom of choice and against the widening of individual experience.

This passage is taken from an Opposition Green Paper, published by the Labour Party in 1972,[9] which went on to recommend that half of all advertising expenditure should be disallowed for tax purposes – a move likely to have achieved social change by depressing the economy and bankrupting both ITV and the press. How this and similar proposals would lead to better programmes and newspapers and more enjoyment and enlightenment for viewers and readers was not spelt out. The overriding desiderata were greater public control, public accountability and internal democracy (or workers' control).

The opportunity to realise these aims came in February 1974 when Labour was returned to power, albeit with a narrow majority of five. Early in March Harold Wilson made known his ministerial appointments. The Ministry of Posts and Telecommunications would disappear. The technical services would go to the Department of Trade and Industry. Responsibility for broadcasting policy would be transferred to the Home Office under Roy Jenkins. A long-time associate of Jenkins, the former journalist John Harris (created Lord Harris of Greenwich), was to be Minister of State at the Home Office with the broadcasting services as one of his special responsibilities. On 10 April Jenkins announced that the government had decided to revive the committee of enquiry, under its previously appointed chairman, with the following terms of reference:

> To consider the future of the broadcasting services in the United Kingdom, including the dissemination by wire of broadcast and other programmes and of television for public showing; to consider the implications for present or any recommended additional services of new techniques; and to propose what constitutional, organisational and financial arrangements and what conditions should apply to the conduct of all these services.[10]

The resurrection of Annan so close to the 1976 deadline necessitated legislation to extend the term of the Independent Broadcasting Act of 1973 for three years until July 1979, together with a similar extension of the BBC Charter. The other members of the committee were named on 12 July, five years before the new deadline. Their report was required in two and a half years. Television's period of prolonged instability had been postponed but was now to stretch for more than six years fraught with politicking.

Within ITV Lord Annan was seen as a good choice for chairman: a worldly academic, politically middle-of-the-road. But the rest of the names inspired no confidence that the report would be fair and balanced. Not a single one of the sixteen members was identifiable as what might be described as an ITV viewer and, to judge an industry with a multimillion pound turnover, fifteen out of sixteen had no first-hand experience of business. Leading campaigners against advertising and the whole ITV system were included. The full list was:

Lord Annan	Provost, University College, London.
Peter Goldman	Director of the Consumers' Association.
Professor Hilde Himmelweit	Professor of Social Psychology, London School of Economics.
Tom Jackson	General Secretary, Union of Post Office Workers.
Anthony Jay	TV writer/producer; former BBC executive.
Marghanita Laski	Novelist, critic, journalist.
Mrs H. M. Lawrence	Hertfordshire County Council Alderman, and member, Stevenage Development Corporation.
A. Dewi Lewis	Chairman, Dyfed Area Health Authority and former grammar school headmaster.
Sir James MacKay	Retired civil servant.

Mrs Charles Morrison	Vice-Chairman of the Conservative party; former Wiltshire County Councillor.
Dipak Nandy	Member of Social and Community Planning Research.
G. J. Parkes	Chairman of Unilever Merseyside Committee.
John D. Pollock	General Secretary, Educational Institute of Scotland; former Chairman of the Scottish Labour Party.
Professor G. D. Sims	Head of Department of Electronics, University of Southampton.
Phillip Whitehead	MP for Derby North; former ITV/BBC producer.
Sir Marcus Worsley	MP for Chelsea.

There were thus three academics (Annan, Himmelweit and Sims), two Members of Parliament (Whitehead and Worsley), two trade unionists (Jackson and Pollock), two political women (Lawrence and Morrison), two representatives of creativity (Jay and Laski), one Welshman (Lewis), one from an ethnic minority (Nandy), one Scottish civil servant (MacKay), one businessman (Parkes) and one consumerist (Goldman). John Pollock also qualified as a Scotsman; Peter Goldman, a former director at party headquarters, as a Conservative; and Dipak Nandy as an academic.

The BBC connection was strong. Tom Jackson was a former governor, Dewi Lewis was Chairman of the BBC's Welsh Appeals Advisory Committee, Marghanita Laski had long been a regular BBC broadcaster, Anthony Jay was a former BBC executive and editor of *Tonight*, and Sir Marcus Worsley and Phillip Whitehead had both worked for the Corporation. Whitehead, who had been a BBC producer and moved on to edit *This Week* for Thames, was the only member with an Independent Television connection.

The secretary to the committee was Jean Goose, an experienced and energetic Home Office official who had been Private Secretary to the Minister of Posts and Telecommunication. By the broadcasters she was judged to be scrupulously fair-minded but as favourably inclined towards the BBC as Dennis Lawrence, the civil servant secretary to the Pilkington Committee who had drafted much of that committee's condemnation of the soul-endangering evils of Independent Television.

The dice were so heavily loaded that a protest seemed called for, despite much doubt at the IBA about the wisdom of such a move. The Chairman wrote to the Minister of State at the Home Office accordingly:

I have been wondering during the past month whether I should write to you about a view which some Members of the Authority have expressed to me about the composition of the Annan Committee. Meanwhile, one

or two journalists (for example, Peter Fiddick in the *Guardian*) happen
to have written about the point which was put to me; and a natural
reluctance to complain about the referee before the game began held me
back. But now I realise that the addition of three new names to the
Royal Commission on the Press may possibly form a precedent for the
addition of a few names to the Annan Committee; and representations
which have probably been made to you from Northern Ireland about the
absence of anyone from that part of the UK might also lead to a fresh
name coming forward; there is also the possibility (as happened with
Pilkington) of present members resigning and being replaced . . .

ITV is always in danger of having no-one who can speak for it with
experience. There tends to be a balance (when broadcasting committees
are being selected) between those who are in principle ready to argue
against the existing structure and those who are ready to defend the
BBC, because they have known and admired it from the inside. So the
IBA, and ITV in general, tend to be crushed by committees of enquiry
between the upper and the nether millstone.

One or two of our members observe in the Annan Committee some
who will, perfectly properly, question on theoretical grounds that which
is; and others who will, equally properly, argue from past experience as
governors or employees of the BBC that in practice the Corporation is
wiser or more sensitive than it may seem to the outside world. We see no
member (apart from a highly committed MP) who can speak from within
of the IBA or ITV. If further appointments are made to the committee,
may I – and other members who will have retired long before Annan
reports – ask that you might bear this point in mind?

Naturally, any disquiet we may feel will not affect in any way our
willingness to do all that we can to help the committee in its task, and to
give it all the information that it needs. Nor is it meant to question the
ability and public spirit of those who have undertaken the job of sitting
on the committee. Those are not in doubt: what worries us is a possible
restriction in the range of interest represented.[11]

The Home Office rebuff to this plea was not unexpected. After a formal
expression of regret at the Authority's feeling of disquiet its reply carefully
missed the point by expressing abhorrence at any suggestion that members
should be representative or have 'direct links with particular interests':

Considerable thought was given to the selection of the membership of the
committee to ensure a balance of experience about broadcasting both from
the broadcasters' and consumers' points of view whilst at the same time
avoiding the situation in which the committee might be open to the
charge that its recommendations had been made by people with a direct
interest in the disposal of valuable public assets. The Home Secretary is

confident that whilst individual members may inevitably hold particular views about the present and future structure of broadcasting, the combined experience and wisdom of the Committee as a whole under Noel Annan's chairmanship could be expected to formulate its recommendations on the basis of reasoned argument rather than on prejudice.[12]

Such sentiments could not disguise the fact that in its composition the committee was a body unsympathetic to ITV, as Pilkington had been, and one which reflected the attitudes and expectations of the government of the day, as in the 1980s the 'free market' complexion of the Peacock Committee was the purposeful choice of the Thatcher administration. 'The composition of the Committee on the Future of Broadcasting was modelled by politics,' its Chairman has written; 'and for good reason. Broadcasting itself has become a political subject.'[13]

Specifically, the inclusion of Phillip Whitehead, the leading public critic of ITV, was seen as a hostile act unbalanced by anyone who might be expected to speak up for Independent Television. Whitehead had been one of those responsible for *The People and the Media*. He was a founder member of The 76 Group. He was a prominent member of The Standing Conference on Broadcasting, one of whose primary objectives was to keep the fourth channel out of ITV hands. His views had been repeatedly expressed and were well known. As a skilled debater and by far the most knowledgeable member of the committee, he would, it was feared in ITV, have little difficulty in persuading the others to his way of thinking and, when the report came to be written, might well be granted the lion's share of the pen.

There were two reasons why Whitehead should not have been a member of the committee. His appointment violated the Balfour Committee principle that those appointed to Royal Commissions or Committees of Enquiry 'should, as far as possible, be persons who have not committed themselves so deeply on any side of the questions involved . . . as to render the probability of an impartial enquiry and a unanimous report practically impossible'.[14] It was also a breach of the convention that sitting Members of Parliament did not serve on such bodies. But Whitehead's political association with the Home Secretary was strong enough to overcome these objections, and Lord Annan and Home Office officials opposed his appointment in vain.

As Leader of the Opposition Edward Heath accepted with reluctance an invitation to nominate three Conservatives. As a pair to Whitehead he included the name of the outgoing minister, Sir John Eden, but this was rejected. Sir Marcus Worsley was selected instead, but a few months later he decided not to stand in the October 1974 general election and left parliament. Official Tory representation was further weakened when Sara

Morrison, a political ally of Heath, resigned her vice-chairmanship of the party on Margaret Thatcher's succession to the leadership.

Lord Annan was more successful in his determination that Whitehead should not be his only member with broadcasting experience. At his insistence Tony Jay was appointed over the heads of the civil servants who argued that, as a freelance earning his living from broadcasting, Jay had a financial interest in what the committee might recommend and was therefore ineligible. A similar argument had been used against all the names of those experienced in advertising put forward by the Authority for consideration by the minister; and the committee lacked even an economist.

Unpaid except for the reimbursement of expenses, the members of this committee on the Future of Broadcasting held regular meetings once a month. There were other meetings and visits, and the amount of paperwork to be studied by the conscientious was prodigious. How much could members be expected to absorb? Fears of another Pilkington were not allayed when they began to visit the companies and revealed an alarming ignorance of broadcasting and an inability to grasp the principles and nuances of company finance. Some were assiduous and learned (as Phillip Whitehead put it) to sift the evidence to find nuggets of truth among all the humbug and special pleading'.[15] Others suffered silent burial beneath a mound of evidence.

Formal submissions were received from 762 bodies and individuals. They ranged (alphabetically) from the Aberdeen Chamber of Commerce; Aberfeldy District Gaelic Choir; Aims of Industry; Albion Cablevision Ltd and Mr David Allen to York Council of Churches; Yorkshire and Humberside Council for Further Education; Mr George Younger MP; Youth Hostels Association (England and Wales) and Yr Academi Cymreig (Academy of Welsh Writers). In addition many members of the public responded to the committee's appeal for their views: 6,000 in the first six months.

In making the case for the defence the IBA and the companies agreed to prepare and submit their evidence separately, and the IBA's advisory bodies, including its National Committees, offered theirs independently. In a 96-page document the Authority produced a clear statement of its principles, policies, practices and plans.[16]

> Independent Broadcasting is unique among broadcasting systems in its way of allowing the forces of the market place, which play a fundamental and valuable part in other social activities, to play their part in broadcasting also, but within a framework of public control.

> The last twenty years have seen public service broadcasting provided without public finance.

> The Authority is ... continuously involved in programme planning as well as having ultimate editorial control of particular programmes in the

schedules ... Its four main functions – the appointment of companies, the control of programme output, the control of advertising, and the building and operation of the transmitters – mean that it is involved in the system, though with a measure of detachment, at all points.

Our wide powers are exercised, and in our view should be exercised, in a way that allows the diversity of the system to be maintained: that is to say that individual companies – their boards, staff and advisers – are given the opportunity to exercise genuine discretion on many issues which involve no conflict with the Authority's own policies. In providing a framework for the production of good programmes, it is important for the Authority to maintain through the system a high general standard of expectation rather than to indulge in continuous harassment of producers.

The Independent Broadcasting services are popular services, but the Authority takes pains to ensure that schedules contain material that is likely to broaden the experience, and indeed enrich the taste, of all viewers.

The Authority prided itself on the creation and development of a plural and regional service: 'a good blend of central strength and local responsibility'. It supported the small regional companies through a policy of differential rental charges and programme pricing: Thames paid an annual rental of £2,250,000; Border paid £1,000; Channel £100. 'The ITV companies' production of local interest material is on average twice as much as that of the BBC regional centres. Moreover, it attracts strong loyalty... ITV audiences for these programmes are greater than those for the BBC – in some regions markedly so.' This strength of the parts also contributed to the strength of the whole: 'It is true that a single monolithic organisation gains some of the advantages of scale; but it also tends to lose that variety of approach which, we believe, characterises the ITV networked product. ITV network programmes often have a regional flavour.'

The expense of regionalism was recognised:

The present structure of ITV results in greater costs than would be incurred by a system that was more centralised and less regional. To achieve savings, however, would entail reducing not simply the number of companies, but also the diversity, and probably the amount, of local programming. The Authority would wish to avoid this and it would not envisage reducing the number of companies unless amalgamations were forced by economic circumstances.

The Authority was at pains to defend its decisions over the 1968 franchises:

Some saw in these changes of structure, and appointments of new companies, a rash leap into the unknown; and in the first two or three years, when the economic climate was difficult, it was easy to say that the Authority had too readily dispensed with some valuable experience which the system had had from the start, and too optimistically injected new elements which found it hard to make performance match promise. Nevertheless, the new companies which came into the system in 1968 have now, in the Authority's view, well justified the original decision to make changes. But the Authority is also conscious of the need for stability; and its recent outline of its plans for 1976–79 has made it clear that it does not envisage a great convulsion of the system every few years.

Steering the correct course between popular and minority programmes was no less tricky than opting for 'development' or 'stability'. As the people's television service ITV had to be both responsive and responsible. 'As democratic governments depend on votes, so programme companies depend on their audiences. It is of the essence of Independent Television that it is broadly responsive to the tastes and interests and wishes of the public.' 'A commercial discipline from the outset put Independent Television in touch with popular style and taste. This was, in its way, a significant democratic development, though there were critics in high places who disliked, and still dislike, this "box office" element in ITV.' Being by definition a mass medium, television is not 'designed to appeal primarily to that small section of the population which brings to bear on broadcasting most spoken and written debate'. 'The task of the broadcaster is to try to provide some programmes for all ... tastes, including not only programmes for decidedly minority tastes but also programmes for a number of majority tastes which do not have many people speaking for them.'

To argue the case for television being responsive does not mean for a moment denying the need for it to be responsible. If the intention of parliament had been simply to provide a service which would give viewers what they already knew that they wanted, there would have been no need to set up an Authority charged with seeing that ITV had social purpose as well as popular appeal. The tradition of serving the public as well as pleasing the public was, fortunately, strongly enough embedded in the British tradition for it to be axiomatic that the new television service set up in 1954 should have among its aims the encouragement of merit, the fostering of quality, and the broadcasting of knowledge and feeling. It has never been the intention that ITV should devote itself entirely, or even predominantly, to undemanding programmes, designed simply to attract the maximum audience all the time, which might have been the outcome of a commercially based system left entirely to its own devices.[17]

The Authority's evidence went on to point out that ITV's programme output of 8,000 hours a year made it one of the largest producers of television programmes in the world. The number of hours of programmes from all sources transmitted by ITV annually was 9,400 – 'about ten times as much . . . as even the most dedicated of busy people can watch'. Failures were candidly admitted: 'That there is some dross in this output is not surprising; it is gratifying that in recent years viewers and critics have recognised much gold.'

The philosophy of the IBA in the Young years is well summarised in these words from its evidence:

The Authority has, then, a particular concern that ITV should, as far as is possible on a single channel, cater for serious interests, introduce people to new and enlightening experiences, and serve individual *and* mass needs, as well as providing a wide range of relaxing entertainment such as the great majority of people expect and require during their limited hours of leisure. Its answer to the old question whether it ought to reflect or to lead the interests and tastes of the public must be that it should do both. The balance it strikes between these two will always be more ambitious than is to be found in a solely commercial service, since the Authority has a special task from parliament to lead; on the other hand, it is always likely to reflect popular taste more plainly than readers of *The Times* and patrons of Covent Garden might wish. The Authority believes in giving the public both 'what it wants' *and* 'what it ought to have'.[18]

Such generalities were succeeded by arguments on specific issues. The proposal originally made by the Pilkington Committee that the Authority should 'collect' the advertising revenue and commission programmes from independent production companies was condemned as 'unlikely to have anything but a deleterious effect on the creative quality of programmes'. As for sharing the licence fee and advertising revenue with the BBC: 'The years of controlled competition between these two systems have seen major improvements in the scope and quality of British television.' 'Moreover, through normal taxation and the levy, ITV since its introduction has contributed about £400 million to the public purse.'

On the question of outside programme contributions the Authority pointed to the greater opportunities which two ITV channels would offer. Meanwhile ITV's schedules, like those of the BBC, included the work of many freelance producers working inside the broadcasting organisations. The idea that it ought to become a programme producer itself was dismissed: 'the Authority would have to set itself up to make programmes on a regular basis: it could hardly make programmes as and when it felt inspired'. Nor was this proposal seen as a practical option for a body with a role as the final arbiter in ITV programme-making, since it would inevitably become 'judge and jury in its own case'.

In referring to 'accountability', the Authority remarked on the almost wilful ambiguity in the current use of the word and on the facile pretext this offered to habitual busybodies to discuss the role and legitimacy of a broadcasting body without ever watching its programmes, which were the chief and proper means whereby broadcasters gave public account of themselves. But if accountability was meant to imply a free two-way flow of information whereby broadcasters explained their actions and decisions while themselves seeking more knowledge and understanding of public reactions and attitudes as a factor in influencing those decisions, then the Authority was fully committed to unremitting efforts towards those ends, principally through the various specialist and regional advisory bodies, direct face-to-face contacts in public meetings, a range of publications and carefully structured research. Yet, desirable as a policy of complete openness might in theory appear to be, the Authority thought that there were some areas which were for good practical reasons better kept confidential: for example, some exchanges with companies over matters of programme detail and, more especially, the interviewing of applicants for programme contracts.

It was consistent with these arguments that the Authority should oppose proposals for a Broadcasting Council. To establish such a Council in addition to the two existing broadcasting authorities would 'either result in a body that had no powers or in one that simply duplicated, to a partial extent, the functions of the members of the Authority'. More detailed arguments against the proposal which had been presented to the minister in 1973 were reproduced in an appendix:

The two most common proposals, for a body analogous to the Press Council or for a 'viewers' voice', would fundamentally alter the structure of Independent Television as laid down by the Television Act and would seriously weaken, if not destroy, the Authority's position as guardian of the public interest.[19]

While the committee was deliberating, supplementary evidence was submitted on a number of specialised topics, sometimes on Authority or company initiative, sometimes in response to a request from the committee. At SCC meetings regular reports were made on encounters with the committee in oral evidence; and some of the supplementary evidence arose therefrom. Extra papers were submitted on the financing of ITV2; the machinery for inter-company programme exchange; company diversification into other fields of investment; details of advertising rate cards; advertising in children's programmes; and company music publishing. The Authority also sent papers on audience research and on its staffing and internal organisation; copies of its observations on the SCNI Report; copies of the Notes of Guidance issued to its programme monitors; a selection from the monthly Intervention Reports by its programme staff; and a list of subjects treated in networked current affairs and documentary programmes from January 1972 to October 1974.

Overall the IBA's written evidence to the Annan Committee was a cogently argued advocacy of its operations and aspirations, in particular its bid for the fourth channel. It deployed a sound defence of its general strategy and served notice to its assailants that any attempt at storming this citadel of power would be strongly resisted by a well-armed garrison.

The written evidence submitted by the ITV companies was defence of a different kind.[20] It took the form of an elegantly produced, easy-to-read instructional manual designed to inform, educate and entertain members of the committee. A collective view was never readily obtainable among the companies, and every word was subjected to scrutiny and approval by fourteen managing directors. (Channel Television, the fifteenth programme company, was not formally involved because it operated outside the jurisdiction of the United Kingdom parliament.)

Only the threat of another Pilkington could have brought the companies to so great a measure of agreement on a common approach and no rocking of the boat. Some with special, mainly national, interests – Grampian, HTV, Scottish, Ulster and Westward – submitted supplementary evidence on those subjects. So, too, did the subsidiaries: ITN (on national and international news coverage) and ITP (on the programme journal). But all subscribed without reservation to the main document, produced under the auspices of ITCA, the trade association; and none of the major companies chose to make individual submissions.

ITCA's Annan Steering Committee, charged with responsibility for assembling the evidence and drafting the document was composed of six company chief executives: William Brown (Scottish), Sir Denis Forman (Granada), John Freeman (LWT), Peter Paine (Tyne Tees), Jeremy Potter (ITP) and Lord Windlesham (ATV), with Mary Lund (General Secretary of ITCA) as secretary. Forman was Chairman of the committee and its master-mind. Freeman and Windlesham were politicians as well as communicators: one had been a Labour, the other a Conservative, minister. Brown, a senior figure in the industry, had a special responsibility for the interests of the nine regional companies. Advertising was Paine's particular expertise. Potter was the scribe.

The opening paragraph was both descriptive and heartfelt:

This is a portrait of Independent Television as seen by the ITV companies themselves. Their diversity is not only geographical. In a federal system each component has its own area of sovereignty, proudly defended, and unanimity is not easily achieved. This submission therefore is not the product of a unitary structure like the IBA or the BBC, but the consensus of fourteen separate interdependent companies differing widely in size, shape, attitude and opinion.

The companies' evidence contained four sections and twelve appendices. A two-page summary of thirty-six conclusions and recommendations was printed at the front for the benefit of those members of Lord Annan's committee who were finding twenty-four hours a day insufficient to reach the end of all the evidence submitted. The first section dealt with the control and finance of television, the second with the making of programmes and their various categories, the third with the companies themselves, both severally and jointly, and the last with advertising and profitability.

The central thesis on which it based its justification of the ITV system was the familiar one of unity in diversity. The special strengths of Independent Television lay, it was claimed, in the effective working relationship which had evolved between small regional companies and the large, nationally oriented – but also regionally based – central ones. This made it possible for viewers in each region to receive not only programmes locally conceived and sensitively adjusted to local concerns but also prestigious major productions comparable with the best of the BBC or any other broadcasting service in the world. Despite a necessary degree of co-ordination, each of the companies in the three groups – central companies, large and small regionals – continued to display its own independent and distinctive character. Profiles of the individual companies were given in support of this claim.

Admittedly the system, with its multiplicity of separate fully-equipped production centres, was more expensive in manpower, money and engineering resources than a simpler centralised system would have been; and the constant switching from one programme source to another to serve varying schedule patterns, to say nothing of different commercials, made great demands on the skills of technical staffs. Yet the outcome of such complexity was the broad well-founded boast that in no other country was there a bespoke local service coupled with network programming to areas with populations as small as those of the Borders or Channel Islands. And, despite the technical complexities, failures affecting the service to the public had amounted in 1974 to no more than 0.01 per cent of transmission time.

It was doubted whether alternative systems would have served the British audience as well. Descriptions were given of the television services in Sweden, West Germany and the Netherlands. Each suffered, it was argued, from disadvantages which ITV had avoided. Because in Britain advertising interests were allowed no say or influence in programme planning or production (and because advertisers had to compete for limited airtime), advertising finance had proved a source of strength, guaranteeing independence from, *inter alia*, central political control. But the alternative proposed by Pilkington whereby advertising revenue would be collected and programmes commissioned by the Authority would have been damaging to standards, because it would have lowered the status of companies from that of creative partners in the planning and provision of a national service to that of 'programme packaging' agents. Moreover, it could be shown that under the existing system programme expenditure was not directly related to advertising income. In the ten-year period from 1965–1974, programme costs had almost trebled while net advertising revenue had risen by 82 per cent and profits by no more than 18 per cent. When calculated at constant prices, taking inflation into account, profits had in fact fallen. Yet over the same period, when there had been five changes in the levy, the government's share of ITV income, taken as levy and by normal taxation, had varied between 62 and 88 per cent and was currently over 75 per cent.

The genesis and development of ITV programmes in four different genres were described in the ITCA evidence: Thames's historical documentary series, *The World at War*; LWT's drama series, *Upstairs, Downstairs*; Granada's light entertainment series, *The Comedians*; and, as an example of the regional news magazine, an edition of Tyne Tees's *Today at Six*. One appendix listed major honours and awards gained in national and international festivals and competitions during the years since 1968. Another offered testimony to the companies' serious practical involvement in the cultural life of the community, at national and local levels, by listing the many ITV benefactions, through the Television Fund and otherwise, to a wide range of educational, artistic and scientific enterprises and institutions.

'In examining the broadcasting services of any country,' ITCA informed the committee:

> there are two questions which are likely to take precedence over all others: Who is to control broadcasting? How is it to be financed? ... What will matter most to most people, however, will be the nature and quality of mainstream television. And this in turn will be decided by the quality of the people who control television, by the degree of freedom that they enjoy, by the nature of whatever restraints are imposed upon them, and by the provision of money to finance the best possible service.

The companies recorded that after Pilkington:

> the Authority began to exercise firmer control in the balance of programming ... Axes fell on quiz games ... Regulations on what could and could not be scheduled and at what time of day grew like Jack's beanstalk, a growth which has flourished luxuriantly into the 1970s ...

Few would dispute that during those years ITV became more serious, more worthy, and introduced many new programmes of quality. The gains were not, however, all on the side of the viewer. It took a public outcry to restore *Crossroads* after its banishment from the London area ... Something of the élan of the early years with their innocently popular programmes departed.

The Authority had made no criticism of the companies in its evidence: they were, after all, the Authority's choice and under the Authority's thumb. The companies, on the other hand, did not feel inhibited from airing a public judgment on the regulating body. They expressed both praise and dissatisfaction.

The nature of the control exercised by the IBA over the programme contractors has never been fully understood by the public nor by parliament .. It would be wrong to assume that, once a company is appointed, the IBA accepts changes in management or shareholding without question. There have been several occasions when it has intervened decisively.

But such acts of control were 'seldom manifest': 'mutual trust rests on confidentiality'.

In the experience of the companies the IBA has performed its public obligations towards them with fairness and firmness. Some companies have particular areas and items of complaint, but all would acknowledge that the IBA acts disinterestedly, always in what it believes to be the best interests of television, and has generally been right. For this it has received little recognition.

ITCA's preference was for an Authority which was strong in policy but left the execution of that policy to the companies. Controversial programmes should be transmitted first and debated afterwards:

The IBA's interpretation of its role is open to question ... on the grounds that it takes too detailed an interest in the minutiae of scheduling and programming. The voluminous canon of restrictions imposed upon the companies' programme staffs, the close supervision of schedules, the counting of every minute, all combine to act as a brake on good programming. The practice and procedures of vetting some programmes before transmission are unsatisfactory to some companies.

This last criticism was a reference to the Authority's interventions in *World in Action*, *This Week* and other current affairs programmes. Within the IBA's Programme Division, so far from being resented, it was greeted as proof that the Authority was doing its job properly. Less well received at 70 Brompton Road was the companies' quip that 'the IBA has a tendency to equate merit in programming with lack of audience appeal'.

Scheduling was a well-trodden battle ground. ITCA claimed that 'a straitjacket of rules inhibits the schedulers' freedom to set out the day or the week in a way that will best suit the majority of viewers. This is a road which has been paved, perhaps almost blocked, with good intentions'. To this the Authority later responded that there were sound, public-service reasons for its elaborate regulations and restrictions on matters such as the showing of feature films and TV movies.

Company criticism extended to the composition of the Authority. The companies wanted direct viewer representation and the appointment of some members with broadcasting experience. In the wake of Lord Hill's interventionist chairmanship they wanted it made clear who was the chief executive: Director General or Chairman. Hill was accused of having 'unbalanced the system and blurred the lines of responsibility'. 'The role of chief executive should be more clearly defined and the companies should have more regular and direct access to the Authority itself.'

Among ITCA's general conclusions were: 'The strength of the Independent Television system lies in its regional structure and its diversity. No reforms are justifiable which would impair these.' 'Competition between two broadcasting systems has raised standards and is an important safeguard against political interference in broadcasting. It also helps to ensure for those who work in the industry a genuine choice of employment.' 'Separate sources of finance are fundamental to the competitive independence and to the diversity of the broadcasting systems.' 'A second ITV channel is necessary if ITV is to provide a fully comprehensive service and cater better for minorities. If launched in a period of buoyant revenue the companies could finance it.'

The companies recommended rolling contracts similar to those of the radio companies, 'with due safeguards against inadequate performance'. They asked for no change in the machinery of control over advertisements, which, 'both in copy and presentation, is the most effective in existence'. Wasteful duplication between ITV and the BBC was deplored: there should be agreement on the principle of alternation in the coverage of regularly recurring major events; engineering research should be combined; there should be a single audience measurement system; and the future of schools broadcasting should be examined jointly.

Criticism of the BBC was avoided. In-fighting would have been an unwise tactic when the whole system was under threat. At the time of Pilkington the BBC under Sir Hugh Greene had been at war with ITV; at Annan time the two organisations stood in the dock side by side, jointly charged with the offences of duopoly and non-accountability – crimes for which some witnesses were demanding the death sentence. In his autobiography Sir Ian Trethowan, the BBC's Director-General from 1977 to 1982, was to write of the BBC and ITCA evidence in terms unthinkable during the BBC's Greene period:

> We bombarded the committee with about a dozen different pieces of evidence, all of them very thorough but none very original in their forward thinking. They were, truth to tell, dull to look at, and dull to read. The ITV companies, by contrast, produced just one report, well-written and most attractively and stylishly presented.[21]

Without collusion, except for a last-minute *sub rosa* exchange of final drafts, the IBA and ITCA submissions dovetailed happily. There was common ground about the general nature and aims of ITV, but divergence on a sufficient number of points to deprive critics of ammunition with which to revive damaging accusations of a cosy relationship.

What both partners in ITV were, above all, united in emphasising was the need for a second Independent Television channel. The long-available but unused fourth channel was the most glittering prize which a recommendation by Lord Annan's committee might persuade the government to unwrap and award. In this ITV was to be, initially, disappointed. The question raised by its influential adversaries was whether the IBA was a fit body to control one channel, let alone be entrusted with two.

It was far from reassuring that one of the pieces of evidence most hostile to Independent Television (and to the BBC) came from the political party in power. The Labour Party's booklet *The People and the Media* was produced expressly to 'assist the committee in its deliberations'. It proposed a government-funded Communications Council, which would keep all the mass media under permanent review, and a Public Broadcasting Commission, which would collect and allocate all broadcasting finance and, although not making programmes itself, would lay down guidelines for programmes and scheduling. The BBC licence fee (a regressive tax) would be phased out. All transmission facilities would be taken into public ownership. Programme-making would be carried out by dispersed programme units, possibly organised through two corporations, each having one national and one regional channel.

While acknowledging that under the existing system 'the international repute of British broadcasting is deservedly high',[22] the main concern of the authors of *The People and The Media* was with what they detected as a growing concentration of power over the mass media. In this they were reflecting an obsession on the political left, not with the quality of programmes or their appeal to viewers, but with big business, excessive profits, high ratings, the evils of advertising and – as evidence of 'a media carve-up' – press shareholdings in ITV companies. In an attack on 'unbelievable profits' they looked back to 'the bonanza days' of ITV's first twelve years and pointed out that only small parts of those profits had been ploughed back into the industry, so that in the lean years, like 1970, expenditure on programmes had had to be cut. 'Previous profits had got safely away.'[23]

Swingeing indictments came too from The Standing Conference on Broadcasting (SCOB), which was financed by the Acton Society Trust, a Rowntree charity. Among its fifty members were twenty-seven academics, six trade union officials and ten former employees of ITV or the BBC. In Phillip Whitehead and Hilde Himmelweit it had two members appointed to the Annan Committee. At a press conference held in May 1975 to launch the first instalment of its evidence, *Broadcasting in the UK: An Agenda for Discussion*, the chairman claimed that SCOB was the only body, apart from Mary Whitehouse's NVALA, giving evidence without having a vested interest. Since the trade unionists represented workers in the industry and some of the former broadcasters in the group were known to their ex-colleagues as drop-outs, this claim was greeted with some incredulity.[24]

In its evidence SCOB described itself as 'an independent organisation to analyse, articulate and promote the public interest',[25] but public interest in its views was not widespread. A public meeting in Birmingham was attended by nine people, two of them from the IBA; in Glasgow the audience was ten.[26] Nevertheless, according to Lord Annan, the committee thought its views of such interest that 'we ought to try to discover their reasons in depth for reaching these conclusions',[27] and SCOB was accorded a two-day hearing 'bending the Annan Committee's collective ear'.[28]

Among SCOB's recommendations were central funding (a majority view); 'open disclosure and discussion' in the interests of public accountability; wider access to the screen; greater participation of the public and workers in the industry in broadcasting policy and management; and a school of research and analysis. Some of its detailed evidence formed a valuable stimulus to thought, but the impact was blunted by internal differences and failure to reach a collective view. Even the all-important definition of the purpose of broadcastng was a subject of disagreement. The choice of 'to improve and maintain the intellectual, moral, political and cultural health of society' was attacked in an addendum by one member (Dr Jay Blumler) as authoritarian. Dr Blumler's alternative was 'to nourish the roots of citizen choice in political, social, moral, intellectual and cultural matters'. From both definitions the world 'entertainment' was noticeably missing. Mr and Mrs Viewer were not to be permitted much fun while their citizen-choice roots were being nourished.

Despite some gaps of this kind between high-mindedness and the real world, SCOB represented the most serious intellectual challenge to the broadcasting establishments. It was pressing for the entire structure of broadcasting to be recast; its voice was heard within the Annan Committee itself; and its public campaign aganst 'the straitjacket of duopoly' was skilfuly conducted to attract support where it mattered most: among decision-makers, not viewers.

The IBA returned fire by criticising SCOB for shaky arguments, assertions unsupported by evidence (e.g. that the existing system had 'manifestly failed') and for 'snide misrepresentations' such as the demonstrably untrue statement that the IBA, ITCA and BBC submissions to Annan 'have presented their present policies as the best possible under any circumstances'. The Authority detected confusion between SCOB's apparent desire for greater central control and its calls for freedom and diversity, and it found the statement 'commercial television and radio have reflected the national trend towards concentration' running in the face of facts. To recognise the diversity of ITV and the even greater diversity of Independent Broadcasting since the launch of Independent Local Radio would have spoiled SCOB's argument.

As the conflicting evidence insisted, the fundamental issue to be determined was the proper method of controlling broadcasting in a democratic society. Although politically inclined to the left, the membership of the committee reflected the moderate Labour leadership's desire for reform, not the revolutionary zeal of those behind *The People and the Media* and SCOB. Indeed the confidently predicted appointment to the committee of Anthony Smith, the most influential of the radicals – who lobbied persistently through SCOB and other pressure groups for a Communications Council (to make broadcasters socially accountable) and a National Broadcasting Centre – was rumoured to have been personally vetoed by the Prime Minister.[29]

Lord Annan himself, keen to redress the balance of his committee and the overkill of Pilkington, was favourably predisposed towards ITV,[30] and he quickly became alarmed at the implications of SCOB's proposals, running along similar lines to those of the Labour Party, to replace the existing broadcasting authorities with Smithish institutions: a National Broadcasting Council which would become in effect a permanent Annan Committee, advising the minister on policy and political, social and technical options, and a National Broadcasting Commission with a full-time membership which would assume executive control. Under this scheme only a third of the commission's members would be appointed by the minister. Two-thirds would be representatives of trade unions, political parties and programme-makers', viewers' and listeners' organisations and could be mandated by the bodies which appointed them.

This, Lord Annan judged, would be disastrous for broadcasting and disastrous for parliamentary democracy. The minister's powers would be eroded, and the duopoly would be replaced by a monopoly representing political and special interests.

Such a commission would have set up its own secretariat, which would have usurped the duties which the civil servants who advise the minister are there to perform. The vast powers which were to have been given to the commission – powers to schedule and control hours given to current affairs, education and political broadcasts – would have led it inescapably to interfere in individual programmes and, far worse, to impose its own decisions upon output.[31]

'Good broadcasting, like freedom, comes from the separation of powers. But freedom was not the concern of some of those who put forward these proposals.'[32] To Lord Annan the scheme was sinister and SCOB's claim to have solved the conflict between editorial independence and public accountability by this contrivance transparently dishonest. Backed by the weight of the broadcasters' evidence, he was to carry his committee with him in resolutely opposing the destruction of the liberal tradition in British broadcasting.

15

ANNAN: REPORT AND DEBATE

The *Report of the Committee on the Future of Broadcasting (Chairman: Lord Annan)*[1] was presented to parliament on 23 March 1977 and published the following day. Give or take a week or two, the committee had met its two-and-a-half-year deadline. It presented, in 522 pages, the findings of the most thorough investigation ever undertaken into British broadcasting, and made 174 recommendations. No one could question the conscientiousness, industry and dedication with which the committee's members had reacted to the high level of public interest in their work.

The report's balanced approach and lucid and urbane tone owed much to the Chairman, who – a journalist as well as an academic – drafted it himself in sparkling prose studded with *obiter dicta*. He had led a team not of his own choosing through forty-four full meetings, twenty-eight group meetings, 750 submissions, 23,000 letters, twenty-five days of oral evidence, thirty-six days of visits to broadcasting organisations, overseas forays to Canada and West Germany (and by individual members to the United States, France, Israel and New Zealand) and months of reading, researching, thinking and writing, involving, according to his computation, 17 stone of paperwork per committee member. Under its terms of reference nothing was to be regarded as sacrosanct, and in forming its conclusions on strategy for the future of broadcasting the committee had been brought by adroit chairmanship to an unexpected degree of apparent unanimity.

ITV's critics had scored some successes. The notion of 'a straitjacket of duopoly' had won acceptance and future developments were to be placed under new authorities. Most woundingly, ITV was not to be granted the long-sought ITV2 which would enable it to compete with the BBC on equal terms. The committee wanted different kinds of programmes and different sources of editorial judgments and believed that 'we can get these only if new channels and services are not spatchcocked into old Authorities'.[2]

The new bodies were to be a Local Broadcasting Authority (to control radio and cable television), an Open Broadcasting Authority (to supervise the fourth channel), a Public Enquiry Board (to hold public hearings every

seven years and other enquiries as required), a Broadcasting Complaints
Commission and a Telecommunications Advisory Committee (with a
wider remit than the Television Advisory Committee, which it would
replace).

In a retaliatory submission to the Home Secretary ITCA waxed sour
over the straitjacket of duopoly and these proposed new statutory bodies.
Of the first it claimed:

> Few have been aware of the existence of this sinister garment. What
> degree of discomfort it inflicts by comparison with similar garments, such
> as the straitjacket of tripoly and the straitjacket of oligopoly, is not made
> clear. Certainly any discomfort caused by the straitjacket of monopoly
> was never brought to the public notice by those who so resolutely
> opposed the birth of ITV. At the time of Beveridge the cast of mind
> which now sees virtue in plurality was ardent in defence of monopoly.[3]

In the second it detected a paradox:

> In proposing these new bodies the committee is making a notable
> contribution to the growth of bureaucracy. All of them will have to be
> staffed, housed and administered, and the report prudently makes no
> attempt to estimate the total cost to the taxpayer. Yet elsewhere the
> report admonishes the BBC for its top-heavy administration and 'swol-
> len bureaucracy'. . . It is paradoxical, too, to suppose that the creativity
> and enterprise which the committee is concerned to foster will be stimu-
> lated by the appointment of a further series of governing bodies.[4]

But, mercifully, the committee had been dissuaded from favouring the
critics' version of accountability. It agreed that broadcasters must be
accountable, but:

> they are made so through the broadcasting authorities who are them-
> selves accountable to parliament. In our view this chain of accountability
> should not be broken or replaced. In the main it has borne the strain well
> of reconciling the freedom of the broadcasters with the rights and duties
> of parliament. Any attempt to weld on intermediate bodies – a Broad-
> casting Council, a Communications Commission, a Tribunal of Taste –
> would weaken the chain of accountability and would usurp the proper
> functions of parliament and the broadcasting authorities.[5]

It was firmly stated as the unanimous opinion of the committee that an
executive Broadcasting Commission superimposed on the IBA and the
BBC would be insufferable,[6] and one member went further in insisting on
recording her disagreement with the proposal to establish a Public Enquiry

Board. In a sharp Note of Dissent Marghanita Laski wrote: 'A majority of the committee having decided against recommending a Broadcasting Council, I find it imprudent to recommend a structure that could so easily become one and will clearly be under constant pressure to be made into one.'[7]

On this clouded subject of the conflict between editorial independence and the principle of accountability, highlighted in the evidence of those who were demanding both more freedom for broadcasters and tighter control through a Broadcasting Council, the committee published an analysis by Mary Warnock (a member of the IBA) distinguishing between the *accountability* of an Authority to parliament and its *responsibility* to the public. This found the notion that everyone can be accountable to everyone else ludicrous.[8]

How then to meet the protest of Mary Whitehouse's NVALA over 'the sense of frustration and powerlessness which many people feel in relation to broadcasting' and the Labour Party's complaint that 'although it is conducted within a broad framework of public accountability, broadcasting in Britain today is controlled by closed and almost autocratic institutions'?[9]

Except among broadcasters it was generally accepted that these cries from the heart, coming from both right and left, were fair comment. ITV was no readier than the press to admit mistakes and apologise, and the BBC's arrogance was notorious. 'Much of the evidence we received was particularly critical of the BBC, whose attitude was described as not only cavalier, but aggressive and arrogant. Several people sent us letters from the BBC and we thought these epithets justified.'[10]

One answer was the recommended Broadcasting Complaints Commission. Because there was said to be no public confidence in complaints bodies which were both judge and jury in their own cause, the IBA's Complaints Review Board and the BBC's Programmes Complaints Commission were to be replaced by this independent body, and the Authorities required to publicise its adjudications. ITV's responses raised no objection to this, although ITCA expressed reservations about the new commission's quasi-judicial status and what it described as complainants being allowed 'two bites at the cherry'. The committee believed that complainants should not be required to make a choice between public vindication and legal redress. The ITV view was that preliminary quasi-legal proceedings would inevitably prejudice any subsequent action at law: complainants should therefore either waive their legal rights or exercise them by going directly to the courts to sue for damages.

The Public Enquiry Board proposal was more coldly received by ITV, but the committee's idea of public hearings at which the public could make its views known found cautious acquiescence; although the IBA thought the committee wrong in being sceptical about the value of advisory bodies such as its General Advisory Council and in underrating the place of research, which 'can be a more reliable indication of public opinion and of public wishes than is likely to be obtained from public meetings and hearings'.[11]

The editorial freedom which parliament allowed the broadcasters was applauded in the report as precious and what the public wanted. 'The evidence we received reverberated with the plea that the government, politicians, and indeed quasi-political organisations, should not be permitted to control broadcasting directly.'[12] The moralists too were warned off. The Reithian concept on which the BBC had been so majestically built was rejected. It was urged on the committee by the Nationwide Festival of Light, which believed it to be the role of broadcasting not only to inform and entertain but also 'to give a positive lead – to reinforce or set a standard of morality and social behaviour which will uplift and strengthen, not debase and corrupt'.[13]

The committee's response to this ideal, which would once have been accepted without question, is symptomatic of new thinking in the 1960s and 1970s. Annan stamped with its seal of approval the direction in which the BBC had been moving since Reith's day:

Too often those who advocate such a policy seem to suppose that social and moral objectives could be formulated, agreed, and then imposed on the broadcasters. No doubt they can in totalitarian countries. They cannot here. We do not accept that it is part of the broadcasters' function to act as arbiters of morals or manners, or set themselves up as social engineers.[14]

The qualification that broadcasters should not be indifferent to morality and the good of society could not disguise this decided rebuff to the lobbyists headed by Mrs Whitehouse and Lord Longford and to the ghost of Lord Reith, who after likening ITV to bubonic plague ('which, to those who knew anything about bubonic plague, seemed to be going rather far') had so despaired of the BBC under his successors as to be driven to offer ITV his services 'in any capacity'.[15]

In the context of social change Lord Annan's committee believed that broadcasters should mirror the multiplicity of attitudes and opinions within a pluralistic society, reflecting 'divisions within society, divisions between classes, between the provinces and London, between pragmatists and ideologues'.[16] It was the disappearance of a consensus which had resulted in attacks on some programmes from opposite points of view: they were thought to be biased or obnoxious either because they were too challenging or because they were not challenging enough. But in a fragmented society it was possible for broadcasting to become a uniquely unifying force.

Yet another division was between the standards of broadcasters and those of their audiences. Bad language in programmes was the greatest single cause of complaint by the public, but dramatists saw no reason to pay heed. Bad manners too offended. 'I do not want to see the Prime Minister insulted in my lounge' was typical of the views conveyed to the committee in thousands of letters from the general public, but insulting prime ministers, whether Wilson, Heath, Callaghan or Thatcher, was part of some broadcasters' stock in trade. The BBC, in particular, did little to hide its contempt for politicians. Even Lord Hill, a politician appointed Chairman by Wilson to keep the Corporation's disrespect within bounds, 'saw no reason why the BBC should have been expected to apologise for a wisecrack in a satire programme to the effect that if you can see the Prime Minister's lips moving you know he is lying.'[17]

But Annan was at one with the broadcasters in believing programmes to be more important than control: 'We do not agree with those who have suggested that the concern of the Pilkington Committee for good programmes was somehow misconceived and that the real issue is the control of broadcasting ... Any broadcasting service must be judged by the quality of its programmes.'[18] The comments of one distinguished broadcaster and former Chairman of the Authority, Lord Clark (of *Civilisation*), were quoted with evident approval:

He believed that the amount of good that British television has done was remarkable. Far from having debased popular taste, he thought it had always been a little bit ahead of it. It had enormously widened people's horizons; it had increased their knowledge of the world and of nature, and even whetted their appetites for art and ideas. It had produced works of dramatic art and familiarised people with great works of literature, which they never would have read for themselves but for seeing the television adaptations. It had alleviated solitude: relieved boredom; and had made us laugh. Nor did Lord Clark support many of the arguments against television, which are frequently advanced, mostly by intellectuals. To the argument that television prevents reading and mental effort, he asked how many intellectuals who so complain read rather than gossip after dinner? It is said that it breeds fantasy and triviality. True; but what is gossip but fantasy about people? Why was it considered wrong for people to identify with the characters in drama and soap operas but reasonable for them to do so with characters in novels? It is said that television popularises half-truths. Yes; but popularisation is a benefit to society and interviews bring people to life, voice discontent, and expose those politicians who evade even half the truth.[19]

In other passages in the report, too, there was welcome praise for broadcasters, who had grown accustomed to nothing but carping:

It is one of the achievements of British broadcasting that programmes are regarded as hand-made products produced by individual craftsmen and not as articles of mass production.[20]

Those of us who have discussed the matter abroad, or with those who have lived in other countries and can make comparisons, are struck by how often we met with the response: 'The British have the best broadcasting in the world.'[21]

At a time when the art of governance in Britain – the reconciliation of public interest with intitiative, productivity and satisfaction in one's job – has fallen into some disrepute, the achievement and success of the British system of organising commercial broadcasting ought not to go unrecorded.[22]

In our view the network does a good job in providing entertainment and excitement. What is more, the TV companies now produce programmes which in prestige and intrinsic worth in a particular range are the equal of the BBC's output.[23]

There is one department in which the commercial sector has by common consent surpassed the BBC. That is in the presentation of news.[24]

As with diversity, the committee found virtue in competition – provided it was *regulated* competition. The spur of competition between ITV and BBC had been an incentive to good programme-making and 'we want to ensure that any new broadcasting system includes competition, not just for audiences but for variety and excellence'.[25]

These bouquets were accompanied by some brickbats. For example: 'Much of ITV's output seems settled in well-worn grooves, safe, stereotyped and routine in its production. Scheduling is often apprehensive, rather than daring.'[26]

Although the IBA had developed an impressive philosophy of commercial broadcasting and 'an addition to the theory of public service broadcasting', it 'remains in the public eye a curiously subfusc body'.[27] Its evidence was 'too bland'. It was still 'too ready to let the financial interests of the companies take precedence over the public interest' and 'too unwilling to place obligations on the companies which they can well afford to undertake'.[28]

Indeed, despite the accolades and Lord Annan's subsequent statement that the IBA (unlike the BBC) worked well and was about the right size,[29] members of the committee seem to have taken a hearty dislike to the Authority. They recommended not only that it should not be allowed a second channel, that its responsibility for local radio be removed and that it should not be given any responsibilities in the development of cable and satellite television, but also that it should interfere less with individual ITV programmes and should even suffer the humiliation of being renamed the Regional Television Authority, thus restoring to the BBC its status as the sole national broadcasting organisation.[30]

This last was an odd proposal. It appeared to stem from long-standing resentment of Independent Television as a choice of name. 'The title "Independent",' said the report, 'has always been open to misinterpretation and implies that the BBC is in some way the State-controlled broadcasting system, whereas the commercial companies are totally independent of the State.'[31]

Within ITV it was felt that the committee might just as well have recommended that the BBC be renamed, on the grounds that the word 'British' implied that it was a State system and that ITV was not British. In fact, the name Independent Television was originally chosen to proclaim that the new service would not be another call on the public purse: it would be financed independently and would be run independently of the then broadcasting establishment (the BBC). And as ITCA enquired:

If the IBA is to be re-christened the Regional Television Authority, what is to be the new name of ITN which provides national and international news coverage? Logic would appear to dictate the abolition of ITN under the new dispensation, but the committee is full of praise for it, acknowledges its superiority over BBC news and wishes it to have opportunities to expand on a fourth channel. Indeed the case against a wholly regional designation is so strong that the committee torpedoes its own recommendation in the very act of making it: ITV's 'regional services would, as now, be part of the national network which the companies would continue to provide' (para 7.9). In para 11.20 it is conceded that ITV's service 'should include high-quality networked programmes including national and international news from ITN'. What, then, is the purpose and effect of the recommendation? If this kind of ITV is to be described as regional, would the committee so describe the House of Commons, which is composed entirely of regional representatives with both regional and national responsibilities?

From the very beginning ITV has been a federal structure which provides both national and regional television, and no alteration in organisation or balance of programming is proposed. The implications of ITV's formal loss of national status seem hardly to have been considered. What of the five major companies which have been brought into being and staffed expressly to provide national as well as regional programmes? The committee appears not to have understood that the networking role of these five central companies in the provision of programmes for the whole nation is an essential feature of ITV's structure and service: without it the regional effort could not be sustained.[32]

This proposal to belittle ITV by downgrading its title and public image was taken by the companies as only one example of bias by a non-commercial, predominantly anti-commercial body against a commercial system. The intelligentsia was at its old games of despising trade and interfering with the pleasures of the common people – from the loftiest of motives, of course. ITCA riposted that the difference between commercial ITV and non-commercial BBC was not one of public service, which both provided:

> The difference is one of background, belief and attitude. If the BBC is seen as the Establishment of broadcasting, ITV represents the broader appeal of populism, and the committee has clearly had difficulty in bringing itself to identify with the needs and tastes of the majority of viewers. ITV is the favourite channel of most of the population: what the report reflects is a minority view of broadcasting. Significantly, Part IV (Programmes) contains sections on News and Current Affairs, Access and Party Political Programmes, Educational Broadcasting, Religious Broadcasting, The Arts in Broadcasting and Films on Television, but none on Drama, Comedy and Light Entertainment. Yet it is entertainment and comedy and the narrative strength of ITV in drama, drama series and serials which mean so much to the ordinary viewer: a good story and laughter and relaxation. Some of the report's recommendations read like the advice of a group of public-spirited men and women whose attitude towards television, and whose need for it, are very far removed from those of the man and woman in the street.[33]

Another critic of the report pointed out that it contained more references to classical music than pop, more references to Glyndebourne than *Coronation Street*, and that while it showed great concern for small minorities like Gaelic speakers (75,000), Open University students (50,000) and remote rural communities it was less sympathetic towards large minorities like the elderly and the inner-city dweller. Professor Jeremy Tunstall's critique was entitled *Annan in Wonderland*, and he condemned the report as a policy disaster.[34]

Yet the committee's statement of general principles, its analysis and evaluation of broadcasting issues and their presentation in the report won admiration, including the Authority's, even if this was tempered by strong reservations about some of the recommendations, particularly those which did not follow logically from the evidence which the committee had so assiduously assembled. These, and some significant errors of fact such as the inaccurate account of LWT's birth-pangs, suggested that Lord Annan had in the end been unable to overcome the prejudices of some of his members, and that the Authority had been right to protest to the Home Office before the start of play.

The proposal for an Open Broadcasting Authority to run a fourth channel to which none of the rules about impartiality and good taste would apply, and for which no financial arrangements were thought out, seemed to the IBA little more than a perverse nursing of illusions despite the evidence. The vision of a band of free communicators unshackled by authority and untrammelled by normal constraints may have been a noble ideal but, to all except the small lobby which had promoted it with such fervour, it appeared patently impractical and a dead duck politically before the ink was dry on the report.

The nub of the Authority's general reaction was expressed in these terms in its formal response to the Home Secretary:

> We welcome the committee's stand against any increased central control of broadcasting and the importance it attaches to editorial independence. We acknowledge the force of much that it has to say about relations between broadcasters and public, even though we doubt the value of the institutional proposals which it goes on to make. We share wholeheartedly the committee's desire for an increased range of choice in viewing and listening, and for greater diversity in the provision of programmes; but we do not share its belief that such increased choice and diversity is to be achieved by increasing the number of regulating bodies. The committee has a fine vision of good broadcasting, but we believe that the structures it proposes for realising that vision are unsound. We do not agree that broadcasting should be categorised into three neat sections – national, regional and local . . . To put broadcasting into compartments, under additional public authorities, is not to make for greater diversity of programming. To turn away from the benefits of complementary scheduling on the fourth television channel is to narrow choice. And if new bodies are set up on a weak financial basis, the result will be to defer the achievement of the committee's aims.[35]

All in all, ITV received the report with more relief than dismay. It was not the threatened end of the world as they knew it. It was a gallimaufry of the acceptable and the deplorable.

On the one hand, the committee adopted ITV's own recommendation that ITV and BBC should not simultaneously show the same sporting events with the same pictures but should agree on alternating coverage, to which the BBC was resolutely opposed.[36] On the other, there was a proposal that the IBA should contribute to the cost of collecting the BBC's licence fee from which ITV gained no benefit, financial or otherwise.[37]

The committee's disappointing rejection of rolling contracts was,[38] it seemed, based on an imperfect knowledge of the nature of the rolling contract system and a misunderstanding of the evidence offered on this subject. For ITCA the need to avoid periodical convulsions, with their threat to employment, was a valid argument in its favour. The IBA recalled the arguments of the White Paper of March 1971 in which the case for rolling contracts had been deployed in relation to Independent Local Radio. Experience had confirmed the view advanced then that three-year rolling contracts provided an effective method of supervising programme contractors. Rolling contracts by no means implied automatic renewal. They could include break-points at which contractors might be warned that they would have to face competition if they failed to achieve improvements in programme performance.

On the other hand, the committee brusquely dissolved one broadcasters' nightmare: the then intellectually fashionable notion that every citizen had a right of access to the medium. On a 24-hour-a-day, 7-day-a-week open-to-all channel, the committee calculated, every adult citizen would receive an allocation of less than a second a year. Broadcasting was a form of publishing, not a mass conversation, an aerial Hyde Park Corner or the equivalent of a witenagemot.[39]

The committee's recommendation on press holdings called for the existing statutory requirements to be retained, but went further in judging it undesirable for any single newspaper to have more than 10 per cent of the voting shares in an ITV company or for the total press interest in any company to exceed 25 per cent.[40] ITCA argued in response that to impose such restrictions would damage both press and television; that, where it existed, association between the two had benefited both and that the feared threats to editorial independence had proved illusory. In fact the benefits were largely one-sided. Since the start of ITV the companies had been required to accommodate press shareholdings in order to protect the press from the full financial consequences of competition for advertisement revenue, and the press had reaped rich dividends. Latterly, though, these holdings (sometimes foisted on unwilling consortia as a condition of a contract award) had come under attack as sinister concentrations of media power.

The IBA pointed out that the provisions of the Act referred to in the recommendation did no more than require corrective action by Authority or Home Secretary in the event that newspaper holdings of an unspecified amount in any ITV company produced results contrary to the public interest; and that so far no one, not even the Annan Committee, as its report recorded, had found any evidence of any attempt by press interests to influence the political or programme policies of the companies in which they held shares. In 1967 the Authority had taken steps to widen press representation and in some companies it was substantial. In terms of voting shares press interests accounted for 62 per cent of Southern, 43 per cent of Border and ATV, 33 per cent of LWT and Anglia, 29 per cent of Channel, and 25 per cent of Scottish. The Authority maintained that the proposed limitations on press holdings would damage these seven companies for no good reason. Unless the law were changed, it would not wish to require any reductions further to those already carried out in the case of Scottish Television.

The best directed and most damaging broadside against the report was delivered by the Authority's Director General in a Granada Guildhall lecture (on 27 June 1977). Harping on the theme summarised in the heading of a leader in *The Times*, 'Fair audit but weak blueprint', he identified 'the real weakness of the report, though it is often brilliant about the present and the past', as its essential hollowness about the future, where its proposals were illogical, naive, ill-thought-out and unrealistic. About the potentially immense consequences of technical change it was either ignorant or lacking in interest.

Sir Brian Young charged that, when faced with opposing points of view, the committee had recorded and emphasised both and then sat on the fence: 'Many viewers told them how much they disliked being offended; a few producers told them how much they disliked interventions designed to prevent such offence. What did the committee do? Did they say that the broadcasting authorities must favour one side or other in this quite central conflict of interest? Not a bit of it. They wagged their finger roundly at the broadcasters for offending; they then wagged their finger roundly at the authorities for intervening.' They stated problems but offered no solutions. They put faith in words rather than action. In Young's view a Public Enquiry Board would merely prolong the war of words. It would go on and on listening to cries and counter-cries of 'too popular – too mandarin', 'too permissive – too restricted'.

He foresaw that the committee's proposals for the fourth channel would mean the further indefinite postponement of a service for which the viewer had already been waiting ten years. ITV2 had been perfunctorily dismissed in favour of an uneconomic OBA channel for only two stated reasons: first, because it would be too popular; secondly for the precise opposite – because it would be too elitist. The proposal for a Local Broadcasting Authority to take over responsibility for local radio would delay the introduction of new stations and more than double the cost to the taxpayer.

'This committee,' he declared, 'dances most elegantly on the point of a pin – angels in their power of expression, but angels also in their tendency to fly above the real world, and their ability to dance on ... both horns of a dilemma.' He noted with pleasure that it had judged the Authority to be right in resisting proposals made by Pilkington and the Select Committee on Nationalised Industries and consoled himself with the prospect of some further committee of enquiry applauding its resistance to Annan's.

In his responding Granada lecture the following week Lord Annan conceded some of Sir Brian's points; other he deflected by flattery:

We record that commercial broadcasting has emerged better organised and better disciplined and is now a system which should command respect and admiration not merely for the product but for the way the ends of commercialism have been caged and the virtues of competitiveness and profitability encouraged ... We judged that the commercial system was so effective and efficient that we took pains to keep it that way.

He argued that ITV did not suffer from the bloated bureaucracy which afflicted the BBC and would be saved from the same fate by not having ITV2, since 'part of the BBC's bureaucratic troubles have come about as a result of running two channels'. (But the committee had not recommended the detachment of BBC2.)

Broadcasting is, as Lord Annan remarked elsewhere,[41] a subject which particularly excites the intelligentsia, and he and those on his committee had risen to the occasion and produced a report which dealt perceptively with most of the issues affecting the broadcasting media, whatever its inadequacy as a route map for exploring the future. But the subsequent glosses put upon the report by some members, not least by the Chairman himself, tended to expose the defects and diminish its stature.

To those puzzled how such admirably firm emphasis on the central importance of the creative process and the climate of free-ranging diversity in which it could flourish came to be coupled with proposals for a proliferation of institutionalised bureaucracies, the answer was soon forthcoming. The contradictions and illogicalities were revealed as resulting from internal dissension and trade-offs between members of the committee with opposing views. Some of the recommendations which the Chairman felt it his duty to promote so busily in public he had argued against in committee, and he was by no means displeased when they were rejected by government.[42]

In conclave at Churchill College, Cambridge the committee had reached a general agreement on four governing principles: flexibility (which signalled the end of the ITV/BBC duopoly) and diversity (which meant that broadcasting should reflect the pluralism of society), and editorial independence and public accountability (the conflict between which had somehow to be resolved). Thereafter the only way forward was by compromise. Phillip Whitehead pressed for a Broadcasting Council which would be, in effect, a permanent Annan Committee undertaking continuous assessment of the broadcasters, as advocated by Anthony Smith and Dr Jay Blumler of SCOB. Unable to carry the majority with him, he accepted a deal under which the rest of the committee (Miss Laski alone standing firm) agreed to recommend a Public Enquiry Board for Broadcasting instead. The argument then moved on to what powers this board should enjoy: strong (Whitehead) or weak (Lord Annan, who was quite as much opposed to the board as Miss Laski but judged it tactically wiser to negotiate for emasculation). The struggle was hard but in the end futile because neither the Labour government nor its Conservative successor was minded to act on the recommendation.[43]

It was Whitehead alone on the committee who had a prepared scheme to put the principles of flexibility and diversity into practice. He had been a leading participant in debates on the future of broadcasting for the past ten years, and his clear-mindedness and eloquence carried the day to the exclusion of other options put forward by the secretariat. The multiplication of Authorities and the denial of an ITV2 were his victories.[44] In another of his objectives – the break-up of the BBC into two corporations, one for television and one for radio – he was blocked by a vote of six in favour and ten against.[45]

Crises were averted by Sara Morrison's mediation, but Whitehead's dominance of the committee's thinking led the Chairman into concessions which he came to regret. He regretted the inflationary proposals for new bureaucracies and the bias against the IBA. He regretted the sop to Professor Himmelweit in the recommendation to exclude advertising from children's programmes, made despite doubts on the committee about the value of the published research on this subject, including her own. Above all he regretted that the committee had funked industrial relations. To his mind the malpractices of the trade unions in television were as scandalous as the misconduct of the print unions in Fleet Street and this omission in the report was shameful. But fear of upsetting powerful trade union leaders caused widespread suppression of the truth in the 1970s, and he believed that this was a topic on which no official body could at that time comment frankly and freely.[46]

The report was introduced in the House of Lords on 19 May 1977 in a highly critical speech by the Labour peer, Lord Willis, better known as Ted Willis, the script-writer of *Dixon of Dock Green* (1953–75) and President of the Writers' Guild of Great Britain. As a man of the people and a broadcasting professional, Lord Willis disapproved of interfering academics and ignorant amateurs. He noted the report's smug announcement that two members of the committee knew something about television from the inside.

> My reaction to this was, Big Deal! I find it absolutely bizarre and incredible that this should be regarded as something to boast about – an inquiry into television for the next 15 years and we have two whole experts on it – in fact, two and a half perhaps! So I welcome the remarks in Chapter 5 which relate to this. Speaking of appointments to the broadcasting authorities the committee ask that 'from the list sometimes referred to as the Great and the Good, the Prime Minister ... may choose some of the Lesser and Better'. I put my hands together and raise my eyes to Heaven or 10 Downing Street, and say, 'Oh yes please'.[47]

As for academics, Lord Willis had nothing against Sir Michael Swann (Chairman of the BBC and former Vice-Chancellor of Edinburgh University) or Lady Plowden, or even Lord Annan, but he thought one could have too much of a 'lofty, high-table attitude', and he castigated Marghanita Laski's expression about giving the masses bread and circuses as 'breathtaking in its arrogance and condescension': broadcasting needed to be rescued from the academic mafia and from 'this pestilence of professors who keep telling us how to run our business'.

On such occasions the Lords could muster an array of experts unrivalled in the Commons. Even with Lords Bernstein and Grade abstaining, speakers in the debate on Annan included two former Authority Chairmen (Lords Hill and Aylestone), the then Chairman's husband (Lord Plowden), a former member of the Authority (Lady Burton), the current Chairman of the ITCA Council (Lord Windlesham), a former member of the ITA General Advisory Council (Lord Parry), a director of Scottish Television (Lord Taylor of Gryfe), a former BBC talks producer (Lord Boston of Faversham), the largest cable television operator (Lord De La Warr) and Lord Annan himself. Most expressed praise for Lord Annan and gratitude for the readability of his report, but the weight of opinion was opposed to the more controversial of the recommendations.

Lord Taylor thought the notion of giving the IBA a regional title 'quite childish. If anything, the name which could be justified for commercial television might be the National Broadcasting Service, or the Free Television Service, because it is not supported by any fee-paying or any government subvention. In fact, it is the other way round – we subsidise the government.' He took the opportunity to explode the myth that an ITV franchise was a money-printing licence. Scottish Television had paid no dividend to its shareholders for three years in succession. Out of every £5 earned by the company, he wanted their lordships to know, £4 went to the Exchequer. Lord Aylestone, too, used the occasion to complain of the government making off with ITV's money. The Treasury had taken more than £12 million which the IBA was holding in reserve for technical development.

Lord Annan told their lordships that his report was 'a massive vote of confidence in the British way of regulating broadcasting. At the end of our labours we had no doubt that British broadcasting was the best in the world and it is the object of our report to try to keep it that way'. The message which he bore from his committee was: 'Keep the basic principles of broadcasting in the United Kingdom as they are, but please look to the future and provide new outlets, new chances, new initiatives if you want to keep the standard as high as it is now.'

In the Commons, four days later, the list of members wanting to take part in the debate was so long that the Speaker had to make a plea for speeches to be kept short. Television was a subject on which everyone in public life had an opinion to express.

There was bipartisan satisfaction at the generally clean bill of health given to the constitutional arrangements for broadcasting. On behalf of the government the Home Secretary (Merlyn Rees) endorsed the committee's first and major recommendation: 'Broadcasting services should continue to be provided as public services and should continue to be the responsibility of public authorities. These should be independent of government in the day-to-day conduct of their business.' For the Opposition William White-law found Lord Annan's belief that his report should be read as a massive vote of confidence in British broadcasting encouraging. The idea of an Open Broadcasting Authority he found original and imaginative but financially flawed, and he expressed 'little doubt that ITV2 is the only means of using the fourth channel reasonably quickly, as the committee itself accepted, and on a sure financial basis'.[48]

Julian Critchley, Chairman of the Conservative Party Media Committee, who had been in disgrace for criticising the report's recommendations before they were published (on the basis of a leak in the *Observer*), wished to amend his verdict on them from 'a dog's breakfast' to '*fritto misto* in which one finds a large amount of squid'. They were both conservative (in acknowledging the merits of the present system and leaving it largely un-touched) and radical (most plainly in hostility towards ITV and advertising). He detected a 'high-minded disapproval of commerce and advertising'. The report was written 'by those who rarely watch the box for the rest of us who do'.

Sir Paul Bryan, a director of Granada Television, thought that the committee had got itself into a great muddle over duopoly and plurality:

> The sort of people who used to be pro the BBC monopoly are now anti-duopoly and pro-plurality. Having at the outset settled for plurality, they soon came up against the road block of the BBC, which clearly cried out for pluralisation but could not be dismembered for fear of what foreigners would think. Then they came to the ITV, which with its fifteen companies plus the ITN is in itself a model of plurality. This should have pleased them very much; but then most of them found that they had a deep-down feeling against commercial broadcasting, so ITV had to be demoted to regional status.

It fell to Whitehead, the only member of the Annan Committee in the House, to defend its proposals against general attack. He confessed that at times during the debate he had felt as though he was 'taking on single-handed the whole House and all the parties in it'. This, it seemed to him, was because the committee had dared to offend both halves of the duopoly. But 'like it or not', he warned the House, 'developments in our society will take broadcasting in the direction that Annan foretells' – towards 'pluralism, more outlets, different ways of seeing and hearing, more innovation'. When he resumed his seat, the next speaker (Wyn

Roberts) claimed to be left in some doubt whether the so-called Annan Report might not more accurately be described as the Whitehead Report.

At the time of the report's publication the Home Secretary had issued an open invitation to all interested parties to submit comments. These should be received, he said, on or before 1 July in order that they might be carefully considered and the necessary legislation prepared in good time. In the following weeks the public debate which had preceded the committee and continued as a background throughout its life developed a renewed intensity. There were lectures, conferences, speeches, broadcasts and articles in the press in which the pros and cons were hotly argued. The more publicity-minded members of the committee were regular participants, Lord Annan himself the star performer: for the Chairman of an official enquiry his openness in public was unparalleled. In the course of all this argument the government was presented with a fair measure of convergence of views over the ends to be attained and a wide measure of divergence over the best means of reaching them.

Its timetable was thrown out of gear by division within its own ranks. At the Home Office Lord Harris and his officials dismissed the recommendation for an Open Broadcasting Authority to run the fourth channel as an absurdity and drafted alternative plans for an ITV2, one closely following the companies' proposals, the other the IBA's. They were supported by the Treasury, which was strongly opposed to the funding of a new television channel by government subvention at a time of severe strain on the public sector borrowing requirement. But a copy of the draft was leaked to Phillip Whitehead from the cabinet office and he lobbied Shirley Williams, Tony Benn and other reformist ministers to reject it.[49] As a result a cabinet committee was set up under the chairmanship of the Prime Minister (James Callaghan) to reconsider the OBA and other issues, including the control and possible dismemberment of the BBC.

The future of broadcasting thus fell out of professional and informed hands into those of a small group in a government which was far too preoccupied with rising inflation, a sterling crisis, wages policy, industrial unrest and Scottish and Welsh devolution to have studied the Annan Report fully, let alone given any proper consideration to the mountain of evidence on which it was perched. It was July 1978 – more than year after the debate in parliament – before the government was able to cobble its differences together and publish its revised intentions. Its White Paper, *Broadcasting*, represented a more than usually unsatisfactory compromise between irreconcilable views.[50]

On the credit side, the government announced its acceptance of Annan's four principles and followed Annan in rejecting the apparatus of control advocated in *The People and the Media*, agreeing that a council or commission charged with supervising the broadcasting authorities would constitute either a return to monopoly or a separation of responsibilities and would threaten the broadcasters' independence. In the belief that direct discussion between broadcasters and the public would achieve greater accountability, it endorsed the Annan recommendations for public hearings. It also endorsed an independent Broadcasting Complaints Commission (although wavering on the legal waiver). It associated itself with Annan's general praise of the IBA's performance and the 'marked improvement in the quality of Independent Television programmes over the past decade', as well as its success in establishing a popular Independent Local Radio service so quickly; and it was more logical than Annan in its plans for the IBA's future in the light of that record.

The government considers that the IBA should continue to be responsible for a television service and a local radio service provided through programme contractors; and that it should assume responsibility for the supervision of cable television and radio services, including community cable services and pilot schemes of pay television, and such other services as hospital and university services. In view of these responsibilities it would not be appropriate to change the name of the Authority to 'Regional Television Authority', as recommended by the Annan Committee. The Authority will therefore continue to be known as the Independent Broadcasting Authority.[51]

But the White Paper also ran in the face of Annan in a different – and ominous – direction. It was, as the IBA's Director of Television observed, 'a haystack stuffed with weapons'.[52] Despite Annan's evidence of a universal dislike and distrust of any direct government involvement in broadcasting, a whole range of new government appointments and interventions was proposed. In an exuberance of placemen and quangos the BBC was to have three Management Boards half of whose members were to be government-appointed, and the government would also appoint all the members of the new Open Broadcasting Authority, National Broadcasting Councils for Scotland, Wales and Northern Ireland, the Broadcasting Complaints Commission and the Broadcasting Technical Advisory Committee. The IBA would be required to consult with the government on several matters previously under its own control: the proportion of foreign material screened, the use of warning symbols for feature films, joint directorships of broadcasting and newspaper companies, joint directorships of ITV and ILR companies, and even on the composition of ITCA's Programme Controllers Group.

In adopting the Annan recommendation of a new kind of Authority to run the fourth channel, on which much of the next stage in the development of television was said to hinge, the White Paper lamely proposed the establishment of an Open Broadcasting Authority so drastically modified as to eliminate nearly all its *raison d'être*.

Broadcasting contained little which could be greeted with enthusiasm in Independent Television circles. The collective comments of the companies were almost entirely critical;[53] those of the Authority more tactful.[54] But there was small likelihood of early legislation. The future of broadcasting was not a priority in the government's programme, and no Broadcasting Bill was forthcoming before the general election in May 1979 brought it down and the incoming Conservatives consigned Labour's White Paper to the waste-paper basket.

The new Home Secretary was William Whitelaw, identified by the BBC's Director-General as a Wykehamist masquerading as an Etonian.[55] His impersonation of a typical member of the squirearchy camouflaged a sharp mind and exceptional political nous; and as its senior member after the Prime Minister he carried the weight in cabinet which his predecessor had lacked. Only when Whitelaw had finished weaving a wary way through the minefield sown in his path by years of controversy could the Annan chapter be closed at last.

What then was the outcome? The political structure of television remained unchanged. The new Authorities were not to be. The duopoly survived into the 1980s. An independent Broadcasting Complaints Commission was established, but with limited terms of reference and few teeth. In yet another piece of political juggling the fourth channel became an ITV channel but not an ITV channel (as described in Chapter 17). Generally, the Annan Committee's achievement lay less in what it recommended than in what it did not – less in its proposals than in what it decisively averted. For the report was a reaffirmation of traditional liberal principles and values. Lord Annan was justifiably proud that it owed more to the spirit of Locke and Montesquieu than to Marcuse and R. D. Laing and the other idols of contemporary thought.[56]

16

PROSPERITY AND
UNCERTAINTY, 1976–80

Independent Television's strength in programmes and finance during the latter half of the 1970s was greatly assisted by the non-occurrence of contractual upheavals scheduled for 1974 and rescheduled for 1976, so that while Lord Annan and his committee were examining the state of television and mapping its future, ITV and its viewers enjoyed the blessings of a reprieve. The wounds inflicted by the surgery of 1967 had healed and the next operation on a healthy patient was postponed until 1980. In a peak period of programme-making and a trough in the nation's fortunes ITV provided the public with some comforting distraction from the impact of a faltering economy.

Programme success was the theme of the IBA's Annual Reports for 1975–76 and 1976–77. These years inherited, consolidated and developed the advances made between 1972 and 1974. The screening of so many good programmes at a time when the Annan Committee was preparing its report was politically helpful and noted as such at the time, but the supposition that this was a contrived coincidence is unrealistically complimentary to production planning procedures, even though the IBA reported:

> Distinction was brought to the 1975–76 programme schedules both in drama and in factual programmes. Dramatic serials of note were *Edward VII*, *The Stars Look Down*, *Clayhanger* and *Upstairs, Downstairs* which reached its triumphant conclusion during the year. Other productions which in different ways broke new ground were *The Nearly Man*, *Shades of Greene* and *Rock Follies*. The latter half of the year saw also a variety of single plays, from *The Naked Civil Servant* through *Willow Cabins* by Alan Plater to Terence Rattigan's *In Praise of Love*. Notable among factual programmes were several editions of *World in Action* dealing with the 'nuts and bolts' of Britain's economy, the *Decision* series, which examines how fundamental decision-making is arrived at, and another Granada programme in which journalists represented the type of arguments lying behind the cabinet's decision to assist Chrysler. The year saw a growth in the amount of regional material.[1]

On 22 September 1976, the anniversary of the inauguration of its programme service from the Croydon transmitter, ITV came of age. Some self-congratulation seemed appropriate. 'As we stand waiting for the curtain to go up on the next act (or perhaps a new play),' wrote Denis Forman, Chairman of Granada Television, in a retrospective article:

> there is one thing about ITV which would surely strike an Annan from Mars as a paradox. The BBC (Public Service Broadcasting as they often remind us) is strong in Public Service Comedy, some of it very broad and very good, Public Service Olympics and sport in general, and with shows such as *It's A* (Public Service) *Knockout* and the like. Contrariwise, ITV, once so clearly the People's Television, has the edge perhaps in popular history as treated in the excellent *The World at War*, and in contemplating our economy-ridden society in programmes like *Weekend World*, *The State of the Nation* and *Decision*. The admirable influence of the IBA has pushed ITV towards *gravitas*; reaction to the early years of ITV has pushed the BBC towards the box-office. Some of us, not from Mars, who look at ITV a lot tend to wonder on occasions whether it has not moved a shade too near to Top People's Television. But then, as we have discovered over the past 21 years, to provide good television for all of the people all of the time is not an easy thing.[2]

Forman looked forward hopefully to a time when those responsible for broadcasting – in government and parliament, at the IBA and on advisory boards and committees of enquiry – would be television literates, brought up as children in a world in which television was recognised as a normal part of life, not as some strange beast from outer space.

At a celebratory dinner in the City of London's Guildhall Merlyn Rees, the newly appointed Home Secretary, confessed that he had been among those who opposed the commencement of ITV because they believed that the pressures of the market would have a debasing effect on broadcasting standards. He and his Labour colleagues, he admitted, had been wrong. 'Independent Television is now a familiar and much loved feature of our national scene and, rather than being a debaser of programme standards, it has often been the pace-setter.'[3]

For special praise he singled out *News at Ten*, introduced in 1967, and the arts programmes *Tempo* and *Aquarius*. As an MP with a Yorkshire constituency, he had been impressed by the achievements of Yorkshire Television in serious drama and by its indictment of social neglect in the documentary *Johnny Go Home*. As Secretary of State for Northern Ireland, he had become aware of the value of Ulster Television's encouragement of community work against a background of violence. He acknowledged ITV's contribution to the balance of payments through the sale of its programmes overseas. This had totalled £14 million the previous year – five times more than the cost of ITV's imports.

But complacency was discouraged by his reference to 'the timely chastisement of Pilkington' and his use of the occasion to air some personal views on ways in which the service might be improved. Coverage of economic and industrial affairs, he believed, could and should pay more attention to what was good. In news and current affairs programmes, the mechanics and techniques of the medium too often took precedence over the information to be conveyed about serious political issues. He did not think that enlightenment of the viewer was best achieved by confrontations between two politicians with precisely opposite and predictable views. Above all, television was still guilty of showing too much violence.

Happily, the coming of age coincided with a noteworthy performance in programme prize-winning. Appreciation by the audiences which ITV existed to serve was gratifyingly supplemented by acclaim from fellow-professionals, and its record in international competitions in 1976–77 was unsurpassed. At the Monte Carlo International Television Festival ITN won the news-reporting prize in successive years: in February 1976 with film of the battle of Newport Bridge in the Vietnam war and in February 1977 with coverage of the cod war with Iceland. In May 1976 *Upstairs, Downstairs* (LWT) achieved a hat trick by winning an Emmy award in Hollywood for the outstanding television drama series for the third year in succession. In June *Opium War Lords* (ATV) won first prize at the American Film Festival (for non-fiction films) and *Johnny Go Home* (Yorkshire) gained the top award in the youth section at the Prix Jeunesse in Munich. In September, in a remarkable double, Thames won both the major awards at the Prix Italia. *The Naked Civil Servant*, which had already won an international Emmy Award in the USA, carried off the drama prize, while *Beauty, Bonny, Daisy, Violet, Grace and Geoffrey Morton* was judged the best documentary film. Thames followed this unique triumph by winning the 1977 Prix Italia in the music category with a production of Benjamin Britten's *St Nicholas Cantata*. Also in 1977 *Upstairs, Downstairs* won a fourth consecutive Emmy; *The Muppet Show* (ATV) beat opposition from twenty-nine other countries to take the hotly contested Golden Rose at Montreux for the world's best light entertainment programme; and an adaptation of Dickens's *Hard Times* (Granada) won the top television award at the New York International Film and Television Festival.

These expensively produced and expertly crafted programmes were the high fliers, but it could hardly be claimed that all ITV's productions rated superlatives. Comedy was one acknowledged area of weakness. The failure rate in this programme area was high. During the three-day Consultation with producers, directors and scriptwriters of comedy and light entertainment held by the IBA in the autumn of 1976 Denis Norden criticised the programme-makers for seeking instant success at the expense of developing characters and situations; for predictability and cosiness; for underestimating the audience's intelligence; and for shying away from the controversial.[4]

This was among the verdicts of the Annan Committee delivered a few months later:

> Commercial television has the defects of its merits. Its routine productions are on a lower level than those of the BBC, even in comedy, light entertainment and sport, the popular areas in which it might be expected to excel. But within the one channel, ITV output cannot be expected to have the range which the BBC can provide on two channels and ITV has come a long way since the Pilkington Committee reported. They have made a real attempt to widen the subject matter of their programmes, particularly in current affairs and documentaries, and the best of their programmes are as good as any made by the BBC. We agree with the ITCA that quality is not to be confused with conventional respectability, and it is not another production of Shakespeare or a lavish, though over-simplified, historical series which will persuade its critics that ITV is giving a good service. What ITV needs to do is to give greater depth and subtlety to their popular series.[5]

A similar point had been made by Jeremy Isaacs, Director of Programmes at Thames and later the founding Chief Executive of Channel Four. In an IBA Lecture in 1975 he applauded ATV's *Antony and Cleopatra*, Southern's *Figaro*, LWT's *Akenfield* and Granada's *The State of the Nation*, but disapproved of ITV being judged by these and other prestigious 'special events'. Isaacs did not want praise for them to distract ITV from what he judged its proper business – 'making every television programme we do as good of its kind as we can, seeking always to improve the general level, taking our audience with us, a little further each time'. What mattered no less than prestige was the quality of the programmes which he taunted the Authority with regarding as 'distressingly popular', programmes which carried its 'positive loathing seal of disapproval' and drew audiences in millions and tens of millions: *Crossroads*, *Coronation Street*, *Opportunity Knocks*, *This is Your Life*, *Love Thy Neighbour*, *Benny Hill*, *Sale of the Century*, *The Golden Shot*.[6]

The Authority's own criticisms at this time embraced 'a need for continued attention to be given to the composition of the schedule on Saturday evenings, to the perennial problem of light entertainment and comedy, and to the quality and scheduling of imported material. The policy of repeating previously-shown material is also a matter for continuing care.'[7] Another perennial problem was the single play, always the most difficult form of television to sustain. There continued to be fears for its future. It was harder to establish with audiences than series or serials, but essential if the live theatre was to be extended into television, and as an outlet for new writing. A commitment to single plays was proclaimed, and more were promised.[8]

In September 1977 a consolidated set of guidelines for programme-makers was published by the Authority for the first time and printed as an appendix to its 1976–77 Annual Report.[9] *Television Programme Guidelines* was a collation of the myriad decisions of principle taken in discussion between Authority and companies over the previous twenty years. It covered all the problem areas which called for special care by producers and directors – good taste, violence, privacy, impartiality, sponsorship, politics, crime, defamation, charitable appeals, contempt of court, etc. The objective throughout was to meet the requirements of parliament with minimum impairment of the creative freedom of programme-makers.

In the late summer and early autumn of the previous year the heavier emphasis on worthiness in programme schedules, prompted partly by a nervous anticipation of Annan's report, had provoked a protest from the paying customers. Advertisers complained that ITV was failing to deliver audiences of the expected and required size. On such occasions the Authority's role as a financially disinterested arbiter in the composition of schedules shielded the companies from commercial pressures, and no rescheduling occurred in response to the advertisers' demands. The matter was resolved when audience figures rose to their customary level in the run-up to Christmas.

The contribution of advertisers to the system was growing annually. The change in 1974 from income-based to profit-based levy had taken immediate effect in a higher level of expenditure on programmes despite dire financial forecasts. During 1975, as recounted in Chapter 5, the companies were in a state of agitation over falling advertisement revenue and rising costs, not least a 14 per cent increase in rentals payable to the Authority during the extended contract period from July 1976 to July 1979; which was, however, substantially less than the rate of inflation, to which rentals were contractually linked.

But by March 1976 the predictions of impoverishment made so forcefully twelve months earlier were confounded when revenue rose by 20 per cent over the previous year's level. By the end of the financial year in July confidence was fully restored. The volatility of ITV's revenue, and the companies' inability to forecast it accurately, were in marked contrast to the BBC's pegged but assured income from the licence fee. In 1975–76 ITV's revenue jumped to £209 million; in 1976–77 to £267 million; in 1977–78 to £335 million. By 1980–81 it had reached £525 million.

Substantial sums were invested in programme production, and the claim that sufficient funds were available for the companies to finance a second channel was justified. In the good years their monopolies yielded a higher revenue than could reasonably be expended on a one-channel service, and the circumstances of the time – government-enforced dividend and wage restraint, and a disincentive to make any sizeable investment in fixed assets because of the impossibility of planning for future development – meant that the companies were retaining large reserves. Because there was no point in making more programmes than the schedules could accommodate, they thought it prudent to employ part of them in other businesses.

Thus one consequence of prolonged uncertainty about the future of ITV and each of its companies was a burgeoning of the practice of diversification. 'Taking money out of television', as it was apt to be called, raised some delicate issues not only for companies whose experience was confined to television but also for the Authority whose approval had to be obtained under the terms of the television contracts. It was not the IBA's policy to block such ventures as long as it could be satisfied that a company's television operations would suffer no ill effects from a diminution of funds or a diversion of board and management attention and effort. Diversification was seen to be useful if it helped the making of programmes in bad times as well as good.[10]

The companies claimed the right to do what they pleased with their own money when all commitments had been honoured and all dues paid. They argued that:

> Not only is advertising revenue subject to extremes of fluctuation, but by the very nature of the franchise normal business expansion cannot take place by extending the basic operation. Companies therefore have to seek opportunities in allied fields and in other areas where they have the appropriate expertise. The benefits of such extension of activity arise from the creation of a stronger asset base, a broader operation and greater stability. As a result, a company can more easily raise further capital as required by loan or rights. The surplus and the net cash flow from the diversified activity are likely to be more stable than those from the volatile business of a programme contractor.[11]

In 1976 the Authority adopted the working rule that not more than one third of a company's capital employed should be used for the purposes of diversification. It insisted, however, that any decision to diversify must be, first and last, the responsibility of the company, which

> has to bear in mind that if some call for capital expenditure on the television side arises for which, through some miscalculation, adequate reserves are not available, it must be ready, if necessary, to call up additional funds from the shareholders or from other sources acceptable to the Authority; it has also in due course to convince the Authority – probably in competition with other applicants – when it makes an application for a new programme contract, that it is financially and managerially in good shape to continue to hold a television contract.[12]

The Annan Committee received complaints on this subject, mostly from trade unions, and its report recorded the defence of the practice given in oral evidence by company spokesmen.[13] Television, they said, was founded on diversification – out of newspapers, cinemas, theatres. Why should they now be treated differently from all other companies? If in the good years high profits were injected into normal programming, this would lead to over-production and over-manning which would in turn lead, in the bad years, to just the cutbacks and redundancies which the unions were anxious to avoid.

The committee concluded, by a majority, that there was no good reason for interference, and the Labour government, in its White Paper of July 1978, declared itself satisfied that the powers of the IBA were adequate to protect the system from harm on this account.[14] But others continued to question the desirability of television companies investing in opticians, book and diary publishing, fine art dealing, funeral parlours and safari parks. 'If advertising revenue drops, you sell off a few zebras and camels. Is that what happens?' enquired one MP disingenuously.[15]

When Lord Aylestone retired from the chairmanship of the Authority at the end of March 1975 his valediction included the words: 'It will be a good year for British broadcasting when an Annual Report does not have to describe the past year as one of unparalleled activity.' By this yardstick Lady Plowden, his successor, who presided over the coming-of-age and occupied the chair until the end of 1980, was denied any good years. Her reign was a period of moneyed unease. For no less than six years, while the Annan Committee sat, while the government pondered on its findings and while contractors for the 1980s were being selected, the companies were left wondering whether there was life after 1981.

The expiry date for the *status quo*, postponed for three years until July 1979, had then, as a further interim measure, been delayed for another two and a half until the end of that year. On the first occasion the Authority offered all the fifteen programme companies contracts on new terms; on the second, an extension on existing terms. All its offers were accepted despite the thick fog which blanketed the future and which was not to be dispersed until the end of the decade. The satisfactions of creative achievement and enviable prosperity ensured that no company ever voluntarily surrendered its contract or spurned the offer of a new one, however loudly it grumbled at times.

During the post-Annan period the Public Accounts Committee subjected the Authority's administration of the levy to close scrutiny, and the House of Commons decided that the time was ripe for its Select Committee on Nationalised Industries to hold another enquiry into Independent Broadcasting.

This committee's interrogation proved a happier experience than that of the previous Select Committee,[16] and the differences in tone and conclusions between the two reports are indicative of the higher standing achieved by Independent Broadcasting. Six years after the harassment of 1972, the Authority's examination by Sub-Committee C under the chairmanship of Sir Donald Kaberry between March and July of 1978 passed off without rancour.

The major television issue of the day – ITV's claim to the fourth channel – was, so to speak, *sub judice* because the government's White Paper was in the course of preparation; and the committee was severely critical of the prolonged delay in its publication. Its attention was therefore directed mostly towards Independent Local Radio. The insatiable political appetite for argument about television had been temporarily assuaged by the enormity of the Annan experience. The written and oral evidence taken by this Select Committee from the Authority and other relevant bodies – which included the Home Office, ISBA and the IPA on behalf of the advertisers, ABS and ACTT on behalf of the unions, and the IBA's own General Advisory Council – was more of an updating of facts: on finance, engineering, labour relations, advertising control and programming.

In programming the Authority was able to point to changes made in the planning and composition of schedules and to the advances in reaching new audiences and developing the regional system made possible by the removal of government restrictions on hours. Its review in 1972 had concluded that the service would not be improved by making radical alterations to the respective functions of the five central and ten regional companies. Instead, arrangements were made to facilitate the inclusion of programmes from the ten in the network schedules. As a result, following derestriction, networked programmes from those companies had more than doubled in three years: from 165 hours in 1970–71 to 350 in 1973–74. By 1976–77, after a further three years, the figure had risen to 450 – an average of about 8½ hours a week – and significant numbers of these programmes were being scheduled during the main evening viewing period. Further growth, said the Authority, was dependent upon the unblocking of a new log-jam in the ITV system, which only another ITV channel would make possible.

The committee's report, published in October 1978, was welcomed by the Authority as stimulating and useful.[17] It recommended that the Authority be permitted to proceed with the expansion of ILR and with engineering for the fourth television channel. It approved IBA policies in television, declaring its conviction that the structure established was a good and efficient one. It endorsed IBA plans for public consultation in the competition for new television contracts and it recommended that the Authority be established by statute without arbitrary limitation on its span of existence.

This last proposal, which would have overcome the need for short-term extensions, was not adopted by government. Nor did the Home Secretary

concur with the committee's view on the advisability of broadening the
membership of the Authority to reflect the additional responsibilities
resulting from its transformation from ITA to IBA. Any significant
increase in Authority members beyond the current number of eleven was
thought likely to make it too large to function effectively as a collective
body.[18]

The appointment of this Select Committee was suspected of being little
more than a government pretext for the postponement of decisions about
Annan's recommendations. The government's delayed White Paper was
published after the committee's hearings but before its report. While ITV
was awaiting the latter and preparing its responses to the former, an attack
was launched from another quarter. The BBC's finances were in poor
shape, and its Chairman, Sir Michael Swann, made a public demand for
parity of misery between the two organisations. He called on the government to raise the levy and make ITV poorer because no increase was being
permitted in the BBC's licence fee. It was hard to detect a genuine concern
for the public interest in this campaign for ITV's impoverishment unless
one was convinced that anything which was good for the BBC must be
good for the viewer; but the BBC lobby was, as always, powerful, and
ITV's indignation soon turned to alarm as the sympathy of parliament was
enlisted.

On 28 November the IBA's Chairman, Lady Plowden, had a meeting
with the Home Secretary to urge him to resist what she described as an ill-
considered proposal. In a letter on the following day she wrote:

I was concerned by the thought that MPs' concern about the BBC
licence fee might lead to suggestions that the money ITV spends on
programmes should be reduced. I believe that the public would regard
such action as a punitive and unnecessary way of diminishing their
pleasure. Moreover, I think the campaign is based on a false premise –
that ITV is too rich. It was understandable that, in their attempt to get a
licence fee increase which would last them over three years, the BBC
should have fostered this notion: but to base action on it during this
coming year (when the BBC's revenue will be more than twice what it
was in March 1975) would be unjust to ITV programme-makers and to
the public.[19]

Despite its growth in money terms ITV's revenue was, Lady Plowden
pointed out, not very different in real terms from what it had always been.
There had been highs in 1972–73 and 1977–78 and lows in 1970–71 and
1974–75. At constant prices (adjusted to take account of inflation in
accordance with the retail price index since 1967–68) revenue from
advertisements had brought in no more than £75.8 million, £86.7 million,
£95 million and £110.5 million in progressive rises over the previous four

years, and these figures appeared in their true light when compared with those for 1968–69 (£94.9 million) and 1972–73 (£104.5 million).

The BBC was complaining of losing staff to commercial companies which could offer better salaries, but these were not always ITV companies. During the past year the two-way flow of staff between the two broadcasting organisations had been even, the Home Secretary was told. The BBC was complaining of being outbid in a contract for covering football matches, but the price offered by ITV had not been extravagant and, in any case, many more exclusive contracts for sporting events were held by the BBC than by ITV. Ceefax was more lavishly funded than Oracle, and so was the BBC's parliamentary radio operation by comparison with ILR's; ITV had been outbid by the BBC in the purchase of feature films, most extravagantly in *The Sound of Music*; and many BBC productions were less tightly budgeted than ITV's.

This was an unedifying squabble and particularly regrettable in the light of the restraint shown by both organisations in evidence to Annan, when ITV had not had a bad word to say about the BBC and the BBC had concentrated its hostility on attempting, understandably, to deny its rival a second channel. The counter-measures taken on this occasion succeeded in neutralising the BBC offensive because ITV had a strong case ably deployed by an energetic Chairman. At this time 80 per cent of its surplus after programme expenditure was going into the public purse in corporation tax and levy, so that in addition to a popular, high-quality, self-financing television service the public was receiving an annual bonus of about £100 million.

However, a general affluence was reflected not only in the availability of funds for diversification but also in ITV's salaries and production costs. Both were markedly higher than the BBC's, in spite of the Corporation's 'swollen bureaucracy'. Owing to regional variations the programme output of ITV on one channel was much the same as the BBC's on two: for the year to the end of March 1980 the BBC's amounted to 11,358 hours, while ITV's for the calendar year 1980 was 11,312. But at a total cost of £262.5 million for its television services the average cost per hour of a BBC programme was approximately £23,000. The nearest comparable figure for ITV – for costs, excluding sales costs, during the year which ended in October 1980 – was £407 million, giving an hourly average of £36,000. At £13,000 or a level of 56 per cent above the BBC's, this excess could not be wholly accounted for by a much praised but uneconomical regional system which necessitated sixteen separate production units.[20] The network was burdened with onerous labour costs through acquiescence in the 'custom and practice' of union restrictions and overmanning, and whereas strikes saved the BBC money they cost ITV dear.

Still, it could be argued, the system gave value for money in the quality and range of its programmes. During the winter of discontent of 1978–79, when the weather was atrocious and rubbish lay uncollected and bodies unburied, the public stood in desperate need of comedy, light entertainment and escapist drama, and ITV obliged with well-established favourites like *George and Mildred*, *Mind Your Language* and *Robin's Nest* and newcomers such as *3–2–1* and *Bless Me, Father*. *The Kenny Everett Video Show* was enjoyed by some and not at all by others, while the ever-popular Bruce Forsyth was savaged by the press for *Bruce's Big Night Out*.

In drama series *Lillie* and *Edward and Mrs Simpson* captivated large audiences drawn towards other periods of time, and those with a taste for action and adventure continued to be thrilled by *The Sweeney* and *The Professionals*. In a determined effort to attract more viewers to the single play, it was scheduled and packaged in groups under generic titles: *The Sunday Drama*, *Saturday Drama* and *ITV Playhouse* (on Tuesdays). Playwrights whose works appeared on ITV screens during 1978–79 included Alan Bennett (six plays), John Bowen, Melvyn Bragg, John Braine, Ian Curteis, Ian Kennedy Martin, David Mercer, Brian Phelan, Alan Plater, Ken Russell and William Trevor.

Escapism and drama were balanced by news, current affairs and documentary programmes which during 1978–79 covered troubles in Northern Ireland, Uganda and Iran; guerrilla warfare in Eritrea; the Chinese invasion of Vietnam; terrorism in West Germany; the kidnapping and murder of the Italian Prime Minister; sanctions-busting in Rhodesia; an air disaster in Zagreb; mass suicide in Guyana; the trial of dissidents in Moscow and of Jeremy Thorpe in Minehead; the neutron bomb; letter bombs; opium addiction; oil pollution; sexual inequality; the plight of the unemployed, of the low-paid, of small businesses, of the British steel industry and of the economy generally; devolution referenda in Scotland and Wales; relations between government and unions; and the strikes by transport and public service unions. There was no excuse for ITV viewers to be ill-informed about the world in which they lived.

In sport the focus was on football. A move by London Weekend Television to secure for ITV exclusive rights to televise League football matches for the next three seasons might have resulted in a more rational approach to avoidance of the duplication of major sporting events – an objective long sought by ITV and recommended by Annan. Initiatives by the Office of Fair Trading and an infuriated BBC aborted what was condemned as a breach of agreement and hailed as Snatch of the Day. The BBC regained its share of Football League matches in return for an agreement to minimise simultaneous coverage of future major sporting events, specifically the 1980 Olympic Games and the 1982 World Cup competition.

In arts programming Southern Television contributed three Glyndebourne operas: Verdi's *Falstaff* and Mozart's *Don Giovanni* and *The Magic Flute*, the last of which was screened on the Saturday of Christmas week and attracted an audience of 2 million. LWT's *South Bank Show*, edited and presented by Melvyn Bragg and still the network's flagship in the arts, began its remarkable run of awards at the Prix Italia, winning the top international prize for music programmes three times in four years. In 1978 it was awarded for the ballet *MacMillan's Mayerling*; in 1980 and 1981 for Tony Palmer's biographies of modern British composers: Benjamin Britten (in *A Time There Was*) and William Walton (*At the Haunted End of the Day*). In 1980 ITV won another Prix Italia when Thames's *Creggan* (about Northern Ireland) was judged the year's best documentary.

The continuing prosperity of the system, coupled with the continued restriction of domestic production to a single channel, led the Authority to negotiate with the companies new arrangements to replace its long-standing control over the quantity of foreign material shown in programmes in accordance with the statutory requirement for 'proper proportions' of material of British origin and performance.[21] The regulation that the programme running time of foreign material should occupy no more than 14 per cent of any company's total transmission time, averaged over six months, had been subject to qualifications. British Commonwealth material was counted as British; foreign programmes approved by the Authority as of outstanding educational or cultural value were exempt; and special allowances were made for programmes co-produced with foreign organisations (calculated according to what percentage was foreign and what home-produced).

Under this arrangement the proportion of British-made material on ITV (programmes, promotions and advertisements) had amounted to some 84.5 per cent. In July 1978 it was decided to raise this in stages to a target of at least 86 per cent of home-produced material by the summer of 1979. The distinction between 'foreign' and 'Commonwealth' was abandoned, both categories being reclassified as 'overseas', but overseas material of outstanding educational or cultural value remained exempt.[22] The aim was to strengthen the domestic flavour of the schedules and increase the scope for UK creative talent without dispensing with the benefits of popular and distinguished programmes from overseas (but other parts of the EEC now counted as 'British'). This was a matter in which the government displayed an unexpected interest in its White Paper, which was published shortly before the agreement was made known.[23]

The national strikes by transport and other workers during the winter of 1978–79 resulted in cancellations by advertisers unable to move stocks in response to customer's demands. By the end of February this had cost the companies £9 million, but much worse was to follow. In a pay dispute the television unions took the network off the air for nearly eleven weeks from 10 August to 19 October 1979. In this, ITV's longest strike, no management service was mounted, so that viewers were deprived of all programmes and the companies of all advertisement revenue, estimated at between £90 million and £100 million. The strike ended in victory for the unions and a costly humiliation for the company managements, who were obliged under the terms of the settlement to meet vastly inflated wage bills which, however, as the unions insisted, they could well afford to pay. Much of the lost revenue was soon recovered or replaced, but a decline in the companies' high profitability was set in motion by this disruption. 1979 was the worst year for industrial disputes in ITV's history, and the loss of audiences, loss of forward production and loss of the confidence of advertisers seriously damaged its health throughout 1980 and beyond.[24]

Political uncertainty persisted pending a general election. Expected in the autumn of 1978, it was not held until May 1979, when the question mark which had been hanging over the fourth channel for a decade was immediately removed by the newly elected Conservative government. A statement in the Queen's Speech gave notice of its intention to entrust the channel to the IBA, subject to safeguards. In September the Royal Television Society, whose membership embraced both ITV and BBC, assembled for its biennial Convention at King's College, Cambridge. Reflecting the changed political climate for the broadcasting establishment, the sun shone comfortingly throughout the weekend. In a carefully prepared announcement of the government's intentions William Whitelaw, now the Home Secretary, delivered the *coup de grâce* to a *bête noire*:

> I am convinced that not only would the creation of an Open Broadcasting Authority directly dependent on the government for funds be potentially dangerous; it is also unnecessary to achieve what we want. The experience and ability of the IBA if used to the full, the money, equipment and skills of the ITV companies and the talents of the independent producers, can be harnessed to provide a different and worthwhile service on the fourth channel.

But already other clouds were gathering on the horizon. The whole future of conventional terrestrial broadcasting appeared to be menaced by the imminence of a multiplicity of channels distributed via cable systems and satellites in the sky. In the USA, it was reported, some of the 4,000 cable TV services were already offering subscribers a choice of as many as seventy-two channels. How many satellite transmissions under overseas control would be receivable in Britain, and how soon? What impact would be made on viewing habits by the confidently predicted sales of millions of video cassettes and video discs? The title of the Annan report had been *The Future of Broadcasting*, but John Freeman told delegates that it might just as well have been called *Institutional Television: The Final Ten Years*.

Whitelaw's speech at Cambridge relieved the industry of a heap of frustrations. In the ensuing months the future shape and face of television were progressively unveiled. On Thursday, 18 October the Home Secretary lunched with the Authority for an informal exchange of views, and on the following afternoon members set out for a working weekend in seclusion at a hotel and former hunting lodge: Great Fosters, Egham. Amply briefed by IBA staff, they were able to take a series of decisions not only about the fourth channel – its structure, finance, advertising, programmes and launch date – but also about the post-1981 ITV contracts. General contract specifications, the names of areas and allocation of transmitters, the duration of contracts and the problem of redundancies in the event of changes of contract were all discussed. Decisions were taken that there should be no new daytime contract in London and that the teletext contracts would be linked to ITV's.

On 29 October the Authority announced its engineering plans for the fourth channel. On 12 November its provisional proposals for the channel itself were published.[25] On 24 January 1980 it made a public statement on the contracts under which the ITV companies would operate from 1 January 1982.[26]

The January statement included the news that the Authority had not bowed to political pressure to split the Midland region into two. Instead it would be a single 'dual region', embracing East Midlands and West Midlands, and a similar two-in-one arrangement was to be applied in South and South-East England. These were modelled on the existing link between Wales and the West of England, which was to continue as a dual region. But the item which attracted most public notice and comment was the announcement that the Authority 'would be prepared to consider applications' from those interested in providing a breakfast-time television service nationwide.

The Authority had spent much of 1979 'taking the public mind' about the ITV service and how its viewers would like to see it develop. It had set out its proposals for canvassing public opinion early in the year (in *ITV: Future Contracts and the Public*), held 250 public meetings, invited written comments and undertaken a major research survey whose findings were published in November. What had been learned in these public relations exercises was all taken into account in the terms of the announcement in January.

On 6 February 1980 the government, not bothering with another White Paper, published its Broadcasting Bill without further ado. After the lengthy process of debate and amendment in parliament this Bill became law on 13 November as the Broadcasting Act 1980. It was to be followed within a year (on 30 October 1981) by the Broadcasting Act 1981 which consolidated the Independent Broadcasting Authority Acts of 1973, 1974 and 1978 and the Broadcasting Act 1980.

The 1980 Act extended the Authority's function for fifteen years until 31 December 1996 and gave the Home Secretary discretion to make a further extension of not more than five years by statutory instrument. Television company contracts could run for not more than eight years. They might be extended, but had to be readvertised when the maximum period was reached. The Act gave the Authority powers to proceed with a second television service.

The Broadcasting Complaints Commission was a survivor from the previous government's White Paper. Annan's advocacy of this independent body had commanded general support. Its function under the provisions of the Act was to consider and adjudicate on complaints of unfair treatment in television and radio programmes or of 'unwarranted infringement of privacy' in connection with programmes.[27] The precise wording of this clause was the subject of much debate at the committee stage of the Bill and the government was forced to concede the addition of the word 'unwarranted' as an amendment. The broadcasters were required to pay for the commission and publish its findings as the commission directed.

In general, a consensus view was expressed in the provisions of this urgently required Act of settlement, and it brought kudos to the Home Office as well as relief to broadcasters. Meanwhile the long-drawn-out process of competition for new contracts brought no recognisable benefit to viewers. As Lord Thomson of Monifieth, the new Chairman of the IBA in 1981, wrote:

> TV companies are sometimes accused of window-dressing during a franchise period and of producing programmes especially to impress the Authority. Apart from the fact that IBA members would be unlikely to be taken in by such tactics, in my experience the reality is rather the reverse. One of the disadvantages of the whole franchise process as established now by successive parliaments and governments over a quarter of a century is that (while it lasts) there is a considerable distraction and diversion of energy from programme-making to corporate survival.[28]

Although ITV was flattered by further international awards, 1980–81 was not, by and large, a distinguished programme year. The inadequacy of its programmes for children, for example, was plain for all to see and there was no disagreement between the Authority and companies on the need for an improvement. The target for most complaints from viewers during 1980 was the suspension of *Crossroads* during coverage of the Moscow Olympics. But it was drama-documentaries which attracted the most damaging criticism. However laudable the intention and skilful the writing and direction, programmes which blended fact and fiction were peculiarly vulnerable to misunderstanding on the part of viewers and to charges of inaccuracy or even deception.

Into this bastard category fell STV's *A Sense of Freedom*, a play about a real criminal, Jimmy Boyle, written by himself while in prison, and Granada's

Invasion, a dramatic reconstruction of events during the Russian occupation of Czechoslovakia in 1968 which made ingenious use of transcripts of actual meetings and the recollections of a member of the Dubček government. These were programmes made with care for the facts, but *A Man Called Intrepid*, an acquired series, featured real people, including Sir Winston Churchill and the physicist Niels Bohr, in a portrayal of British intelligence operations during the Second World War which was far removed from historical accuracy.

Most controversial of all was *Death of a Princess*, an ATV programme about the execution of a young member of the Saudi royal family and her lover for adultery. This precipitated a crisis in diplomatic and economic relations between Britain and Saudi Arabia. To prevent its transmission, the Saudi royal family made strong representations, at first to the company and then to the British government. But Lord Windlesham, ATV's Managing Director, resisted pressure from former colleagues in government and stood firm on the principle of editorial independence.

The IBA, whose senior staff had previewed and approved the programme, was not approached and did not intervene. It could have saved the government embarrassment and averted the threat of a loss of millions of pounds in export earnings if contracts were cancelled in angry retaliation, but an act of censorship at the behest of a foreign government would have caused a storm and created an intolerable precedent. Moreover, it found itself to be without any clear statutory power which would justify banning a programme whose only offence lay in giving offence outside the United Kingdom.[29] The programme's accuracy and impartiality were not in question, and expediency was judged a matter for government, not the Authority.

Creatively, the flaw in *Death of a Princess* was a failure to differentiate between real people and actors playing real people so that the viewer might be aware which was which. Working from a script based on interviews, the programme's presentation of reconstructed scenes was indistinguishable from its filmed recordings of real events. The princesses seen by millions of viewers car-cruising through the desert, to the fury of the Saudis, were in fact actresses.

The programme was screened at 8 p.m. on 9 April 1980, and reports of Saudi displeasure, a possible oil embargo and an apology by the Foreign Secretary were prominent in the press and on BBC news services the following day. The Foreign Office then issued a statement to the effect that Lord Carrington's apology had not been an apology but an expression of regret that offence had been given. According to *The Times*, the Foreign Secretary had explained that the government had no powers to interfere with the editorial content of programmes, still less to ban them. This was widely believed but untrue. Under the law, formal government control over ITV programmes was absolute. As the minister responsible for broadcasting, the Home Secretary – in the words of the Act – 'may at any time by notice in writing require the Authority to refrain from broadcasting any matter or classes of matter specified in the notice, and it shall be the duty of the Authority to comply with the notice'.[30]

Controversy over what had become known as docu-drama or faction was centuries old, as protest over Shakespeare's caricature of King Richard III bore witness. But television, with its audiences in millions, brought a new dimension to the issue. Viewers received news bulletins from the same source and were not necessarily qualified to distinguish between the portrayal of real people in different programme categories. At a time of public regrets by the Foreign Secretary for *Death of a Princess* ('mixing fact with fiction . . . can be dangerous and misleading') and a public apology by the Authority for *A Man Called Intrepid* ('dramatic licence should not lead to a travesty of the truth'), it was not surprising that the 1980 Broadcasting Bill should contain a provision allowing complaints about unfair treatment in television and radio programmes to be made even on behalf of the dead, who had never been so privileged under the libel laws.

The government probably had the reputation of Sir Winston Churchill mostly in mind when this provision was framed, but according to the Chairman of the BBC (George Howard) it quickly became known as 'the Richard III clause'. In a House of Lords debate it was foreseen that much historical dispute would result. The government therefore tabled an amendment, and the Act as passed stipulated that complaints on behalf of the dead might be entertained by the new Broadcasting Complaints Commission only in the case of programmes broadcast within five years of their death.[31] Shakespeare and all classical historical drama were thus rescued at the eleventh hour.

Financially the system at the end of the decade was in good heart but precariously balanced. More and more money was still coming in and more and more was still being spent every year on programmes. In 1980 programme expenditure amounted to 69 per cent of the companies' income. Fifteen per cent went to the government in corporation tax and levy and 5 per cent to the Authority in rentals, leaving 11 per cent to cover the provision of other services, depreciation of assets and after-tax profit (required for new equipment, dividends and reserves).[32] Costs were high and rising: between July 1978 and October 1980 they increased by 87 per cent; income by 58 per cent. The pay settlement which ended the 1979 strike was particularly costly in an industry which remained obstinately labour-intensive despite labour-saving developments in technology; and before the start of the new contract period even more staff would have to be recruited: for increased local programme origination, for the new dual regions, and for programme production, sales, research and technical operations for the fourth channel.

The overall financial state of Independent Television during the second half of the 1960s and throughout the 1970s is illustrated in outline by the table below. This shows, for each financial year from end July to end July, net advertisement revenue and profit before interest and tax, both actual and as adjusted (in accordance with the Department of Employment's index of retail prices) to take account of the high rate of inflation prevailing during the period.[33]

	NAR (£ million)	NAR at constant prices (£ million)	Profit before interest and tax (£ million)	Profit before interest and tax at constant prices (£ million)
1964–65	82.6	82.6	19.0	19.0
1965–66	82.0	79.2	16.8	16.2
1966–67	89.9	84.6	19.1	18.0
1967–68	94.9	85.0	18.3	16.4
1968–69	98.6	83.7	9.6	8.1
1969–70	93.3	74.6	6.0	4.8
1970–71	101.4	73.7	17.5	12.7
1971–72	120.4	82.2	26.9	18.4
1972–73	149.5	94.1	32.5	20.5
1973–74	150.3	81.7	22.5	12.2
1974–75	160.7	75.2	15.1	7.1
1975–76	209.3	80.1	27.9	10.8
1976–77	266.9	88.9	35.3	11.8
1977–78	334.9	100.8	47.5	14.3
1978–79	395.1	108.3	47.7	13.1
1979–80	428.8	99.1	39.0	9.0
1980–81	525.4	107.1	41.2	8.4

These figures do not lend support to the frequently reiterated assertion that the level of ITV's (post-levy) profits was excessive. Over the seventeen years income from advertisements multiplied more than sixfold and profits more than doubled. But in real terms advertisement revenue grew by 25 per cent and profits fell to less than half. The use of other base years, such as 1967–68, would produce different adjusted figures, but the general picture would not be greatly dissimilar. In the 1970s it could no longer be said (though it often was) that an Independent Television contract was a licence to print money or that the industry was the most lucrative in Britain since the building of the railways. Shareholders would have done better to invest in real estate.

17

THE EMPTY CHANNEL

After the rapid welcome of colour television into millions of homes the single most beneficial development waiting to be bestowed on viewers was the introduction of a service on the fourth channel. Available but unused, it remained throughout the 1970s the empty room of British broadcasting. The nation was said to be unable to afford it at a time of economic stringency, and years were wasted in bitter wrangling over its character and control between advocates and decriers of an ITV2. In all, the saga of the fourth channel ran for more than twenty years before its first programme appeared on the screen on 2 November 1982.

Between 1954 and 1968 the Authority was continually looking forward to introducing a second service provided by different contractors. Competition within ITV was seen as implicit in the doctrine which had led to the first Television Act. Its authors regarded monopoly in broadcasting as wrong, both socially and politically, and the destruction of the BBC monopoly in television as only a first step.

In 1962 two White Papers endorsed the Pilkington conclusion that there would be scope for another ITV channel in due course. In the following year the Conservative government reiterated its intention to allocate the fourth UHF channel to a second ITV service unless financial or other obstacles proved insurmountable, and it was the expectation of this which led the Authority to extend existing contracts from 1964 to 1968. Lord Hill, the Chairman of the ITA, looked forward to a launch in the autumn of 1965 'if all goes well'.

It didn't and in December 1966 the Labour government's White Paper on Broadcasting announced that no allocation of frequencies for such a service would be authorised for at least three years. There were said to be higher national priorities than a fourth television channel and competing claims to be considered – from the Open University, for example.[1]

When the three years had passed, informal discussions between the ITA and the Post Office were not pursued in view of ITV's financial difficulties at the time, but in 1970 the Authority and its Policy Committee considered the question again, and from that time until the early 1980s the minutes of

Authority and SCC meetings are replete with accounts of the running debate on whether and when and how the goal of a second ITV service could be achieved.

In a revolution in ITA thinking during 1969 and 1970, 'complementary' displaced 'competition' in the vocabulary of the argument. This had been made feasible by an amendment to section 13 of the 1963 Television Bill, which would have required the Authority to ensure that 'where reasonably practicable' any second set of programmes was supplied by different contractors. After amendment the Act merely stipulated that the Authority should 'ensure that, so far as possible, the same kind of subject-matter is not broadcast at the same time in the different programmes'.[2]

The Authority's conversion to complementarity was a volte-face by Sir Robert Fraser, the Director General. His long allegiance to a belief in the virtues of the competition was swayed by financial realism and the recognition that institutional considerations must yield to a structure which promised a better programme service. With studios and staff duplicating those of ITV1, a competitive ITV2 would have to incur the full costs of a separate service, and ITV1 areas and contractors would have to be merged to make new, financially viable units. This would cause disruption to the system and to the careers of those working in it. Even with mergers, there would be difficulty in creating areas which could support two competing companies everywhere, and some amalgamations would result in areas of little relevance in social and geographical terms. Regions like Northern Ireland would never be able to support two competitive services. And how was the avoidance of programme clashes between competitive services to be secured, as the Act required?

It was anticipation of a new committee of enquiry following the appointment of Lord Annan in May 1970 which brought ITV2 off the shelf and into a whirlpool of public controversy. At an SCC meeting on 13 May the Director General asked the companies for figures of additional costs, assuming that ITV2 would follow the example of BBC2 and broadcast in main viewing time for some 35 hours a week and that there would be a standardised network schedule with programmes produced predominantly by the five central companies as in ITV1.[3]

In July an influential article in the *Sunday Times* by the producer Peter Morley argued that a two-channel ITV, totally complementary, would release the system from the ratings tyranny of a single commercial channel and enable ITV, for the first time, to provide the service required of it by parliament. The deadlock between quality and profitability would be broken.[4] But in October the report of the National Board for Prices and Incomes found 'the financial outcome to be too problematical for us to recommend the opening of a second commercial service'.[5]

In May 1971 the Authority established a working party of Authority staff and ITV company managing directors to examine the practicalities of a second service supplied by the existing contractors and paid for by advertisement revenue collected by them. In June Brian Young, the new Director General, widened the debate by addressing an open letter to all ITV employees and to organisations and unions within the television industry. He invited ideas and proposals on the shape and character of a second ITV service, subject to two assumptions: first, that ITV2 should be complementary to ITV1, not competitive, and secondly that 'since ITV's total audience share could not be dramatically increased, some way needed to be found of financing two services from an amount nearer to present income than to twice that income'.[6] His letter was followed by a Consultation held on the Authority's premises on 2 November 1971.

At that meeting it was generally agreed that 'complementary' should not be interpreted as meaning one wholly serious and one wholly popular channel: if all minority programming was concentrated on one channel it would become a ghetto. Other forms of complementarity were advocated: experimental as opposed to traditional. A plea for ITV2 to become 'a sanctuary of worthwhile programming' led to an argument over the distinction, if any, between a sanctuary and a ghetto. Speakers stressed the desirability of common junction points (programmes beginning and ending at the same time at key points in the two schedules), the cross-trailing of programmes between channels, cross-channel repeats and the carry-over of some outside broadcasts from one channel to the other to facilitate long non-stop coverage. Decision-making about content and balance – the 'mix' – was identified as a major problem. Should the Authority be given greater powers or exercise more forcefully those which it possessed already, or would too close an involvement in the details of scheduling detract from its credibility as a regulatory body and reduce the companies to the status of programme factories?

After these and further deliberations on its own and with its advisory committees the Authority submitted to the Minister of Posts and Telecommunications on 30 November 1971, and published on 8 December, a 24-page document, *ITV2*, arguing strongly for a second service to permit ITV to broaden its range of programming, and for an early decision so that transmissions could begin in 1974. This set out the advantages of:

a system where the two Independent channels are as little as possible in rivalry with each other for the same part of the audience at the same time, although there would be competition on an equal footing with the BBC. If this complementarity between ITV1 and ITV2 is achieved, then a wider range of the total audience will be provided with programmes which are to their taste. A self-supporting single service will always find it difficult regularly to screen programmes which it is known will attract a relatively small proportion of the total audience. With two services, the planners will be able to accept the possibility of small audiences at peak

time with equanimity, since the audience for the alternative ITV programme need not be substantially affected. There is therefore, in a
complementary situation, an inherent, rather than an imposed, reason
for scheduling programmes in the way that serves the viewer best: and
planners are not tempted to place programmes of mass appeal against
each other, since that would diminish the following of both.[7]

A regular and reliable supply of programmes from large production
teams was seen as essential to the success of the new service:

The success of Independent Television has owed much to its use of the
five central companies in such a way that they could plan ahead and
bring major resources to bear on creative ideas, with the knowledge that
the resulting programmes would be assured of a showing throughout the
network. This means that much of the time on ITV1 is shared out and
scheduled by these five major suppliers in a way that facilitates planning
and production ... The results of this general impetus to good
programme-making should be seen in ITV2 as well as in ITV1, if the
second service is not to be a rag-bag; a large number of the programme-
makers who could provide good material to fill a second channel
worthily are already working for these five central companies.[8]

The desirability of wider access to the channel was recognised nevertheless. Other sources to be tapped for programmes catering for a greater
variety of viewers' interests were the regional companies and freelance
producers and even amateurs: 'particular sections of the community who
feel that they have some special message or viewpoint'. The paper hinted
at an equal division of airtime between the central companies, the regionals and freelances, subject to the availability of acceptable programmes; and control was to be in the hands of a new programme planning
board with an Authority chairman, not in those of the ITV1 Programme
Controllers Group. An appetising list of new programme opportunities
was enumerated.

But the government did not rise to the bait. On 9 December the
minister, Christopher Chataway, told Lord Aylestone privately of his
belief that the channel could not be allocated before 1976 without a
measure of general agreement; and this did not then exist. Press comment
on the ITA proposals was mostly critical, and on 13 November an 'open
conference' had been held at the Central Polytechnic of London organised
by the Free Communications Group and *Time Out* magazine as an
alternative forum to the Authority's Consultation.

The objective of the TV4 Campaign, born at that conference, was to
rouse public opinion against an ITV2. On 10 December it presented the
minister with a vituperative memorandum, *Opportunities for the Fourth*

Channel, which was published four days later. A statement from the group, reported in *The Times*, denounced the ITA plan for displaying 'an arrogant and bland ignorance of the needs of the public – the needs for access, accountability and participation on the part of the people who watch (and finance) television'.[9] This campaign stimulated an immense amount of media coverage, but probably of more effect on the government was an Early Day Motion on the Order Paper of the House of Commons from five Labour MPs on 2 December calling for the channel to be kept out of the hands of the present contractors and for a public enquiry. It attracted more than a hundred signatures.

The following August the report from the House of Commons Select Committee on Nationalised Industries[10] recommended that 'no decision should be taken on allocating the Fourth Channel except as part of a general review embracing the broadcasting needs of the country'.[11] Although the minister had told the House that the Authority's case was very well argued,[12] the committee was unimpressed, and it thought the 'principle of equity' propounded by Sir Hugh Greene in a Granada Guildhall lecture in 1972 (that ITV should in fairness have two channels because the BBC had two) 'a poor argument for the allocation of a scarce national resource'.

In his White Paper of March 1973 the new minister, Sir John Eden, prevaricated on the allocation of the fourth channel and invited views. The response was an abundance of submissions, which the Home Office later deposited in the laps of the Annan Committee. In a Further Submission the Authority welcomed the opportunity to counter the arguments of its critics:[13] that there was no public demand; that ITV2 would mean more of the same; that there was need of a third force in broadcasting; that it would be wrong to hand the channel over to the present ITV companies; and that the Authority would be incapable of fulfilling its intentions to control it.

'It should be clear ... from all the Authority has said that, in proposing ITV2, it is not seeking to see the existing programme companies made more powerful or more prosperous,' this submission declared defensively. Earlier emphasis on the crucial role of the five central companies as substantial programme providers had been tactically imprudent and the Authority now stressed its willingness to accommodate contributions not only from independent producers but also from a National Television Foundation as proposed by one of its principal critics.

In evidence to the Annan Committee the following year the IBA drew attention to its 1971 and 1973 submissions and, in briefly restating the case for a second ITV service, dealt severely with the spoiling tactics of the BBC, whose memorandum to the minister had asserted that the existing 'delicate and subtle balances' protected programme standards. According to the BBC, 'the present system depends on there being a rough balance of audience between the one ITV channel and the two BBC channels': an

ITV2 would therefore narrow viewers' choice because the BBC would be obliged to maintain its share of the audience by substituting popular programmes for minority ones. The IBA's response to this threat was pained:

> The claim that the BBC has to have roughly half the television audience is one of the premises on which the memorandum is based, and not a conclusion reached from any arguments put forward. The Authority does not feel that the claim provides the basis of a valid argument against a second ITV service, or that it justifies the BBC's declared intention to respond to ITV's widening of programme choice for the viewer by reducing its own range of programmes.[14]

Not surprisingly, Sir Charles Curran, the BBC's Director-General, saw no need at all for a general fourth channel. Outside Wales, he believed, it would be best handed over to the Open University (whose programmes the BBC was anxious to off-load).[15]

How to finance the channel was less of a dilemma for the Authority than for others: 'It is not a precondition for ITV2 that it should necessarily be self-supporting from its own income, without any transfer of revenue from ITV1.' Although this would cause some loss of revenue to the Exchequer through a reduction in levy payments, the Authority did not believe that a more economical means of establishing a fourth channel service of comparable quality could be found – a belief which was unquestionably correct and never seriously challenged.

Yet how, on financial grounds, could any government in the 1970s be persuaded to authorise another television channel of any kind? That was the biggest obstacle to be overcome. The IBA's initial submission to the Annan Committee, written in August 1974 and looking forward to the launch of an ITV2 not before 1979, was made public in December 1974 in the same week as Britain's worst-ever trade figures were published, sterling reached its lowest-ever level, the latest retail price index showed inflation running at 18.3 per cent and the BBC announced programme cuts because the government was delaying an increase in the licence fee.[16]

On the other hand, television was an exceptionally economical method of disseminating not only entertainment but information and an education service too. Investment in television, it could be argued, was an investment both in provision for the nation's leisure hours and in the enlargement of the public's understanding of the issues of the day. In 1976 the running costs of an ITV2, including amortisation of the initial capital investment, were calculated to amount to less than a pound per person per year. The additional cost to viewers – in sets, aerials and licence fees – would have been nil.[17]

The ITV companies' memorandum to the minister in 1973 was reprinted in their evidence to Annan.[18] Here they pointed to the deprivation of the public denied an available channel, the inequity of two-against-one com-petition and the under-utilisation of creative talent in an industry confined to a single-channel service. They pointed to the fact that a new channel would need programmes, production facilities and finance, and argued that only ITV could supply all three. To meet the strong demand for innovation and the inclusion of talent from outside the system they too adopted the suggestion of a National Television Foundation to encourage new kinds of programmes in all categories, particularly the experimental, and proposed an Independent Programme Board, which would initiate and schedule programmes, to serve the interests of outside programme-makers and groups who would not normally have access to the medium. The com-panies promised to 'divest themselves of a significant proportion of airtime and programme responsibility', including periods of peak time, and hand it over to those independent bodies.

This endeavour to overcome the opposition by meeting it more than half-way failed. ITV was thought too powerful already, and it was still suspected, despite all protestations and specific declarations of intent, that ITV2 would inevitably mean 'more of the same'. This phrase joined the boo words 'duopoly' and 'straitjacket' as the battle slogans of ITV's adversaries. The dark suspicion was even entertained that, despite all forecasts, the companies would somehow contrive to make money out of it; and if there was to be a profitable second commercial channel, fair play surely demanded (as some graphically expressed it) different snouts in the trough.

The idea of a National Television Foundation had been in currency since the spring of 1972. Ignoring the diversity within ITV, it was based on the proposition that broadcasting was a form of publishing confined in Britain to two large publishing houses whose authors were for the most part their own employees and who imposed their own house style on all their publications. Anthony Smith, the media guru and evangelist whose brain-child it was, had worked as a member of the BBC's *Tonight* team before becoming a research fellow of St Antony's College, Oxford. He visualised that the Foundation

would supplement existing broadcasting by broadening the input, by allowing anyone to bring a project to it, whether an independent programme-maker with a finely worked-out plan, neatly costed, or a firm, organisation or individual with merely a well-argued complaint that some issue was failing to get across to the public. The Foundation would then play a kind of impresario role, merely by allocating resources to some, but fitting producers, writers, technicians to others who arrived only with an idea, a grievance, a cause.[19]

Neither Smith nor his band of articulate supporters was willing to be beguiled by the companies into swallowing their compromise proposal. As

Peter Fiddick expressed it in the *Guardian*, which became a valuable mouthpiece for Stop ITV2 campaigners: 'The point of his NTF idea is that it should run an entire channel free from traditional pressures of scheduling and ingrained thinking about how programmes "should" be made and when to show them. A limited number of hours within a channel that is by itself run in harness with ITV1 is by no stretching of words the same idea.'[20]

But ITCA's evidence to Annan went on to argue that an ITV2 was needed for the sake of ITV itself, because a static industry's arteries harden:

> Growth has served to keep the youthful industry of television vigorous. As yet there has been little wastage through retirement, either in ITV or among the majority of BBC television staff. Lebensraum for new talent will be hard to find for a further decade or more until the pioneers of the mid-1950s begin to reach the age of sixty. To date there have been periods of swift expansion at intervals of four or five years: the recruitment for ITV from 1955 to 1960, for BBC2 in 1964, for the new Yorkshire contractor in 1968, for the extension of hours in 1972. Such periods facilitate the recruitment of new talent. They also offer an opportunity for the best talent to leapfrog over those older or less able. The next growth point must be ITV2.[21]

This point was emphasised in supplementary evidence in November 1975, submitted to silence a whispering campaign that, because of the cost, the companies did not really want a second channel, whatever their public stance: 'It has been said that ITV is ambivalent about a second channel. It is not. The ITV companies want a second channel and they would like it allocated now.' Not only was it impossible to provide a truly comprehensive service to viewers on a single channel, but 'ITV's creative staffs require wider opportunities for experiment and innovation: if these become available through ITV2 the quality of ITV1 will be raised.'[22] This enthusiasm in public did, however, conceal some real apprehensions about the impact of any second channel on the system, particularly among the smaller companies.

In its report on this complex issue the Annan Committee noted that successive governments had rejected proposals for a second commercial channel, mainly for economic reasons, and that an obstacle had been rolled in the path of a nationwide service by government acceptance in principle of the recommendations of two technical committees – Crawford and Siberry – that the channel should be reserved in Wales for programmes in the Welsh language. It noted the zeal of the IBA for a second channel which would broaden its range of broadcasting and permit a larger number of serious programmes in peak viewing periods. It acknowledged that the companies had made a formidable case in excellently presented evidence. But it then drew a distinction between those 'who wanted an ITV2; those who wanted a channel for education; and those who wanted a service for minorities' – a false distinction because the IBA and ITCA proposals for ITV2 contained provisions for minority and educational interests.

On grounds of cost, time and the integration of transmitter sites, the committee accepted the necessity of making the IBA responsible for transmitting the new channel. But, guided by Phillip Whitehead, nudged by the BBC and told by Anthony Smith that 'awarding a new channel or even a substantial part of it to the IBA and the companies would damage broadcasting irreparably', it was persuaded that an ITV2 would be an undesirable 'confirmation of the duopoly' and result in an undesirable 'battle for ratings' with the BBC. 'Although ITV2 might begin by providing programmes for minorities, the result of intensified competition would be that fewer programmes for minorities would be provided both by ITV and by the BBC.' New initiatives would be stultified, and 'if the new service were the responsibility of the IBA, how could the IBA resist pressure from the television companies to allow them a considerable share of peak programming time on the new channel?' Moreover, ITCA's 'programme plans for the channel suggested to us that the service would be aimed primarily at those most able and most likely to provide themselves with other sorts of entertainment, information and education'.[23]

The idealist concept of an Open Broadcasting Authority which the committee was led to adopt as an alternative was described as 'a new kind of broadcast publishing'. It was almost the National Television Foundation under another name. The Authority was to be a Non-Authority, not responsible for the content of its programmes except for compliance with the laws of libel, obscenity and incitement to crime, disorder or racial hatred. There would be no requirement of balance, either politically or in the mix of programmes. The only safeguard against bias, propaganda and indoctrination by political extremists would be the loss of its remit through government intervention (strongly deprecated elsewhere in the report). The OBA would broadcast educational programmes, including those of the Open University (which alone would have filled most of the proposed airtime), programmes from individual ITV companies and, especially, the work of independent producers. The untrustworthy IBA was to be confined to the role of engineer.

When the report was published this novel idea caught the public imagination and was hotly debated. In the 1977 Fleming Memorial Lecture[24] Lord Annan reiterated the committee's belief that, no matter what the IBA and the companies said, an ITV2 would provide more of the same – 'and if we get more of the same it will mean worse'. The battle for the ratings would intensify if a second channel enabled ITV to take 60 per cent of the audience. The BBC would then hit back. 'Their knives would be out and they would match comedy with comedy, quiz show with quiz show, or put on a crime series to knock out an arts programme. So far from there being more choice for the viewer, there would be less choice.'

This was an argument which could be deployed only by overlooking the companies' repeated assertions that they wished to run a complementary channel and that it was not even in their financial interests to produce more of the same to supplement ITV1, since this would merely serve to depress advertisement rates. New advertising money to finance the channel would come from offering advertisers different, specialised audiences. It was strange, too, that Lord Annan should object to the scheduling of a crime series against an arts programme. Could he really be saying that giving the viewers more choice meant choice between one arts programme and another?

But the most crucial question still not satisfactorily answered was: How could an OBA channel be financed? From 'a variety of sources', said the report. These were to include the sponsorship of programmes, block advertising, longer advertisements, a government grant, and leasing time to the IBA. But sponsorship had long been taboo in a Britain fed on stories from the USA of such horrors as a sponsoring gas company insisting on all mention of gas ovens being removed from a programme on Nazi extermination camps. OBA rules governing sponsorship and advertising were therefore to be strict – indeed so strict as to make them singularly unattractive to would-be sponsors and advertisers. Thus it seemed as though the taxpayer would be footing nearly the whole bill unless the IBA magnanimously came to the rescue. In his Fleming Lecture Lord Annan was commendably frank on the subject:

> How precisely the channel was to be financed was a matter on which we disagreed. One source must be advertising. But we did not all agree on how much or of what kind. For instance, I, and one or two others, thought that the simplest and most economical way of introducing ITV to the fourth channel was for the Open Broadcasting Authority to hand over its channel for so many nights a week at certain specified times to the IBA. The time would be leased in blocks to the IBA who would see that the programmes that were scheduled were financed, as ITV is now, by spot advertising. Others on our committee wanted the Open Authority to purchase or commission programmes direct from the ITV companies. One of our members envisaged the ITV companies having the channel leased to them.

The Home Secretary (Merlyn Rees) was among those who were noticeably cool towards the OBA: 'The biggest question mark that hangs over the proposals seems to me to be their financial viability' and 'another aspect of the proposals with which I have found some difficulty is the suggestion in the report that the requirement that controversial matters should be treated with due impartiality might be relaxed'. For the Opposition, William Whitelaw was also sceptical about the financial provisions and thought that, since the new channel was to be so reliant on ITV anyway, it would be 'sensible to start the other way round, by giving the fourth channel to the IBA and the ITV companies, which can immediately provide the necessary equipment and skills, but on conditions that would meet the committee's main purpose'.[25]

In the ensuing debate in the House of Commons Julian Critchley seized on a passage in the report which ran: 'It is not lack of confidence in ITV to deliver the goods which has led us to this conclusion. What could have been easier in the short term and cheaper in terms of public expenditure than to have recommended that the fourth channel should have been allocated to the IBA for a second ITV service? It is simply that in our belief an ITV2 will result in worse television services than we have now because the BBC and ITV will engage in a self-destructive battle for the ratings.'[26] Where, Critchley wanted to know, had Annan been all its life? 'Why should a so-called "battle" be self-destructive? Somewhere in the report Lord Annan claims that the advent of ITV has served to improve programmes.'

The ITCA response to Annan analysed the two faces of competition. The committee had distinguished between competition for excellence ('admirable') and competition solely to gain audiences. It approved of rivalry between ITV and the BBC provided that it was regulated, not competition for the same source of revenue, and a contest for excellence which did not degenerate into the 'battle for ratings'. But the companies found this distinction artificial. What of *Upstairs, Downstairs*, for example? Was it to be acclaimed for its high standards or frowned upon for attracting large audiences? Indeed, what programme-maker did not seek to get his work seen by as many people as possible? 'The battle for excellence and the battle for ratings are identical' – as the professionals knew, but amateurs could perhaps not be expected to realise.[27]

In the companies' analysis the point which Annan had missed was that 'the potential harmful effect of competition lies not in its effect on individual programmes but in the composition of the schedule', and the committee's proposals for the fourth channel would do nothing to ensure complementary scheduling, which was the all-important consideration for the viewer. Indeed, to make the fourth channel the responsibility of another Authority would do much to precipitate the conflict which the committee was so anxious to avert, for 'the OBA would have a greater incentive to fight for ratings because its very existence would depend either on a programme subsidy from the taxpayer or on its ability to attract sufficient advertising revenue to meet the cost of its programmes and overheads.'[28]

The companies believed it incontestable that an ITV2 would provide a better service, more choice for viewers, more scope for programme-makers and more jobs in the industry. There was little difference between ITV's and Annan's plans for programmes on the new channel. The essential difference lay in the fact that ITV2 would be complementary to and financially supported by ITV1. Annan had stressed that it was charged with investigating the future of broadcasting and admitted that its OBA was not a practical proposition in the current economic climate. ITCA claimed that only ITV could finance and bring a fourth channel to life in the foreseeable future. The companies were prepared to bear the initial losses, which would admittedly mean an initial loss of revenue to the Treasury too and thus a cost to the taxpayer. 'By the third year, however, it is estimated that increased advertising revenue will be such that the Treasury's take from the levy will be restored to its present level and from then on, with the service established, the taxpayer will be in credit.'[29] The IBA believed that ITV's total audience would need to be increased by no more than one-sixth to make a two-channel service viable, and that a complementary ITV service for minorities would be much more attractive to advertisers than one planned in isolation.

While the companies continued to promote an ITV2 which would cater, through 'a policy of systematic co-ordination', for 'interests which cannot now be accommodated on ITV without depriving the majority',[30] the Authority was subjecting the practicality of the Annan blueprint to a close examination. An internal memorandum noted that neither the proposed freedom of access to the new channel (i.e. the submission of programmes without strict regard for the editorial policy of a channel controller in such matters as content, duration and length of series), nor independence of authorship (the general acceptance of programmes offered from outside the system) could be taken literally. There would have to be selection by some body acting according to some criteria. A more relaxed attitude towards due impartiality and freedom from the requirement of proper balance in the schedules would also need to be qualified by some restraints, as Lord Annan himself had admitted. Thus in practice the most appealing features of an OBA would prove illusory. And although the freedoms from the imperatives of competitive planning and the requirements of complementary scheduling envisaged in an OBA channel might still be achieved despite these necessary selection processes and restraints on programming, it was doubted whether that would be the best means of serving the viewer.[31]

Scheduling, a recondite and little understood process, gradually rose to the surface of the argument. It was the crux of the relationship between an ITV1 and an ITV2, and Lord Annan thought the IBA's refusal to schedule sinister: 'If one lets the companies schedule ITV2, then one says goodbye to diversity in broadcasting'.[32] But the Authority believed it wrong for a controlling body to become too closely involved. Scheduling was not only the end of programme-making; it was also the beginning, because a programme which could not be found a time-slot would be stillborn. An Authority which assumed a role beyond that of consultation, admonishment and final approval would effectively become responsible for all the companies' production operations and the employment of their staff.

Under the IBA's scheme for complementarity between twin channels each would have its own scheduling body, on which the Authority would be represented, while reserving its position as 'even-handed judge'. ITV2's schedulers would include representatives from the central companies, regional companies which were major contributors, independent producers and educational interests. Planning conflicts between the two channels would be reconciled by a small co-ordinating committee under IBA chairmanship before final schedules were submitted to the Authority for approval.

In January 1978 the Home Secretary (Merlyn Rees) met the Authority for an exchange of views. He was accompanied by his senior officials: the Permanent Secretary at the Home Office (Robert Armstrong), the Deputy Secretary (Philip Woodfield) and the Under-Secretary in charge of the Broadcasting Department (Shirley Littler). How could the Annan philosophy be realised except by an OBA, he wanted to know. The Authority, he was assured at length, could and would satisfy Annan's requirements. 'We are not saying, Give the fourth channel to the companies', he was told: 'We are saying, Give it to the IBA.' This response may have been acceptable to the Home Office representatives, but the minister indicated, before leaving, that the content of a White Paper and the timing of legislation were largely out of his hands: broadcasting was a subject on which his government colleagues had views.[33]

In the post-Annan public debate over OBA versus ITV2, ITV had the better of the argument. Home Office and Treasury officials as well as the Opposition were convinced that the OBA concept, despite its superficial appeal, was fatally flawed and that an ITV2 in some form or other was the only practical option on offer. There was also trade union opposition to an OBA, led by ACTT, which foresaw that it would de-stabilise the industry. It says much therefore for the persuasive powers of a small but dedicated band of lobbyists that they contrived to defeat the Home Office's draft proposals for an ITV2 and secure a Cabinet majority in favour of an OBA.[34]

In opting for 'programmes which will say something new in new ways' on the fourth channel, the redrafted White Paper of July 1978 agreed with the Annan Committee that 'a different kind of service requires a new Authority'. An Open Broadcasting Authority would be established accordingly, to 'commission, purchase or otherwise acquire programmes' but 'not itself make programmes; nor will it engineer the channel or transmit the service'.[35]

Although the OBA would operate 'in some sense as a publisher', Annan's recommendation that it should not have a responsibility similar to the IBA's for the programmes broadcast on its channel was rejected. 'The concept of a broadcasting authority requires that the authority should be responsible for the service it provides, and that the service should conform to certain basic requirements, for example that due impartiality is preserved in the treatment of controversial matters, and that nothing should be broadcast which incites to crime or is offensive to public feeling.'[36] The OBA, however, would not be required to secure a 'proper balance' or a wide range of subject matter in its programming.

The government's expectation that this watered-down OBA channel would gradually become financially less burdensome to the taxpayer sounded no more plausible than Annan's and the IBA commented tartly to the Home Office: 'It may be realistic, therefore, to regard the new Authority's need for government support as little short of permanent. As such, it would set a precedent which could constitute, in the long term, a threat to the independence of broadcasting. It would be ironic, if one outcome of a Report on Broadcasting which stressed the need for independence were a grave diminution of independence.'[37]

The complications arising from the creation of a third force in broadcasting were made painfully apparent in the tangled relationship between the OBA and the IBA foreseen in the White Paper:

It will be essential, however, to ensure both that the programmes which the ITV companies provide for the fourth channel are of the kind and quality required and that their involvement with the fourth channel does not adversely affect the service on the existing ITV channel. It will also be necessary to consider how any revenues which the companies might derive from, and any expenditure which they might incur as a result of, any involvement with the fourth channel should be treated for levy purposes. The OBA will be responsible for deciding the arrangements for the allocation of time to, and the provision of programmes by, the Independent Television system. These decisions will have financial consequences: so long, therefore, as the OBA is receiving or is liable to need a government grant, the government will need to be consulted about them. Because the decisions could also affect the quality and balance of programmes provided by the Independent Television companies for the existing ITV channel, for which the IBA is responsible, the OBA will need to consult the IBA about the need for, and the extent of, the involvement of the IBA in the arrangements between the OBA and the Independent Television system. There will in any case need to be discussions about the question of scheduling and the OBA may wish to consider with the IBA the possibility of some complementary scheduling between the ITV channel and the fourth channel. These organisational and operational matters will need to be considered further.[38]

Indeed they would; and since the government invited comments on its White Paper neither IBA nor ITCA was slow to oblige. In dismissing the suggestion that the new channel might become financially viable through advertising support, the IBA recalled the Annan Committee's strong opposition to competition among broadcasters for the same source of revenue: 'The White Paper sows the seeds of a conflict which could well have consequences of a kind to be regretted for broadcasting standards'. The IBA wanted to know what forms of control for advertisements the government was envisaging for the new channel. Would one Authority be expected to reject automatically any advertisement rejected by the other or would there, as the desire for pluralism seemed to suggest, be anomalies, so that advertisements considered unacceptable in the public interest on one channel would be broadcast on the other?

ITCA roundly condemned the OBA as 'patently misconceived' and predicted grave embarrassment for the government if it legislated for a channel which would not work. Since, contrary to the Annan recommendation, the OBA would be responsible for its programmes, it would be just one more Authority – and one spending large sums of public money for which there must surely be higher priorities. 'Programme-makers for the OBA will be subject to the same degree of "censorship" as those working for other channels. On what grounds therefore is it now believed that an OBA service would be different from, let alone superior to, an ITV2?' In its White Paper the government had recognised the need for some complementary scheduling, and this could only be achieved effectively through an ITV2: common junction points were essential; 'programme-making and scheduling are indissolubly linked'.[39]

In believing that a new service required a new Authority, ITCA argued, the government had become confused between Authorities and programme-makers: the function of Authorities was regulatory, not editorial, and diversity in broadcasting did not stem from their proliferation. BBC2 did not require its own board of governors in order to be different from BBC1.

'The fourth channel service will have to compete on its merits for the attention of viewers.' This assertion in the White Paper suggested to the companies that the government was condoning, or even endorsing, the battle for ratings which Lord Annan had so deplored, and ITCA noted that 'a competitive fourth channel appealing to a large enough audience to become financially viable could lead to a narrowing of the range of programming on all channels and, in terms of programme content, is unlikely to be what advocates of the OBA intend. Should the OBA succeed in securing a significant share of advertisement revenue, this could threaten the existence of the smaller regional Independent Television companies'.[40]

Similarly scathing comment had come from Paul Fox, Managing Director of Yorkshire Television, in an article in the *Listener* entitled *The Misbegotten Fourth Child*:

> If the OBA is really to be financed by spot advertising, it will need to make programmes that attract audiences. Inevitably, it will be in competition with ITV and, indeed, BBC1, and we will have the American situation all over again: three channels battling it out for the audience. That, presumably, was not the idea. But as things stand, the government-shaped OBA will not be able to give away its advertising time, let alone sell it.
>
> Somebody, probably in the Home Office, realised that the cabinet committee's proposals for financing the OBA were as amateurish as Annan's. Hence the reference to the ITV companies 'providing a significant part' of the fourth channel's output and, indeed, becoming a 'regular' supplier of programmes.
>
> This, if I may say so, is a little presumptuous. In addition to the privilege of paying the Exchequer a levy of more than £60 million a year, we are now being asked to make programmes for our competitor. It is like asking the proprietor of the *Daily Telegraph* to provide a significant part of a new and worthy Fleet Street daily. We are a programme company – not a facility company for other people's programmes.[41]

To what extent the Labour government would have acknowledged the force of all these objections will remain unknown because it fell before drafting a Bill and its misbegotten child landed on the desk of a Conservative Home Secretary. Whitelaw did not look on the OBA with favour; nor, as he had told the Broadcasting Press Guild in a speech in November 1977, did he want the fourth channel to be 'just another channel controlled by the ITV companies'. After a meeting with the Chairman and Director General of the IBA arranged by Julian Critchley, the counter-Whitehead Conservative back-bencher, he had become convinced that the right solution was an ITV2 with safeguards; and by safeguards he meant above all a channel not dominated by five companies which were generally regarded by politicians as already over-powerful. With his mind already made up, his decision was announced in the Queen's Speech within days of the Conservative victory at the polls in May 1979.

During a summer of renewed ferment intensive discussions were held between the Authority and the Broadcasting Department of the Home Office, when the Authority discovered that it was expected itself to formulate what had by then become the 'strict safeguards' on which the government was insisting. This it did in a hand-delivered letter to the

Home Secretary towards the end of July. The same IBA requirements which governed control of ITV1 were to be applied to ITV2. Programming was to be complementary. 'A due proportion' of programmes on the new channel was to be contributed by organisations and persons other than the ITV1 companies: the Authority was anxious not to be committed to any specific quotas which might prove 'inappropriate and inflexible'.[42]

But the battle was still not over. The Channel Four Group, a new alignment of ITV's critics and dissidents, was developing into the most effective of all the fourth channel pressure groups. It even plotted successfully a reversal of ACTT policy, so that the union's annual conference adopted an emergency resolution urging that the majority of the new channel's programmes should be drawn from 'a variety of British independent production sources other than the existing ITCA companies'.[43]

Another threat to 'ITV2 with safeguards' lay in the possibility of intervention by the new Prime Minister. Margaret Thatcher's commitment to the merits of competition was already well known, and the advertisers' lobby for a competitive channel had strong support among Conservatives. Within the government their cause was being pressed by Sally Oppenheim, the Minister for Consumer Affairs. But in committee, which he himself chaired, Whitelaw carried the day, aided by Treasury calculations about the effect of the different channels proposed on the size of the ITV companies' levy payments. Ever since its introduction in 1963 Conservative policy on television had been influenced by 'the levy factor'.

The proposals for the fourth channel announced by the Home Secretary in his address to the Royal Television Society at Cambridge on 14 September were based on those advanced by the Authority in July. There was to be no new Authority, he confirmed, and no competition for advertisement revenue because this would 'inevitably result in a move towards a single-minded concentration on maximising the audience for programmes'.

The new channel was intended, said the Home Secretary, to provide new opportunities for creative people in British television, to find new ways of serving minority and specialist audiences and to add greater satisfactions to those already available to viewers, as the Annan Committee had recommended. The ITV companies were not to be allowed to dominate programme planning and scheduling. The Authority's 'due proportion' of programmes from outside the existing system had become 'the largest practical proportion'. More significantly, 'complementary', the cornerstone of a decade of thinking, was discarded. Instead, the new service would be required to be 'distinctive'.

The price for the companies' retention of their monopoly in the sale of advertising time was to be payment for the new channel out of the proceeds. Apart from advertisements, the channel would offer a single national service with no regional variations except in Wales; but there the Home Secretary dropped a bombshell. Instead of a single ITV/BBC Welsh-language service, as heralded in the Conservative election manifesto for Wales, two separate channels, one ITV, the other BBC, would carry programmes for Welsh-speaking audiences. Economy was the reason for this volte-face: a single Welsh-language service would be unattractive to advertisers and bound to need government financing to support it.[44]

With the main burden of the Home Secretary's speech variously interpreted – was this or was this not 'ITV2 by the back door'? – and with legislation still to be drafted, lobbying by the interested parties ran on unabated. At the end of September the IBA held a Consultation, attended by some 150 independent producers, under the neutral chairmanship of Harold Evans, editor of the *Sunday Times*. On 10 October an Open Letter to the Home Secretary from the Channel Four Group was published in the *Guardian* signed by nearly 400 individuals and organisations. This endorsed the spirit of Whitelaw's address and demanded the incorporation of a number of principles in legislation: guaranteed finance for the channel, a management board answerable to the IBA, an independent programme controller free to schedule without reference to the ITV companies, and 'a majority of the channel's programmes to come from sources wholly independent of the BBC and ITV'.

At the beginning of October the companies submitted to the Authority a lengthy memorandum on important aspects of what they still regarded as ITV2. This supplied data requested by the Authority and contained a proposed structure for the new channel, an outline programme schedule, estimates of audience ratings, revenues and costs and a recommendation on the selling of airtime. They wanted the channel controlled by a trust which would be jointly owned by the IBA and themselves.[45]

October became a crowded month of internal consultation, and discussion at an SCC meeting failed to paper over substantial cracks in what had formerly been common ground. The companies still believed in complementary scheduling and its unattainability without some restriction on the new channel's independence, which by now the Authority was committed to protecting. Differences of opinion on the naming of the channel were not merely semantic. The IBA had come to accept the view of the politicians that 'ITV2' did not imply the 'new things in new ways' demanded by Annan.[46] Confusingly, by distinguishing the new IBA channel from ITV, abandonment of the name suggested that the companies, and not the Authority itself, were 'ITV'.

It was the advertisers who derived least satisfaction from Whitelaw's address. The companies therefore commissioned from Professor Harry Henry, an acknowledged expert in the field of media research, a monograph which sought to demonstrate that competitive selling would not be to the advantage of the advertiser. Henry argued that despite lack of competition television advertisement rates in real terms had shown a downward trend since 1961, and that a fourth channel operated on complementary lines was likely to be more beneficial to advertisers in providing advertising coverage, frequency and concentration in the form in which they most needed it. This argument carried weight, and IPA and ISBA representations during October concentrated on the undesirability of the companies employing the same individuals to sell airtime for both ITV and the fourth channel. They proposed a jointly owned, separately manned sales force.

A background paper prepared by its staff for the Authority meeting and 'discussion weekend' towards the end of the month set out the changes in the Authority's thinking since its ITV2 submission in 1971. These were noteworthy:

> The original concept of a service provided by the programme contractors – either on the same lines as for ITV1 or through a wholly-owned ITV2 Limited – has evolved into the proposal for a service that would be provided by a subsidiary of the Authority, with contributions coming from the ITV programme contractors among others. The principle of complementary and non-competitive services has remained firm; the contribution expected from independents has been given greater emphasis, and the role envisaged for the central ITV companies has diminished.[47]

In negotiation with the government the Authority urged legislation in general terms to allow itself room for manoeuvre in changing circumstances, but the Home Office was insistent that the Bill should specify certain matters in detail: how the programmes were to be obtained and financed; who would provide them if the IBA did not do so itself; who would have the power to call on the ITV companies for funds, and on what basis; and how the funds were to be accounted for.[48]

The Fourth Channel : The Authority's Proposals, a provisional, but full, statement of policy written by Colin Shaw, Director of Television, reached the Home Office on 1 November 1979 and was published on the 12th.[49] It was the fourth version of the IBA's plans for the fourth channel in eight years. The Authority now proposed to distance itself from the running of the channel. It intended to establish a company with its own board of twelve to fourteen directors, eleven of whom would be non-executive and chosen by the Authority. There would be an independent Chairman and Deputy Chairman. The other members would broadly represent those who were likely to provide programmes for the service, but they would not be delegates. Four would come from the ITV companies, which would be providing some of the programmes and all the money. The Authority would not itself be represented on the board, which would enjoy a substantial measure of independence: subject to the Authority's ultimate control through the appointment of members of the board, approval of schedules, co-ordination of planning and scheduling between its two channels, and the establishment of an annual budget.

The new channel was to have its own distinctive character but not to be an exclusively minority service, and ITV was not to move in the direction of an unrestrained search for popularity: 'Rather, we would see the present mix on ITV's single channel continuing, while the fourth channel roughly reversed that mix with about two-thirds of its programmes addressing sections of the audience who want something particular or who want something different. In the remaining one-third there would be programmes intended to appeal to larger audiences, though often in a style different from that of some popular programmes now seen.'[50]

The overriding concern of the programme controller would be with the quality of programmes, said the Authority in these proposals, and there should therefore be no quotas or rights to contribute. However, a possible pattern at the start would be 15–35 per cent of the output coming from independent producers, 25–40 per cent from the major ITV contractors, a further 10–20 per cent from the regional contractors, up to 15 per cent from ITN, and 5–14 per cent from foreign sources. Subject to three exceptions, it was not intended to prescribe particular categories of programme. The exceptions were a regular provision for ITN news bulletins and programmes providing background information on matters of public interest; at least an hour a week of programmes recognisably religious in aim; and educational programmes amounting to about 15 per cent of the channel's output.[51]

Forecasts of the annual budget suggested £60–80 million at 1979 prices, and this would be raised from the ITV companies as a fourth channel subscription, divided between them in roughly the same proportion as their IBA rentals. On the issue of the sale of airtime the IBA remarked that it had always been clear that a channel of the kind now envisaged was not compatible with competitive selling. The advertisers' problems stemmed mainly from the necessary limitation on the amount of airtime available, and a separate sales force would produce only the appearance, not the reality, of competition. The Authority had formed the opinion that, on balance, it was preferable for time on the fourth channel to be sold by the ITV companies in their own regions, but it was prepared to consider safeguards: for example, that there should be no conditions relating the sale of ITV airtime to the sale of fourth channel airtime; that there should be no linked discounts between the two channels; and that separate rate cards should be published. To alleviate discontent among the paying customers, the Advertising Liaison Committee was proposed.[52]

This prescription did little to appease the advertisers; while the Channel Four Group suspected that the companies had not been put down firmly enough in some of the proposals. The Group took exception to the idea of a representative board. It believed that the four ITV members would somehow rule the roost, although a minority. It wanted discrimination in favour of independent production. Sceptical about the Authority's repeated assertion that the fourth channel would not be treated as a junior partner to ITV in scheduling, it regarded complementary scheduling as a positive drawback. The companies, on the other hand, while accepting that they would not be calling the tune, disliked the prospect of forced payments to a piper who might become a permanent pensioner. Experience had taught that an economic recession could cause problems enough for one advertising-financed channel, let alone two.

All through December 1979 and January 1980, while the new Broadcasting Bill was gestating in Whitehall, the IBA listened to a dissonance of voices critical of almost every aspect of its plan. To clear a few minds about the logic behind the Authority's thinking on financing the channel, Sir Brian Young composed an algorithm. It was published in *Broadcast* on 12 December and is reproduced here overleaf.

When the Bill was finally published on 6 February 1980 it was seen that the Authority's proposal had carried the day, winning unqualified approval from a government weary of the skirmishing. Indeed it almost seemed as though the Authority had been co-opted into the Home Office to frame the legislation for what was now described non-committally as 'Service 2'. During the ensuing debates the advertising lobby won only a minor amendment: that any complaint received by the Authority about the conduct of the companies in the sale of airtime, together with any action taken by the Authority, should be recorded in its Annual Report. The independent producers, lobbying mainly but not exclusively through the Channel Four Group, gained no concessions at all; but their most important demands in preventing the ITV companies from dominating the channel had been met.

Unexpectedly, a third lobby was so successful that it got a whole section of the Bill rewritten. This was Plaid Cymru and its sympathisers, irate at the jettisoning of the separate fourth channel for Wales promised in the Conservative Party manifesto. In an outbreak of civil disobedience in the Principality transmitters were raided, programmes were blacked out and more than 2,000 viewers refused to pay licence fees. In April Gwynfor Evans, the 68-year-old President of Plaid Cymru, took the extreme step of threatening a hunger strike. Unless the government changed its mind, he announced, he would begin starving himself to death in October.

The threat of what might happen if this Welsh Gandhi were to die was taken seriously. Uncharacteristically the government decided to capitulate. In September, by which time the Bill had reached the Lords, the Home Secretary announced that amendments would be tabled to provide for the establishment of a separate Authority to supervise a single-channel Welsh television service, with the proviso that there could be a return to the two-

The IBA's Fourth Channel Algorithm

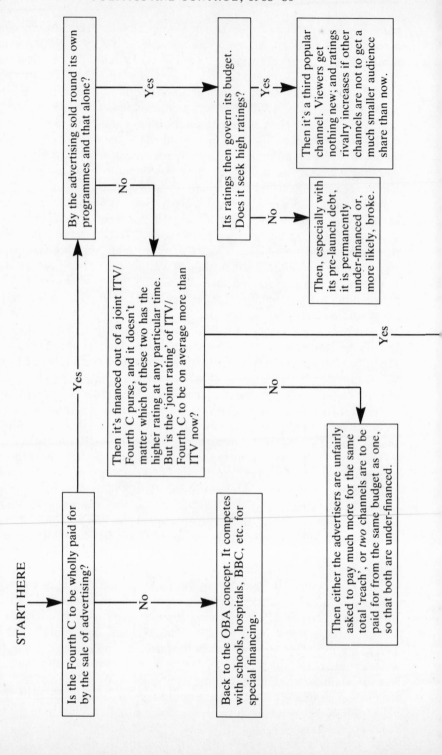

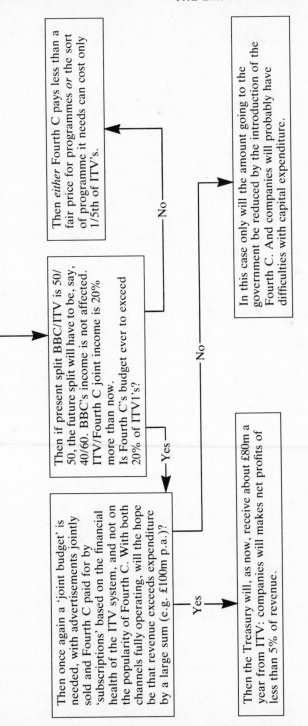

The flow chart reads (rotated 90°):

Then once again a 'joint budget' is needed, with advertisements jointly sold and Fourth C paid for by 'subscriptions' based on the financial health of the ITV system, and not on the popularity of Fourth C. With both channels fully operating, will the hope be that revenue exceeds expenditure by a large sum (e.g. £100m p.a.)?

— Yes →

Then the Treasury will, as now, receive about £80m a year from ITV: companies will makes net profits of less than 5% of revenue.

Then if present split BBC/ITV is 50/50, the future split will have to be, say, 40/60. BBC's income is not affected. ITV/Fourth C joint income is 20% more than now.
Is Fourth C's budget ever to exceed 20% of ITV1's?

— Yes →

— No →

Then either Fourth C pays less than a fair price for programmes or the sort of programme it needs can cost only 1/5th of ITV's.

— No →

In this case only will the amount going to the government be reduced by the introduction of the Fourth C. And companies will probably have difficulties with capital expenditure.

channel scheme should it later prove to be in the best interests of the Welsh language and of broadcasting in Wales generally. In making this, its second volte-face on the subject, the government was decisively influenced by a strong deputation to the Home Secretary and Secretary of State for Wales led by Lord Cledwyn of Penrhos who, as Cledwyn Hughes, had been MP for Anglesey, Secretary of State for Wales and Chairman of the Parliamentary Labour Party.

The amendments tabled by the government in the Lords provided for a Welsh Fourth Channel Authority (Chairman and four members) to be appointed by the Home Secretary. This new Authority would be responsible for supplying the IBA (which would be the transmitting body) with all the programmes broadcast on the Fourth Channel in Wales. It was expected to obtain its programmes from, first, the BBC (which was under a duty to supply – free of charge – the same amount of Welsh-language programming as it would have broadcast under previous arrangements); secondly, the local ITV company, which had to supply a 'reasonable proportion' of Welsh-language material on commercial terms; and thirdly, independent producers. A 'substantial proportion' of the programmes had to be in Welsh, and between 6.30 and 10 p.m. mainly so. When Welsh programmes were not being broadcast, the programmes, although formally supplied by the Welsh Authority, would normally be the same as those being broadcast on the Fourth Channel elsewhere in the United Kingdom.

This unforeseen body was not to be a little brother of the ITV companies or of the new Fourth Channel company. The IBA was not to be its parent. Its functions were envisaged as acquiring and scheduling rather than producing programmes, but it was to be autonomous, with a direct responsibility for complying with the statutory requirements in respect of all the programmes which it supplied to the IBA for transmission.

Revenue for the service would come from the IBA, which would collect it in the form of rentals from all the ITV companies. The amount had to be agreed between the IBA and the new Welsh Authority or might be fixed by the Home Secretary in default of agreement, and the Home Secretary might take it into account when fixing the rate of the levy on television companies' profits. Advertisements would be sold by, and the revenue accrue to, the local ITV company.[53]

Thus, under these hastily contrived arrangements, the ITV companies everywhere in the United Kingdom were obliged to pick up a share of the bill. With its income assured by law, Sianel Pedwar Cymru (S4C) was granted the means to survive where, in different circumstances nearly twenty years earlier, Teledu Cymru had failed.[54] Because of the very small size of audiences for Welsh-language programmes it was foreseen that this would be the most expensive television service per viewer in the world, and among the government's own back-benchers it was felt that a language and culture in peril could have been rescued without this unique departure from the pattern of British broadcasting and what the Home Secretary freely admitted to have been abject surrender to blackmail.

The Broadcasting Bill which established the two fourth channels received the Royal Assent and became law on 13 November 1980. Its passage was anticipated by the Authority which, to avoid further delay, had already announced the names of those who were to take charge of its 'TV4' subsidiary company. The constitutional proprieties were observed by describing the board initially as a panel and its members as consultants.

Edmund Dell, a former Labour minister and city financier, was appointed Chairman, and Sir Richard Attenborough, the film actor, producer and director, Deputy Chairman. The members formed a judicious blend of ITV2 men and OBAites: William Brown (Managing Director, Scottish Television), Roger Graef (independent producer and director), Dr Glyn Tegai Hughes (the BBC's National Governor for Wales 1971–79), David McCall (Chief Executive, Anglia Television), Sara Morrison (a member of the Annan Committee), Anthony Smith (Director of the British Film Institute), Anne Sofer (member of the Inner London Education Authority), Brian Tesler (Managing Director, London Weekend Television) and Joy Whitby (Head of Children's Programmes, Yorkshire Television). They were appointed for three, four or five-year terms at fees of £2,000 a year.

The IBA had been keen to bury hatchets and signal a coalescence of opposing views by inviting Anthony Smith and Sara Morrison to join the board. The companies were doubly disappointed. Of the four, originally five, members to be drawn from their ranks, one was not a chief executive as they had expected. The composition of the board was the clearest confirmation that the companies were not to be in the driving seat.

Towards the end of September the board chose the Chief Executive who was to be responsible for the Channel Four Television Company's operations and programme output. Jeremy Isaacs was appointed for a five-year term from 1 September 1981. It was a widely expected and welcomed appointment which assured the success of the new enterprise. Aged 48, Isaacs had spent his working life in television. He had produced *This Week* for Associated Rediffusion, *Panorama* for the BBC and *World at War* for Thames, where he had been Director of Programmes from 1974 to 1978. His private submission to the Ministry of Posts and Telecommunications in 1973, outlining what form the fourth channel should take, had proved an uncannily accurate prediction.[55] In August 1979 he had taken the opportunity to develop his ideas in the McTaggart Memorial Lecture at the Edinburgh International Television Festival. Channel Four would, he promised on his appointment, enlarge the scope of British broadcasting, increase choice for viewers, encourage innovation and broadcast voices from the widest possible spectrum of opinion; it would release new creative energy and add salt to the porridge.

Paul Bonner, a 'dark horse' contender for the top job, had acquitted himself so impressively when interviewed in competition with Isaacs that he was offered a post which had not been advertised. A BBC career man who had been editor of the Community Programme Unit and Head of Science and Features, he was appointed Channel Controller – Isaac's deputy in all programme matters. The other leading contender for Isaacs' job had been John Birt, the Controller of Features and Current Affairs at London Weekend Television who (with David Elstein of Thames) had also in 1973 submitted a blueprint envisaging a structure for the fourth channel remarkably similar to Channel Four's.

In December the new board, now formally constituted and including Isaacs, received from the Authority a memorandum defining and describing its terms of reference, together with a Programme Policy Statement charging it to serve 'special interests and concerns for which television has until now lacked adequate time'.[56] Fifteen per cent of broadcasting time was to be devoted to educational material, the work of independent producers was to be encouraged, and 'due' contributions from ITV's regional companies were expected. The dilemma of complementarity remained unresolved, the Authority settling for equality of status between its two channels without specifying the mechanism of co-ordination. 'The Fourth Channel,' it declared, 'is intended to complement and to be complemented by the present Independent Television service. Complementarity means two things: the provision of reasonable choice between two schedules, with a number of common junctions; and the co-ordinated use of the schedules in the best interest of the viewer.'

Isaacs interpreted this as permitting Channel Four to draw up its own schedule independently. It would know the basic shape of the ITV schedule and its object would be to offer alternatives, but it would not act slavishly in order to be distinctive. He promised close liaison with ITV's programme controllers and planners. At some points there would be common junctions; at other times ITV junctions would be bypassed to allow Channel Four programmes to vary in length. Any serious clashes could be eliminated by the IBA when the final schedules came to be compared.[57]

When it came on air in 2 November 1982 Channel Four was the first new television service in the United Kingdom since the launch of BBC2 in 1964 and the first to begin with almost full national coverage. Its financing was secure; no new Authority had been created; its programming was under the control, not of the ITV companies, but of its own, independently-minded Chief Executive. It was the major achievement of Brian Young's and Colin Shaw's periods of office. As Young told IBA staff shortly before his retirement at the end of the same month: 'The channel's final structure embodies ten years of IBA thought and action to produce a service designed to enlarge and enrich the one with which we began in 1955'.[58]

But, as another senior figure in ITV, Lord Windlesham, pointed out,[59] the Whitelaw compromise on the fourth channel satisfied none of the interested parties except the IBA: not the Channel Four Group, who still wanted an OBA; nor the ITV companies, who wanted an ITV2; nor the advertisers, who wanted the companies' monopolies broken; nor the Treasury, which was to forgo some hundreds of millions of pounds in levy and taxation for as long as the ITV companies' Channel Four subsidies reduced their profits.

Despite the current vogue for consultation and accountability this was not a subject on which the public mind was taken, but it seems sensible to suppose that a co-ordinated two-channel service in double harness, like BBC1 and BBC2, would have served the viewer better. Fully complementary scheduling by a single body had been central to the concept of an ITV1 and ITV2 which the IBA had advocated for so long, but it was sacrificed to Channel Four's full independence, even though, as the Authority had commented on the Annan report: 'To turn away from complementary scheduling is to narrow choice'.

On the first page of the first volume of this history a former Lord Chancellor – Lord Kilmuir – is quoted as describing the origin and genesis of institutional changes as 'complex, complicated, sometimes long in gestation and sometimes... extremely painful in parturition'. Channel Four was far longer in gestation and no less painful in parturition than Independent Television itself. Swings in the political pendulum twice altered the course of events: first when a change of government in 1964 prevented the introduction of a service based upon direct competition between two sets of ITA-appointed companies operating in the same areas; and again when in 1979 another change of government thwarted the establishment of a third broadcasting authority to introduce a separately conducted service. In the event the duopoly established in 1954 survived substantially intact. Through the compromise of Channel Four the Authority was awarded its second channel, and the companies were fortunate both in retaining their monopolies and in obtaining a new outlet through which to enlarge the scope of their programme-making.

The passage of the Broadcasting Act 1980 was quickly followed by the award of new ITV contracts to run for eight years from 1 January 1982. An account of the events which culminated in the final act of judgment on 28 December 1980 will be found in the next volume of this history.

APPENDIX A

GOVERNMENTS AND MINISTERS RESPONSIBLE FOR BROADCASTING, 1954–80

Party	Prime Minister	Date of Appointment	Postmaster-General	Date of Appointment
Conservative	Sir Winston Churchill	26 October 1951	Earl De La Warr	5 November 1951
	Sir Anthony Eden	6 April 1955	Dr Charles Hill	7 April 1955
	Harold Macmillan	10 January 1957	Ernest Marples	16 January 1957
			Reginald Bevins	22 October 1959
	Sir Alec Douglas-Home	18 October 1963		
Labour	Harold Wilson	16 October 1964	Anthony Wedgwood Benn	19 October 1964
			Edward Short	4 July 1966
			Roy Mason	6 April 1968
			John Stonehouse	1 July 1968
			Minister of Posts and Telecommunications	
Conservative	Edward Heath	19 June 1970	John Stonehouse	1 October 1969
			Christopher Chataway	24 June 1970
			Sir John Eden	7 April 1972
Labour	Harold Wilson	4 March 1974	Anthony Wedgwood Benn	7 March 1974
			Home Secretary	
			Roy Jenkins	30 March 1974
			Merlyn Rees	10 September 1976
	James Callaghan	5 April 1976		
Conservative	Margaret Thatcher	4 May 1979	William Whitelaw	5 May 1979

APPENDIX B

MEMBERS OF THE AUTHORITY, 1954–80

Chairmen	Terms of Office	Background
Sir Kenneth Clark KCB (later Lord Clark OM,CH)	4 August 1954–31 August 1957	Former Director of the National Gallery and Surveyor of the King's Pictures; Chairman, Arts Council of Great Britain
Sir Ivone Kirkpatrick GCB,GCMG	7 November 1957–6 November 1962	Former UK High Commissioner for Germany; former Permanent Under-Secretary, Foreign Office
Rt Hon. Lord Hill of Luton PC	1 July 1963–30 August 1967	Former Secretary, British Medical Association; former Postmaster-General; former Chancellor of Duchy of Lancaster; former Minister of Housing & Local Government and Minister for Welsh Affairs
Rt Hon. Lord Aylestone PC,CH,CBE	1 September 1967–31 March 1975	Former President of the Council and Leader of the House of Commons; former Secretary of State for Commonwealth Affairs
Lady Plowden DBE	1 April 1975–31 December 1980	Former Governor and Vice-Chairman of the BBC; former Chairman of Central Advisory Council for Education (England); President, Pre-School Playgroups Association

Deputy Chairmen	Term of Office	Background
Sir Charles Colston	4 August 1954–9 December 1954	Former Chairman and Managing Director, Hoover Ltd
Sir Ronald Matthews JP, DL (Acting Chairman 1 September 1957–6 November 1957)	3 January 1955–1 July 1959	Chairman & Managing Director, Turton Bros & Matthews Ltd; Director, Thos. Cook & Son
Sir John Carmichael KBE (Deputy Chairman 22 June 1960–29 July 1964; Acting Chairman 7 November 1962–30 June 1963)	8 January 1960–29 July 1964	Former Member of Sudan Government Civil Service; Member, Scottish Gas Board
Sir Sydney Caine KCMG (Deputy Chairman 30 July 1964–30 June 1967)	4 August 1960–30 June 1967	Director, London School of Economics
Sir Ronald Gould	1 July 1967–30 June 1972	General Secretary, National Union of Teachers
Mr Christopher Bland	1 July 1972–31 December 1979	Chairman, Sir Joseph Causton & Sons Ltd; former Member, GLC; former Chairman of Bow Group
Rt Hon. Lord Thomson of Monifeith KT, PC (Chairman from 1 January 1981)	1 February 1980–31 December 1988	Former Chancellor of Duchy of Lancaster and Minister for Europe; former Secretary of State for Commonwealth Affairs; former EEC Commissioner; former Chairman, Advertising Standards Authority

Members for Scotland	Terms of Office	Background
Dr T. J. Honeyman JP	4 August 1954–3 August 1958	Former Director, Glasgow Art Gallery; Rector, Glasgow University
Professor David Talbot Rice MBE	4 August 1958–3 August 1963	Watson-Gordon Professor of History of Fine Art, Edinburgh University

Mr W. Macfarlane Gray OBE, JP	22 January 1964–21 January 1970	Senior Partner, Macfarlane Gray & Co. (accountants); Hon. Sheriff for Stirling and Clackmannan
Dr T. F. Carbery	22 January 1970–31 July 1979	Head of the Department of Office Organisation, Strathclyde University
Rev. Dr W. J. Morris JP	1 August 1979–31 December 1984	Minister of Glasgow Cathedral and a Chaplain to HM The Queen in Scotland

Members for Wales

Lord Aberdare of Duffryn GBE	4 August 1954–3 August 1956	Former Chairman, National Advisory Council for Physical Training & Recreation; Chairman, National Association of Boys' Clubs
Mr J. Alban Davies	4 August 1956–29 July 1964	Director, Hitchman's Dairies; former High Sheriff of Cardiganshire
Sir Ben Bowen Thomas	30 July 1964–29 July 1970	Former Permanent Secretary, Welsh Department, Ministry of Education; Member, Welsh panel of British Council
Mr T. Glyn Davies CBE	10 October 1970–9 October 1975	Former Director of Education for Denbighshire; former Chairman, Schools Broadcasting Council for Wales; Chairman, Welsh Committee, Central Council of Physical Education
Professor Huw Morris-Jones	25 March 1976–24 March 1982	Former member of BBC Broadcasting Council for Wales; Professor of Social Theory and Institutions, University College of North Wales, Bangor

Members for Northern Ireland

Lt-Col. Arthur Chichester OBE, MC	4 August 1954–31 December 1959	Former Clerk of Parliaments, Northern Ireland and Mayor of Ballymena; Chairman, Moygashel Ltd

	Term of Office	Background
Members for Northern Ireland		
Sir Lucius O'Brien	1 January 1960–29 July 1965	Member, Northern Ireland Council of Social Service; Senator, Queen's University, Belfast; Vice President, Trustee Savings Bank Association
Mr D. J. T. Gilliland	8 November 1965–7 November 1970	Solicitor and farmer
Mr H. W. McMullan OBE	11 July 1971–31 January 1974	Former journalist, broadcaster and Head of Programmes, BBC Northern Ireland
Mr W. J. Blease JP (later Lord Blease of Cromac)	10 May 1974–31 July 1979	Northern Ireland Officer of Irish Congress of Trade Unions
Mrs Jill McIvor	1 January 1980–31 December 1986	Former Librarian, Department of Director of Public Prosecutions, Belfast
Other Members		
Sir Henry Hinchcliffe JP, DL	4 August 1954–3 August 1959	Chairman, Glazebrook Steel & Co. Ltd; Director, Barclays Bank; Director, Manchester Chamber of Commerce
Miss M. E. Popham CBE	4 August 1954–3 August 1956	Principal, Cheltenham Ladies' College
Miss Dilys Powell	4 August 1954–3 August 1957	Film Critic, *Sunday Times*
Mr G. B. Thorneycroft CBE	4 August 1954–3 August 1958	Former General Secretary, Transport Salaried Staffs Association
Lord Layton CH, CBE	11 August 1954–3 August 1956	Chairman, *The Economist*; Director and Vice-Chairman, Daily News Ltd; Deputy Leader, Liberal Party in House of Lords
Miss Diana Reader Harris (later Dame Diana Reader Harris DBE)	4 August 1956–3 August 1960	Headmistress, Sherborne School for Girls

Mr T. H. Summerson JP	1 February 1957–3 August 1960	Chairman and Joint Managing Director, Summerson Holdings Ltd; former High Sheriff of Co. Durham
The Hon. Dame Frances Farrer DBE	4 September 1957–3 August 1961	General Secretary, National Federation of Women's Institutes
Mr W. B. Beard OBE	4 August 1958–3 August 1961	General Secretary, United Patternmakers' Association; Member, TUC General Council
Mrs Isabel Graham-Bryce (later Dame Isabel Graham-Bryce DBE)	4 August 1960–3 August 1965	Chairman, Oxford Regional Hospital Board
Mr A. C. Cropper JP, DL	11 October 1960–29 July 1964	Chairman, James Cropper & Co. Ltd; former High Sheriff of Westmorland
Lord Williamson of Eccleston CBE, JP	4 August 1961–29 July 1964	Former General Secretary, General and Municipal Workers Union
Dame Anne Bryans DBE	1 September 1961–31 August 1966	Deputy Chairman, British Red Cross Society
Baroness Burton of Coventry	30 July 1964–29 July 1969	Former Labour MP for Coventry South
Sir Patrick Hamilton Bt	30 July 1964–29 July 1969	Director, Simon Engineering Ltd; Chairman, Expanded Metal Co. Ltd
Professor Hugh Hunt	30 July 1964–29 July 1969	Professor of Drama, Manchester University
Professor Sir Owen Saunders	30 July 1964–29 July 1969	Professor of Mechanical Engineering, Imperial College, London; Dean, City and Guilds College
Sir Vincent Tewson CBE, MC	30 July 1964–29 July 1969	Former General Secretary, TUC
Mrs Mary Adams OBE	1 June 1965–31 May 1970	Former Assistant to Controller of Television Programmes, BBC; Deputy Chairman, Consumers' Association

Other Members	Term of Office	Background
Baroness Plummer JP	1 March 1966–28 February 1971	Widow of Sir Leslie Plummer MP
Baroness Sharp of Hornsey GBE	1 October 1966–30 September 1973	Former Permanent Secretary, Ministry of Housing and Local Government
Sir Frederick Hayday CBE	30 July 1969–29 July 1973	National Industrial Officer, National Union of General and Municipal Workers; Member, TUC General Council
Mr Stephen Keynes	30 July 1969–29 July 1974	Director, Charterhouse Japhet Ltd and Charterhouse Finance Corporation
Professor J. M. Meek	30 July 1969–29 July 1974	David Jardine Professor of Electrical Engineering, Liverpool University; President, Institution of Electrical Engineers 1968–69
Mr A. W. Page MBE (later Sir Alex Page)	1 August 1970–31 July 1976	Chairman and Managing Director, Metal Box Company Ltd
Baroness Macleod of Borve JP	1 March 1971–31 December 1975	Widow of Rt Hon. Ian Macleod MP
Mr W. C. Anderson CBE	30 July 1973–29 July 1978	Former General Secretary, NALGO; former Member, TUC General Council
Mrs Mary Warnock (later Baroness Warnock)	17 December 1973–31 December 1981	Former Headmistress, Oxford High School; Chairman, Committee of Inquiry into Special Education
Professor James Ring	2 September 1974–31 December 1981	Professor of Physics, Imperial College, London; former member and Chairman, IBA General Advisory Council
Baroness Stedman OBE	25 June 1975–4 December 1975	Vice-Chairman, Cambridgeshire County Council; Member, Peterborough Development Corporation Board

313

The Marchioness of Anglesey CBE	1 January 1976–31 December 1981	Former Chairman, National Federation of Women's Institutes; Chairman, Welsh Arts Council
Mrs Ann Coulson	1 January 1976–31 December 1980	Assistant Director, N. Worcestershire College of Higher and Further Education; Member, Birmingham District Council
Mr A. J. R. Purssell	1 October 1976–30 September 1981	Managing Director, Arthur Guinness Son & Co. Ltd
Mr A. M. G. Christopher	18 September 1978–21 December 1983	General Secretary, Inland Revenue Staff Federation; Member, General Council TUC
Mr George Russell	1 August 1979–31 December 1986	Managing Director and Chief Executive, British Alcan Aluminium; Member, Northern Industrial Development Board

APPENDIX C

RULES ON HOURS OF BROADCASTING FROM EACH TELEVISION STATION OF THE ITA

1. Broadcasting hours shall not exceed:
   (i) 53½ hours a week;
   (ii) 8½ hours on any one day.
2. The following broadcasts shall not be taken into account for the purposes of rule 1:
   (i) Religious programmes (namely acts of worship from a church or studio, and other programmes which the Authority, with the advice of the Central Religious Advisory Committee, approves for broadcasting under the terms of Section 3(5)(a) of the Television Act 1964);
   (ii) Ministerial and party political broadcasts;
   (iii) Outside broadcasts, up to a maximum of 450 hours in any one calendar year, of events which are neither devised nor promoted by the Authority, or its programme contractors, of which up to 120 hours may be in recorded form provided they are broadcast within 8 days of the event recorded;
   (iv) Any broadcasting in excess of 8½ hours on Christmas Eve, Christmas Day, Boxing Day, New Year's Eve or New Year's Day, and in Scotland the day following New Year's Day designated a Public Holiday in Scotland;
   (v) School broadcasts (namely:
       (a) programmes created at the request of the Authority's Educational Advisory Council for showing on school days between the hours of 9 a.m. and 4.45 p.m., and
       (b) other programmes, up to a maximum of 25 hours in one calendar year,
       requested or assented to by the Educational Advisory Council on educational grounds, provided such programmes are transmitted on school days between the hours of 9 a.m. and 4.45 p.m. for viewing in schools and other places where children of school age are educated);
   (vi) Repeats, for up to 1 hour daily or not more than a total of 10 days during school holidays, of school broadcasts;
   (vii) Adult Education broadcasts (namely programmes – other than school broadcasts – arranged in series and specifically planned in consultation with and with the approval of the Authority's Educational Advisory Council for the purpose of helping viewers towards a progressive mastery or understanding of some skill or body of knowledge);
   (viii) Broadcasts in the Welsh language;

314

(ix) Broadcast of a half-hour 'parade' of new advertisements, made before 12 o'clock noon on one day a week (other than Sunday).

3. On Sundays:
   There shall be no broadcasting between 6.15 p.m. and 7.25 p.m.; but this rule does not apply to religious programmes as defined in rule 2(i); to appeals for charitable or benevolent purposes; to live outside broadcasts of events which are neither devised nor promoted by the Authority or its programme contractors; to programmes in the Welsh language; to programmes designed specially for the deaf.

4. None of the above rules applies to engineering test transmissions, trade test transmissions, or tuning signals.

5. These rules supersede the previous rules.

6. These rules are subject to any future direction by the Postmaster-General under Section 17 of the Television Act 1964.

28 February 1969

APPENDIX D

THE 1947 AIDE-MÉMOIRE

1. It is desirable that political broadcasts of a controversial character shall be resumed.

2. In view of their responsibilities for the care of the nation the government should be able to use the wireless from time to time for ministerial broadcasts which, for example, are purely factual, or explanatory of legislation or administrative policies approved by parliament; or in the nature of appeals to the nation to co-operate in national policies, such as fuel economy or recruiting, which require the active participation of the public. Broadcasts on state occasions also come in the same category.

    It will be incumbent on ministers making such broadcasts to be as impartial as possible, and in the ordinary way there will be no question of a reply by the opposition. Where, however, the opposition think that a government broadcast is controversial it will be open to them to take the matter up through the usual channels with a view to a reply.
      (i) As a reply if one is to be made should normally be within a very short period after the original broadcast, say three days, the BBC will be free to exercise its own judgment if no agreement is arrived at within that period.
     (ii) Replies under this paragraph will not be included in the number of broadcasts provided for under paragraph 4.
    (iii) Copies of the scripts of broadcasts under this paragraph shall be supplied to the leaders of each party.
    (iv) All requests for ministerial broadcasts under this paragraph shall be canalised through the minister designated for this purpose – at present the Postmaster-General.

3. 'Outside' broadcasts, e.g. of speeches at party conferences which are in the nature of news items, shall carry no right of reply by the other side.

4. A limited number of controversial party political broadcasts shall be allocated to the various parties in accordance with their polls at the last general election. The allocation shall be calculated on a yearly basis and the total number of such broadcasts shall be a matter for discussion between the parties and the BBC.

5. The opposition parties shall have the right, subject to discussion through the usual channels, to choose the subjects for their own broadcasts. Either side will be free, if it wishes, to use one of its quota for the purpose of replying to a previous broadcast, but it will be under no necessity to do so. There will, of course, be no obligation on a party to use its whole quota.

6.   (i) Paragraphs 4 and 5 relate to controversial party political broadcasts on issues of major policy on behalf of the leading political parties. For the ensuing year the total number, excluding budget broadcasts, shall be 12 – divided as to government 6, Conservative opposition 5, Liberal opposition 1. Reasonable notice will be given to the BBC.

    (ii) The BBC reserve the right, after consultation with the party leaders, to invite to the microphone a member of either House of outstanding national eminence who may have become detached from any party.

    (iii) Apart from these limited broadcasts on major policy the BBC are free to invite members of either House to take part in controversial broadcasts of a round-table character in which political questions are dealt with, provided two or more persons representing different sides take part in the broadcasts.

    (iv) No broadcasts arranged by the BBC other than the normal reporting of parliamentary proceedings are to take place on any question while it is the subject of discussion in either House.

7.   Where any dispute arises an effort shall be made to settle it through the usual channels. Where this is not possible, the BBC will have to decide the matter on its own responsibility.

8.   These arrangements shall be reviewed after a year, or earlier if any party to the conference so desires.

6 February 1947

    The above agreement was revised in July 1948, when government and opposition agreed with the BBC that 6(iv) should be construed as:

    (a) that the BBC will not have discussions or *ex parte* statements on any issues for a period of a fortnight before they are debated in either House;

    (b) that while matters are subjects of legislation MPs will not be used in such discussions.

APPENDIX E

THE 1969 AIDE-MÉMOIRE

1. In view of its executive responsibilities the government of the day has the right to explain events to the public, or seek co-operation of the public, through the medium of broadcasting.

2. Experience has shown that such occasions are of two kinds and that different arrangements are appropriate for each.

3. The first category relates to ministers wishing to explain legislation or administrative policies approved by parliament, or to seek the co-operation of the public in matters where there is a general consensus of opinion. The BBC will provide suitable opportunities for such broadcasts within the regular framework of their programmes; there will be no right of reply by the opposition.

4. The second category relates to more important and normally infrequent occasions, when the Prime Minister or one of his senior cabinet colleagues designated by him wishes to broadcast to the nation in order to provide information or explanation of events of prime national or international importance, or to seek the co-operation of the public in connection with such events.

5. The BBC will provide the Prime Minister or cabinet minister with suitable facilities on each occasion in this second category. Following such an occasion they may be asked to provide an equivalent opportunity for a broadcast by a leading member of the opposition and will in that event do so.

6. When the opposition exercises this right to broadcast, there will follow as soon as possible, arranged by the BBC, a broadcast discussion of the issues between a member of the cabinet and a senior member of the opposition nominated respectively by the government and opposition but not necessarily those who gave the preceding broadcasts. An opportunity to participate in such a discussion should be offered to a representative of any other party with electoral support at the time in question on a scale not appreciably less than that of the Liberal Party at the date of this *aide-mémoire*.

7. As it will be desirable that such an opposition broadcast and discussion between government and opposition should follow the preceding broadcast with as little delay as possible, a request for the necessary facilities by the

opposition should reach the BBC before noon on the day following the ministerial broadcast. This will enable the BBC to arrange the opposition broadcast and the discussion as soon as possible.

8. Copies of the scripts of these broadcasts will be supplied to the leaders of the government, the opposition and of other parties when they participate.

9. These arrangements will be reviewed annually.

25 February 1969

LIST OF ABBREVIATIONS

ABC	ABC Television
ABS	Association of Broadcasting Staff
ABPC	Associated British Picture Corporation
ACTT	Association of Cinematograph and Television Technicians
AGB	Audits of Great Britain
AI	Appreciation Index
AMMC	Audience Measurement Management Committee
ARMC	Audience Reaction Management Committee
ATV	ATV Network
AURA	Audience Reaction Assessment
BAFTA	British Academy of Film and Television Arts
BARB	Broadcasters' Audience Research Board
BBC	British Broadcasting Corporation
BET	British Electric Traction Company
BICC	British Insulated Callender's Cables
BMRB	British Market Research Bureau
CRAC	Central Religious Advisory Committee
DAME	Digital Automatic Measuring Equipment
DBS	Direct Broadcasting by Satellite
DDG (AS)	Deputy Director General (Administrative Services)
DDG (PS)	Deputy Director General (Programme Services)
DEC	Disaster Emergency Committee
DG	Director General
DHSS	Department of Health and Social Security
DICE	Digital Intercontinental Conversion Equipment
EEC	European Economic Community
EMI	Electrical and Musical Industries
ENG	Electronic News Gathering
ETU	Electrical Trades Union
FCC	Federal Communications Commission
GAC	General Advisory Council
GEC	General Electrical Company
GPO	General Post Office
HTV	Harlech Television
IBA	Independent Broadcasting Authority
ICPC	Inter-Company Production Committee
ILEA	Inner London Education Authority
ILR	Independent Local Radio

IMF	International Monetary Fund
IPA	Institute of Practitioners in Advertising
IRA	Irish Republican Army
ISBA	Incorporated Society of British Advertisers
ITA	Independent Television Authority
ITCA	Independent Television Companies Association
ITN	Independent Television News
ITP	Independent Television Publications
ITV	Independent Television
JICTAR	Joint Industry Committee for Television Advertising Research
LWT	London Weekend Television
MAC	Multiplex Analogue Components
MP	Member of Parliament
NAR	Net Advertising Revenue
NARAL	Net Advertising Revenue After Levy
NATO	North Atlantic Treaty Organisation
NBPI	National Board for Prices and Incomes
NPC	Network Planning Committee
NTSC	National Television System Committee
NVALA	National Viewers' and Listeners' Association
OBE	Order of the British Empire
PAL	Phase Alternation Line
PAR	Predicted Audience Rating
PEB	Party Election Broadcast
PMG	Postmaster-General
PPB	Party Political Broadcast
PPC	Programme Policy Committee
PR	Public Relations
RAI	Radiotelevisione Italiana
RIBA	Royal Institute of British Architects
SCC	Standing Consultative Committee
SCNI	Select Committee on Nationalised Industries
SECAM	Sequential Couleur A'Memorie
SNP	Scottish National Party
STV	Scottish Television
S4C	Sianel Pedwar Cymru
TAC	Television Advisory Committee
TAM	Television Audience Measurement
TAPE	Television Audience Programme Evaluation
TOP	Television Opinion Panel
TUC	Trades Union Congress
TWW	Television Wales and West of England
UHF	Ultra High Frequency
UK	United Kingdom
UKIB	United Kingdom Independent Broadcasting
USA	United States of America
UTV	Ulster Television
VCR	Video Cassette Recorder
VHF	Very High Frequency
VTR	Video Tape Recorder
YTV	Yorkshire Television

REFERENCES

CHAPTER 1: TELEVISION AND SOCIETY IN THE 1970s

1. *Report of the Broadcasting Committee 1949* (HMSO) (Cmd. 8116) paras. 27 and 33.
2. *Report of the Committee on Broadcasting 1960* (HMSO) (Cmnd. 1753) para. 102.
3. Peter Black, *The Mirror in the Corner: People's Television* (Hutchinson, 1972) p. 186.
4. Sir Hugh Greene, *The Third Floor Front* (The Bodley Head, 1969) pp. 131–32.
5. *Television*, Journal of the Royal Television Society, October 1986 p. 224.
6. *Second Report from the Select Committee on Nationalised Industries* (HMSO, August 1972) Minutes of Evidence para. 320.
7. Asa Briggs, *Governing the BBC* (British Broadcasting Corporation, 1979) pp. 247–248.
8. *Independent Broadcasting* 4 (Independent Broadcasting Authority, May 1975) p. 14.
9. Robin Day, *Day by Day* (William Kimber, 1975) p. 112.
10. *The Times*, 28 February, 30 September and 1 October 1975.
11. Letter in *The Times*, 17 October 1975.
12. See David Glencross, Birt and Jay: A Case for Understanding in *Independent Broadcasting* 8, June 1976, pp. 10–13.
13. Christopher Booker, *The Seventies* (Allen Lane, 1980) p. 164.
14. Merlyn Rees, Speech reported in *Independent Broadcasting* 10, December 1976, pp. 2–4.
15. Arts Council of Great Britain Evidence to the Annan Committee on Broadcasting.
16. Christopher Booker, *The Seventies* pp. 122–23.
17. Mary Whitehouse, quoted in Phillip Whitehead, *The Writing on the Wall* (Brook Productions, 1985) p. 208.
18. House of Lords, Hansard (HMSO) 19 May 1977 Col. 938.
19. IBA Annual Report and Accounts 1979–80 p. 5.

CHAPTER 2: RECESSION AND THE LEVY

1. Volume 2 pp. 364–5.
2. Sir Sydney Caine, *Statement on TV Policy* A Supplement to Hobart Paper 43 (The Institute of Economic Affairs, July 1969).

3. *The Times*, 26 February 1968.
4. ITA Annual Report and Accounts 1968–69 p. 3.
5. ITA Paper 54(69).
6. Ibid.
7. Memorandum Director General to Chairman dated 11 February 1970 ITA File 29/1.
8. *Second Report from the Select Committee on Nationalised Industries* pp. 184 and 187.
9. ITA Annual Report and Accounts 1969–70 p. 2.
10. *Report 156 from the National Board for Prices and Incomes: Costs and Revenues of Independent Television Companies* (HMSO, October 1970) (Cmnd. 4524).
11. Ibid. para. 119 (5).
12. ITA, Note One of *Five Notes for the National Board for Prices and Incomes*, 4 September 1970.
13. Cmnd. 4524 para. 150.
14. Ibid. para. 154.
15. LWT Board Minutes 54/70.
16. ITV's Answer to Aubrey Jones, *Financial Times*, 17 December 1970.
17. ITA Annual Report and Accounts 1970–71 pp. 2–3.
18. House of Commons, Hansard (HMSO) 15 February 1971 Cols. 1211–12.
19. ITCA Statement, 15 February 1971: attachment to ITA Paper 9(71).
20. ITA Annual Report and Accounts 1970–71 pp. 78–79.
21. *Investors Chronicle*, 19 February 1971.
22. ITA Annual Report and Accounts 1971–72 p. 4.
23. SCC Minutes 154(71).
24. ITA Annual Report and Accounts 1971–72 p. 2.
25. SCC Paper 35(71).
26. Letter Director General to Minister of Posts and Telecommunications dated 9 November 1971 ITA File 29/2.
27. Appendix I to IBA Paper 3(74).
28. Ibid.
29. IBA Paper 139(73).
30. Letter Home Office to IBA dated 10 October 1977 IBA File 29.
31. Ibid. Letter IBA to Home Office dated 4 November 1977.
32. H of C 29 March 1974 Col. 815.
33. Appendix II to IBA Paper 41(80).

CHAPTER 3: TROUBLE AT LONDON WEEKEND

1. Lord Hill of Luton, *Behind the Screen* (Sidgwick and Jackson, 1974) p. 45.
2. London Television Consortium: An application for the award of ITA Programme Contract B (London Weekend), 11 April 1967.
3. In conversation with the author, 9 December 1986.
4. *Daily Mirror*, 20 November 1969.
5. Information from Michael Peacock in conversation with the author, 11 December 1986.
6. Letter Michael Peacock to ITA dated 1 May 1969 ITA File 7052/1A.
7. Ibid. Letter Lord Aylestone to Ben Whitaker MP dated 10 June 1969.
8. Ibid. Letter Lord Aylestone to Ben Whitaker MP dated 30 June 1969.
9. LWT Board Paper 21/69.
10. *The Times*, 19 September 1969.

11.   Memorandum Director General to Chairman dated 1 June 1970 ITA File
       7052/1.
12.   Ibid. Note for File by Director General dated 3 August 1970.
13.   IBA Paper 251(73).
14.   ITA File 7052/1.
15.   Ibid. Memorandum Director General to Chairman dated 4 November 1970.
16.   Ibid. Memorandum Director General to Chairman dated 3 February 1971.
17.   Memorandum Director General to Chairman dated 13 July 1970 ITA File
       7052/2.
18.   Television Act 1964 Sections 10 and 12.
19.   Memorandum DDG(AS) to Chairman dated 28 December 1970 ITA File
       7052/2.
20.   ITA Paper 142(70).
21.   LWT Board Minutes.
22.   Letter dated 19 February 1971 ITA File 7052/1.
23.   *The Times*, 25 February 1971.
24.   ITA Paper 21(71).
25.   *New York Herald Tribune*, 22 February 1971.
26.   Letter Brian Young to Rupert Murdoch dated 12 March 1971 ITA File
       7052/1.
27.   *Sunday Times*, 14 March 1971.
28.   ITA File 7052/1.
29.   *Independent Broadcasting* 12, July 1977, Appendix I.
30.   *Report of the Committee on the Future of Broadcasting* (HMSO, 1977)
       (Cmnd. 6753) para. 13.30.

CHAPTER 4: PARLIAMENTARY INVESTIGATION

1.   Edward Heath MP, H of C 3 December 1969 Cols. 1496–1634.
2.   *Second Report from the Select Committee on Nationalised Industries*, Minutes
      of Evidence paras. 967–94.
3.   *New Statesman*, 29 September 1972.
4.   *Second Report from the Select Committee on Nationalised Industries*, Minutes
      of Evidence para. 206.
5.   Ibid. para. 969.
6.   Ibid. para. 757.
7.   Ibid. para. 754.
8.   Ibid. para. 755.
9.   Ibid. para. 758.
10.   *Second Report from the Select Committee on Nationalised Industries*
      para. 148.
11.   Ibid.
12.   Ibid. paras. 59–60.
13.   Ibid. para. 65.
14.   Ibid. para. 70.
15.   Pilkington Report para. 45.
16.   Letter to David Wilson dated 3 October 1972 ITA File 3055/11.
17.   *Observer* leader, 1 October 1972.
18.   Letter to Hon. David Astor dated 2 October 1972 ITA File 3055/11.
19.   *Daily Telegraph*, 28 September 1972.
20.   *The Times*, 28 September 1972.
21.   *Daily Mirror*, 28 September 1972.

22.   *Financial Times*, 3 October 1972.
23.   *Independent Broadcasting Authority (formerly Independent Television Authority)* (HMSO, March 1973 (Cmnd. 5244) p. 7.
24.   Ibid. p. 8.
25.   Ibid. p. 9.
26.   Ibid.
27.   Ibid. p. 16.
28.   Cmnd. 5244.
29.   H of C May 1973 Col. 1538.

CHAPTER 5: THE FRUITS OF DERESTRICTION

1.   Volume 2 p. xiv.
2.   Ibid. p. 365.
3.   Letter dated 14 October 1970 ITA File 3006.
4.   ITA Annual Report and Accounts 1971–72 p. 2.
5.   PPC Minutes 31(72).
6.   IBA Annual Report and Accounts 1972–73 p. 9.
7.   Letter Lord Aylestone to Christopher Chataway MP dated 2 June 1971 ITA File 3006.
8.   GAC Paper 12(73).
9.   IBA Annual Report and Accounts 1972–73 p. 9.
10.   Ibid.
11.   IBA Annual Report and Accounts 1973–74 p. 12.
12.   See Burton Paulu, *Television and Radio in the United Kingdom* (The Macmillan Press Ltd., 1981) p. 262 and pp. 265–69.
13.   *Television & Radio 1977* (Independent Broadcasting Authority/Independent Television Publications Ltd.) p. 63.
14.   Letter Ministry of Posts and Telecommunications to IBA dated 14 December 1973 IBA File 3006.
15.   *The Times*, 8 January 1974.
16.   Letter in *The Times*, 14 January 1974.
17.   IBA File 3006.
18.   *ITV Evidence to the Annan Committee* (Independent Television Books Ltd., March 1975) pp. 131 and 132.
19.   IBA Annual Report and Accounts 1973–74 p. 5.
20.   IBA Paper 72(75) Appendix I.
21.   SCC Paper 7(75) Attachment I.
22.   Christopher Rowley, ITV's Programme Balance 1970–75 in *Independent Broadcasting* 6, November 1975, pp. 2–4.

CHAPTER 6: THE AUTHORITY

1.   ITA Annual Report and Accounts 1970–71 p. 9.
2.   Peter Black, *The Mirror in the Corner*, p. 70.
3.   Lord Clark, *The Other Half* (John Murray, 1977) p. 139.
4.   Peter Black, *The Mirror in the Corner*, p. 135.
5.   E. G. Wedell, *Broadcasting and Public Policy* (Michael Joseph, 1968) p. 197.
6.   Speech to the Manchester Luncheon Club, 17 May 1960.
7.   Ibid.
8.   Ibid. Quoted in Volume I p. 317.

9.   Speech at farewell staff party, 7 October 1970.
10.   Kenneth Adam, *Evening News*, 13 April 1971.
11.   See Chapter 14.
12.   Speech at Chelsea College, 10 February 1977.
13.   The Watt Club Lecture 1983, Heriot Watt University.
14.   BBC radio interview with Robin Day, 9 November 1972.
15.   Harold Wilson to R. H. S. Crossman as recorded in *Crossman's Diaries*
      Volume 2 (Hamish Hamilton & Jonathan Cape, 1976) pp. 442–3.
16.   Letter Lord Aylestone to Bernard Sendall dated 4 March 1980.
17.   Lord Aylestone in conversation with the author, 4 February 1987.
18.   Annan Report paras. 5.23 and 5.24.
19.   Letter Lord Aylestone to Roy Mason MP dated 24 June 1968 ITA File 200.
20.   Ibid. Letter K. Hind to A. W. Pragnell dated 24 July 1968.
21.   Christopher Bland, *Independent Broadcasting* 4, May 1975.
22.   Lady Plowden in conversation with the author, 8 December 1987.
23.   See Chapter 11.
24.   See Chapter 2.
25.   IBA Memorandum dated 9 March 1978 IBA File 3055/11.
26.   Annan Report para. 6.13.
27.   Letter from Sir Brian Young dated 3 July 1980 IBA File 5010.
28.   See Chapter 8.
29.   EBU Statutes, Article 2.
30.   Viscount Whitelaw in conversation with the author, 12 November 1987.

CHAPTER 7: EDITORIAL CONTROL OR CENSORSHIP

1.   Independent Broadcasting Authority Act 1973 Section 2.
2.   IBA Annual Report and Accounts 1974–75 p. 11.
3.   Appendix to IBA Paper 283(77).
4.   IBA Paper 172(72).
5.   IBA Paper 2(73).
6.   IBA Minutes 305(73).
7.   IBA Minutes 310(73).
8.   Letter Brian Young to John Freeman dated 30 April 1973 IBA File 5000/1/8.
9.   IBA Paper 1(73).
10.   IBA Paper 231(80).
11.   Independent Broadcasting Authority Act 1973 Section 8(6).
12.   Ibid. Section 8(7)(c).
13.   Ibid. Section 4(1)(f).
14.   Section 4(2).
15.   Annan Report para. 17.9.
16.   14 November 1973.
17.   IBA Paper 157(77).
18.   Letter dated 15 February 1977 IBA File 5096/2/3.
19.   Ibid. Letter from Lady Plowden dated 29 March 1977.
20.   Letter to the author dated 16 November 1987.
21.   PPC Paper 4(78).
22.   *Sunday Times*, 11 June 1978.
23.   Jeremy Isaacs in conversation with the author, 12 November 1987.
24.   Memorandum DDT to Director General dated 19 June 1978.
25.   *The Times*, 14 June 1978.
26.   H of L 19 May 1977 Col. 961.

27. Volume 2 p. 301.
28. *Independent Broadcasting* 6, November 1975, p. 14.
29. Volume 2 p. 297.
30. H of C 28 November 1974 Col. 780.
31. Annan Report para. 13.11.
32. See Chapter 8.
33. Geoffrey Robertson & Andrew G. L. Nicol, *Media Law* (Oyez Longman Publishing, 1984) p. 390.

CHAPTER 8: WARHOL AND POULSON

1. Letter Brian Young to Mary Whitehouse dated 15 March 1973 IBA File 3078/2/79/1.
2. IBA Annual Report and Accounts 1972–73 pp. 103–5.
3. Court of Appeal Judgment dated 5 February 1973 IBA File 3078/2/79.
4. Letter Lord Aylestone to Lord Shawcross dated 22 January 1973 IBA File 3078/2/79/2.
5. Letter Brian Young to R. H. Mitchell dated 3 April 1973 IBA File 3078/2/79/1.
6. H of C 3 May 1973 Cols. 1495 and 1520.
7. IBA Paper 31(73).
8. *Sunday Times*, 4 February 1973.
9. *Socialist Worker*, 3 February 1973.
10. *Guardian*, 30 April and 30 May 1973; *The Times*, 30 April 1973.

CHAPTER 9: VIOLENCE, SEX AND BAD LANGUAGE

1. Pilkington Report para. 156.
2. IBA Annual Report and Accounts 1976–77 Appendix XIII.
3. Independent Broadcasting Authority Act 1973 Section 5(1)(a).
4. Statement by the Eisenhower Commission on Violence in Television Programmes – Attachment to PPC Paper 6(69).
5. PPC Paper 1(70).
6. Attachment to ITA Paper 55(70).
7. ITA Annual Report and Accounts 1971–72 Appendix IX.
8. IBA Annual Report and Accounts 1972–73 Appendix XIV.
9. *Daily Mail*, 1 September 1973.
10. Portrayal of Violence on Television, *Independent Broadcasting* 3, February 1975.
11. IBA Annual Report and Accounts 1974–75 Appendix XIII.
12. Annan Report paras. 16.19 and 16.20.
13. Information from Joseph Weltman, October 1987.
14. IBA Paper 286(74).
15. Annan Report para. 16.5.
16. Ibid. para. 16.22.
17. PPC Minutes 51(77).
18. IBA Paper 272(77).
19. IBA Minutes 417(77).
20. Stephen Brody, *Screen Violence and Film Censorship* (HMSO, 1977) p. 10.
21. Ibid. p. 125.

22. William Belson, *Television Violence and the Adolescent Boy* (Saxon House, 1978) p. 15.
23. H. J. Eysenck and D. K. B. Nias, *Sex, Violence and the Media* (Maurice Temple Smith, 1978) p. 252.
24. Barrie Gunter, Plenty of evidence, very little proof in *The Listener*, 23 January 1986.
25. Barrie Gunter, Injury Time in *The Listener*, 14 March 1985.
26. *Sunday Times*, 25 February 1973.
27. PPC Paper 5(70).
28. SCC Paper 52(70).
29. SCC Minutes 152(71).
30. Memorandum dated 10 April 1972 IBA File 3057/12.
31. PPC Paper 5(72).
32. PPC Minutes 30(72).
33. Memorandum dated 28 June 1972 IBA File 3101.
34. PPC Paper 6(73).
35. See Chapter 8.
36. Dated 6 January 1973.
37. Letter dated 18 January 1973 ITA File 3057C.
38. Ibid. Letter Brian Young to D. J. Trevelyan dated 20 October 1976.
39. IBA Paper 160(76).
40. IBA Paper 209(76).
41. IBA Paper 252(76).
42. IBA Memorandum: SPO to HPS dated 6 August 1976 IBA File 5096/2/32.
43. Speech at Chelsea College, 10 February 1977.
44. PPC Paper 8(78).
45. Letter dated 18 April 1977 IBA File 3057/2.
46. PPC Paper 2(78).
47. PPC Paper 11(79).
48. PPC Paper 9(82).
49. Annan Report paras. 16.4 and 16.5.

CHAPTER 10: POLITICAL PROPAGANDA

1. Algorithm and notes dated 29 October 1973 – Attachment to SCC Paper 22(73).
2. Party Political Broadcasting, BBC/IBA Paper dated 22 March 1982 IBA File 5012/1.
3. Patrick Gordon-Walker MP – see Volume 1 p. 239.
4. IBA Minutes 287(71).
5. PPB Committee Minutes dated 11 September 1974 IBA File 5012/1/1.
6. ITA Minutes 151(70).
7. Memorandum DDG(PS) to DG dated 29 January 1971 IBA File 5012/1.
8. *The Times*, 10 May 1974.
9. *The Times*, 9 May 1974.
10. *Guardian*, 9 May 1974.
11. Memorandum A. H. Warren to Lord Aylestone dated 3 July 1974 IBA File 5012/1.
12. Ibid. Note for File by Bryan Rook dated 24 May 1972.
13. H of C 23 October 1972 Col. 797.
14. *The Times*, 11 October 1973.
15. Robert Mellish MP to A. H. Warren dated 15 October 1973 IBA File 5012/1/1.

16. *Evening Standard*, 11 October 1973.
17. Letter dated 7 November 1974 IBA File 5012/1.
18. SCC Minutes 198(75).
19. Reported by Alan Watkins, *New Statesman*, 6 July 1973.
20. Memorandum DDG to DG dated 3 December 1980.
21. Letter dated 18 April 1969 ITA File 5012/1.
22. IBA Paper 44(73).
23. PPB Committee Minutes dated 14 January 1974 IBA File 5012/1/1.
24. IBA Paper 59(74) and IBA Minutes 333(74).
25. Speech at London Press Centre.
26. *Referendum on United Kingdom Membership of the European Community* (HMSO, February 1975) (Cmnd. 5925).
27. PPC Paper 1(75).
28. Cmnd. 5925 para. 33.
29. *Studies of the Impact of the Radio and Television Coverage of the EEC Referendum Campaign* (British Broadcasting Corporation, January 1976) IBA File 5013/1.
30. *Daily Telegraph*, 9 June 1975.
31. IBA Paper 50(79).
32. *Glasgow Herald*, 9 February 1979.
33. Letter dated 22 February 1979 IBA File 5012/1.
34. PPC Paper 9(79).
35. IBA Paper 43(77).

CHAPTER 11: ENGINEERING: THE SYSTEM TRANSFORMED

1.   *Evidence to the Committee on the Future of Broadcasting under the Chairmanship of Lord Annan* (Independent Broadcasting Authority, September 1974) para. 250.
2.   Ibid. para. 251.
3.   Television Act 1964 Section 1(3).
4.   ITA Memorandum to the House of Commons Select Committee on Nationalised Industries, November 1971, para. 2.3.
5.   See Volume 2 pp. 324–9.
6.   SCC Paper 2(74).
7.   Howard Steele, The Story of Television 1967–77 in *Television*, Journal of the Royal Television Society, November-December 1977.
8.   Annan Report para 24.23.
9.   SCC Paper 22(68).
10. *Report of the Committee on Broadcasting Coverage* (HMSO, November 1974) (Cmnd. 5774).
11. Pat Hawker, Television Coverage – The Next Steps in *BKSTS Journal*, November 1980.
12. IBA Engineering Progress, September 1978.
13. Crawford Report para. 64.
14. Ibid. para. 62.
15. Ibid. para. 68.
16. IBA News Release, 3 February 1975.
17. Patrick Halliday, The Story behind Teletext in *Television & Home Video*, July 1979.
18. Annan Report para. 25.29.
19. *Independent Broadcasting* 29, June 1981 pp. 18–20.

20.   Annan Report para. 24.24.
21.   Ibid. para. 24.12.
22.   Letter dated 6 August 1976 IBA File 3056.

CHAPTER 12:   DOMESTICATED ADVERTISING

1.   *British Television Advertising: The First Thirty Years* ed. Brian Henry (Century Benham, 1986).
2.   IBA Evidence to Annan paras. 34 and 35.
3.   Brian Young, How Should Broadcasting be Financed? in *Independent Broadcasting* 5, August 1975.
4.   Volume 1 p. 99.
5.   Annan Report para. 12.4.
6.   Ibid.
7.   Anthony Pragnell in *British Television Advertising* p. 326.
8.   *Campaign*, 22 June 1973.
9.   IBA Minutes 316(73).
10.   IBA Paper 172(73).
11.   *Second Report from the Select Committee on Nationalised Industries* Minutes of Evidence para. 993.
12.   Annan Report para. 12.8.
13.   *ITV 1974* (Independent Broadcasting Authority, January 1974) p. 131.
14.   Peter Woodhouse in *British Television Advertising* p. 370.
15.   IBA Paper 147(72).
16.   Peter Woodhouse in *British Television Advertising* p. 363.
17.   Ibid. p. 371.
18.   Television Act 1964, Schedule 2 para. 8.
19.   IBA Paper 182(72).
20.   IBA Paper 242(75).
21.   *ITV 1974* pp. 129–39.
22.   IBA Minutes 327(74).
23.   IBA Paper 194(78).
24.   *Observer*, 10 June 1973.
25.   IBA Paper 118(73).
26.   Annan Report para. 12.4.
27.   Ibid. para. 12.6.
28.   Ibid. para. 12.5.
29.   Ibid. para. 12.10.
30.   IBA Paper 198(76).
31.   ITCA Aide-mémoire to IBA dated 22 July 1976 IBA File 8033.
32.   IBA Paper 147(77).
33.   IBA Paper 111(78) and PPC Minutes 57(78).
34.   SCC Minutes 171(72).
35.   Independent Broadcasting Authority Act 1973, Schedule 2 para. 7.
36.   *British Television Advertising* p. 124.
37.   IBA Paper 219(77).
38.   *British Television Advertising* Appendix M.
39.   *Second Report from the Select Committee on Nationalised Industries* p. 231.
40.   *Tenth Report from the Select Committee on Nationalised Industries* (HMSO, July 1978) Minutes of Evidence para. 573.
41.   IPA 8017, 9 October 1979.
42.   Speech at Royal Television Society Convention in Cambridge, 1979.

43. IBA Paper 219(77).
44. *British Television Advertising* p. 411.
45. IBA Paper 343(80).
46. IBA Paper 84(73); SCC Paper 1(73).
47. IBA Minutes 314(73).
48. SCC Minutes 176(73) and 179(73).
49. Letter A. Graham to R.A. Fleming dated 2 October 1973 IBA File 8059.
50. See Sir Ronald Halstead, The Effect of Television on Marketing in *British Television Advertising*.
51. Ibid. p. 411.
52. Ibid. p. 408.
53. Tim Bell in *British Television Advertising* p. 440.
54. Jo Gable, *The Tuppenny Punch and Judy Show* (Michael Joseph, 1980) p. 32.
55. Ibid. pp. 83–90.

CHAPTER 13: ASPECTS OF RESEARCH

1. Pilkington Report p. 292.
2. William Belson, *The Impact of Television* (Crosby Lockwood & Son, 1967) p. 1.
3. Television Act 1964 Section 24.
4. Roger Beeson, But I Happen to Like Charles Laughton in Newspaper Stories in *Admap*, January 1973 p. 4.
5. IBA Paper 11(76).
6. See Chapter 9.
7. J.M. Wober, TV's Menu: The Viewers' Order in IBA Research Department Working Paper, May 1984.
8. G.J. Goodhardt, A.S.C. Ehrenberg, M.A. Collins and Aske Research Ltd., *The Television Audience: Patterns of Viewing* (Saxon House/Lexington Books, 1975).
9. Annan Report paras. 29.6 and 29.7.
10. IBA Paper 298(75).
11. *Attitudes to ITV: Report on a National Survey* (The British Market Research Bureau Ltd., BMRB/RJG.90338, June 1979).
12. IBA Paper 298(75).
13. Letter to David Wilson dated 18 December 1973 IBA File 9012/8.
14. Annan Report para. 29.15.
15. *The ITV Strike: Its Effect on Sales* (D'Arcy McManus and Masius).
16. Letter to the Rt Hon. Merlyn Rees MP, Secretary of State for the Home Department, dated 14 April 1978 – attached to IBA Paper 100(78).

CHAPTER 14: ANNAN: PREAMBLE AND EVIDENCE

1. Memorandum dated 14 January 1970 IBA File 3056.
2. H of L 25 February 1970 Col. 451.
3. H of C 14 May 1970 Col. 1455.
4. H of L 14 May 1970 Col. 728.
5. Lord Annan, *The Politics of a Broadcasting Enquiry*, The 1981 Ulster Television Lecture delivered at the Queen's University of Belfast on 29 May 1981 (Ulster Television) p. 1.

6.   H of C 15 December 1971 Col. 572.
7.   Tom Carbery, *Independent Broadcasting* 6, November 1975, p. 9.
8.   Ibid.
9.   p. 49.
10. Annan Report para. 1.1.
11. Letter dated 14 August 1974 IBA File 3056.
12. Ibid. Letter dated 30 August 1974.
13. *The Politics of a Broadcasting Enquiry* p. 8.
14. Report of the 1910 Balfour Committee of Enquiry into Royal Commissions
     quoted in *The Politics of a Broadcasting Enquiry* p. 6.
15. Inside the Annan Committee in *New Statesman*, 25 March 1977.
16. *Evidence to the Committee on the Future of Broadcasting under the Chair-
     manship of Lord Annan* (IBA, September 1974).
17. Ibid. para. 37.
18. Ibid. para. 42.
19. Ibid. Appendix III para. 1.
20. *ITV Evidence to the Annan Committee* (ITCA, March 1975).
21. Sir Ian Trethowan, *The Split Screen* (Hamish Hamilton, 1984) pp. 145–6.
22. *The People and the Media* (Labour Party, 1974) p. 11.
23. Ibid. pp. 5 and 6.
24. Peter Fiddick in *Guardian*, 2 June 1975.
25. Broadcasting in the UK: An Agenda for Discussion p. 1.
26. Tom Carbery, *Independent Broadcasting* 6, November 1975, p. 9.
27. *The Times*, 28 June 1975.
28. *Campaign*, 4 July 1975.
29. Phillip Whitehead, Inside the Annan Committee in *New Statesman*, 25
     March 1977.
30. In conversation with the author, 25 March 1987.
31. *The Politics of a Broadcasting Enquiry* p. 10.
32. Ibid.

CHAPTER 15:   ANNAN: REPORT AND DEBATE

1.   Cmnd. 6753.
2.   Speech by Lord Annan at press conference on 24 March 1977 IBA File 3056.
3.   *The Annan Report: An ITV View* (Indpendent Television Companies
     Association, June 1977). p. 7.
4.   Ibid. p. 6.
5.   Speech by Lord Annan at press conference on 24 March 1977 IBA File 3056.
6.   Annan Report para. 4.24.
7.   Ibid. p. 70.
8.   Ibid. para. 4.11.
9.   Ibid. para. 4.10.
10. Ibid. para. 6.13.
11. *IBA Comments on the Report of the Committee on the Future of Broadcasting*
     (Independent Broadcasting Authority, June 1977) para. 2.16.
12. Annan Report para. 4.7.
13. Ibid. para. 3.20.
14. Ibid. para. 3.21.
15. Lord Clark, *The Other Half* p. 138.
16. Annan Report para. 2.26.
17. Ibid. para. 2.28.

18. Ibid. para. 3.23.
19. Ibid. para. 3.16, quoting Richard Bradford Lecture, Royal Institution, October 1975.
20. Ibid. para. 3.25.
21. Ibid.
22. Ibid. para. 13.46.
23. Ibid. para. 11.7.
24. Ibid. para. 11.10.
25. Ibid. para. 4.3.
26. Summary of the Annan Report p. 14 IBA File 3056.
27. Annan Report para. 13.44.
28. Ibid. para. 13.45.
29. Granada Guildhall Lecture.
30. Annan Report Recommendation 47.
31. Ibid. para. 13.43.
32. *The Annan Report: An ITV View* paras. 11 and 12.
33. Ibid. para. 14.
34. *The Media Reporter* Vol. 1 No. 3 1977.
35. *IBA Comments on the Report of the Committee on the Future of Broadcasting* para. 1.3.
36. Annan Report Recommendation 136.
37. Ibid. Recommendation 41.
38. Ibid. Recommendation 55.
39. Ibid. para. 3.14.
40. Ibid. Recommendation 58.
41. *The Politics of a Broadcasting Enquiry* p. 17.
42. Ibid.
43. Ibid. p. 15.
44. Ibid. pp. 12 and 13.
45. *New Statesman*, 25 March 1977.
46. Lord Annan in conversation with the author, 25 March 1987.
47. H of L 19 May 1977 Col. 886.
48. H of C 23 May 1977 Col. 1038.
49. Phillip Whitehead in conversation with the author, 19 November 1987.
50. Cmnd. 7294 (HMSO, July 1978).
51. Ibid. para. 53.
52. Colin Shaw in *Independent Broadcasting* 18, November 1978.
53. *Comments on the White Paper on Broadcasting* (Independent Television Companies Association, 1978).
54. SCC Paper 20(78).
55. Sir Ian Trethowan, *The Split Screen* p. 180.
56. *The Politics of a Broadcasting Enquiry* p. 18.

CHAPTER 16:   PROSPERITY AND UNCERTAINTY, 1976–80

1. IBA Annual Report and Accounts 1975–76 p. 7.
2. *Independent Broadcasting* 9, September 1976. p. 6.
3. *Independent Broadcasting* 10, December 1976, p. 2.
4. IBA Annual Report and Accounts 1976–77 p. 15.
5. Annan Report para. 11.9.
6. *Independent Broadcasting* 4, May 1975, p. 10.
7. IBA Annual Report and Accounts 1976–77 p. 8.

8.   Ibid. p. 14.
9.   Appendix XIII.
10.   Sir Brian Young in evidence to SCNI, 5 July 1978: see *Tenth Report from the Select Committee on Nationalised Industries* Minutes of Evidence para. 987.
11.   Letter from Aubrey Buxton to Lord Aylestone dated 13 March 1975 IBA File 29.
12.   IBA Annual Report and Accounts 1977–78 p. 8.
13.   Annan Report para. 12.47.
14.   Cmnd. 7294 para. 151.
15.   Sir Donald Kaberry, SCNI Minutes of Evidence, 5 July 1978: see *Tenth Report from the Select Committee on Nationalised Industries* para. 991.
16.   See Chapter 4.
17.   IBA Annual Report and Accounts 1978–79 p. 7.
18.   IBA Paper 33(79).
19.   Letter dated 29 November 1978 IBA File 5003.
20.   SCC Paper 5(81).
21.   See Volume 1 Chapter 17.
22.   IBA Annual Report and Accounts 1977–78 Appendix X.
23.   Cmnd. 7294 paras. 114 and 115.
24.   *British Television Advertising: The First Thirty Years* ed. Brian Henry pp. 189, 192 and 193.
25.   Reprinted as Appendix XIV in IBA Annual Report and Accounts 1979–80.
26.   Ibid. Appendix XIII.
27.   Broadcasting Act 1980 Section 18(1).
28.   IBA Annual Report and Accounts 1980–81 p. 5.
29.   Ibid. p. 10.
30.   Independent Broadcasting Act 1973 Section 22(3).
31.   Broadcasting Act 1980 Section 19(4)(a).
32.   *Television and Radio 1981* p. 215.
33.   *ITV Evidence to the Annan Committee* p. 132: table extended by Paul Young, Senior External Finance Officer, IBA.

CHAPTER 17:   THE EMPTY CHANNEL

1.   Cmnd 3169 (HMSO, December 1966).
2.   Television Act 1964 Section 25.
3.   SCC Minutes 145(70).
4.   ITV's Second Channel in *Sunday Times*, 5 July 1970.
5.   Cmnd. 4524 para. 135.
6.   *Report on Consultation on ITV 2*; see *Second Report from the Select Committee on Nationalised Industries* Appendix 6 p. 324.
7.   *ITV 2* (Independent Television Authority, December 1971) para. 17.
8.   Ibid. para. 26.
9.   *The Times*, 9 December 1971.
10.   See Chapter 4.
11.   *Second Report from the Select Committee on Nationalised Industries* para. 106.
12.   H of C 19 January 1972 Col. 479.
13.   IBA Annual Report and Accounts 1973–74 Appendex XIII.
14.   *Evidence to the Committee on the Future of Broadcasting* para. 143.
15.   Sir Charles Curran, *The Fourth Television Network: A Question of Priorities* (British Broadcasting Corporation, 1974).

16. *Independent Broadcasting* 3, February 1975 p. 8.
17. Attachment to IBA Paper 108(76).
18. *ITV Evidence to the Annan Committee* Appendix C.
19. Anthony Smith, *The Shadow in the Cave* (Quartet Books, 1976, first published by George Allen & Unwin, 1973) p. 296.
20. *Guardian*, 13 August 1973.
21. *ITV Evidence to the Annan Committee* para. 97.
22. *The Annan Report: An ITV View* p. 36.
23. Annan Report paras. 15.8, 11.27, 15.32, 15.19, 11.27.
24. 28 April 1977.
25. H of C 23 May 1977 Col. 1037.
26. Annan Report para. 15.32.
27. *The Annan Report: An ITV View* p. 40.
28. Ibid. p. 25.
29. Ibid. p. 24.
30. Ibid. p. 22.
31. IBA Paper 154(77).
32. H of L 19 May 1977 Col. 941.
33. IBA Paper 18(78).
34. See Chapter 15.
35. Cmnd. 7294 para. 15.
36. Ibid. para. 16.
37. *IBA Comments on the Proposals about the Fourth Television Channel in the 1978 White Paper on Broadcasting* (IBA, September 1978) para. 2.
38. Cmnd. 7294 para. 24.
39. *Comments on the White Paper on Broadcasting* paras. 14 and 12.
40. Ibid. para. 16.
41. Paul Fox in *The Listener*, 17 August 1978.
42. Letter from Lady Plowden to the Rt Hon. William Whitelaw MP dated 27 July 1979, IBA File 3200.
43. *Film and Television Technician*, June 1979, p. 12.
44. Home Office news release dated 14 September 1979 IBA File 3200.
45. *ITV2: The Fourth Channel*: ITCA memorandum dated October 1979 attached to IBA Paper 282(79).
46. SCC Paper 18(79).
47. IBA Paper 281(79).
48. IBA Paper 262(79).
49. Reprinted in IBA Annual Report and Accounts 1979–80 as Appendix XIV.
50. Ibid. para. 17.
51. Ibid. paras. 20 and 21.
52. Ibid. paras. 33–8. See p. 206.
53. Broadcasting Act 1980 Part III.
54. See Volume 2 Chapter 8.
55. Stephen Lambert, *Channel Four: Television with a Difference?* (BFI Publishing, 1982) Appendix 1.
56. Reprinted in *Independent Broadcasting* 27, January 1981.
57. CF Paper 42(81).
58. Channel Four – A Message from the Director General, 3 November 1982 IBA File 3205/7.
59. Lord Windlesham, *Broadcasting in a Free Society* (Basil Blackwell, 1980) pp. 72–5.

INDEX

ABC Television
    company, 15; merger with Rediffusion to
        form Thames Television, 18
    contract, ITV, Midlands and the North
        weekend service, 1956–68, 188–9
    engineering, 188
    staff, 18
    studios, 188–9
ACTT, *see* Association of Cinematograph,
    Television and Allied Technicians
AGB, *see* Audits of Great Britain
ATV, *see* Associated Television
AURA (Audience Research Assessment),
    212–13, 214
Abbreviations, 320–1
Aberdare of Duffryn, Lord, 309
About Britain, 76
Adams, Cliff, 209
Adams, Mary, 93, 311
Additional payments, *see* Exchequer Levy
Adult education programmes, 72, 76–7, 250,
    279, 284
Advertising, 192–210
    controls, 2, 104, 193–202, 206; advisory
        councils and committees, 197–8, 199,
        200; amount and length of
        advertisements, 193–4; and children
        198, 201, 216; *Code of Advertising
        Standards and Practice*, 198, 199, 201;
        copy clearance, 199; distinction
        between advertisements and
        programmes to be clear, 193, 199–
        200; government campaigns, 197;
        'legal, decent, honest and truthful',
        195–6; natural breaks, 194–5;
        Principles for Television Advertising,
        198; and religious programmes, 194,
        201–2; unacceptable products and
        services, 195–7
    forms of advertising: commercials, 209–
        10; on consumer advice programmes,
        108; sponsorship, 109, 288; at sports

events, 109; spot advertising, 294;
    subliminal, 198
legislation and recommendation: Annan
    Committee, 195, 200, 201–2;
    Broadcasting Act 1980, 299, 300–1,
    302; Pilkington Committee, 200, 232,
    236
products and services, 195–6, 204–5;
    alcoholic drinks, 200, 209; children's
    toys, 201; cigarettes, 109, 200, 202,
    209; financial services, 198; fireworks,
    201; football pools, 198; medicines,
    198
revenue from, 2, 20, 25–6, 70, 78–82, 101,
    206, 217, 265–70, 277–8; competition
    in sales of advertising, 205; Net
    Advertising Revenue (NAR), 36,
    278; Net Advertising Revenue After
    Levy (NARAL), 36, 50; Oracle, 188;
    rate cards, 202–3; sales, 202–8
Advertising Advisory Committee (ITA/
    IBA), 197–8, 200
Advertising Liaison Committee, 206, 298
Advertising Standards Authority, 206
Advisory councils and committees, ITA/
    IBA ITV, 61, 73, 96–7, 104
advertising, 96, 197–8, 199, 200
appeals on television, 97, 111
audience research, 213–14
educational broadcasting, 73, 96–7, 104
engineering, 58, 67, 181, 222–3
programme planning and policy, 67, 73,
    99, 105, 114–15, 138, 141, 144, 147,
    149, 161, 202
religious broadcasting, 97, 104, 194
Age Concern, 112
Agnew, Spiro, 122
Aide-mémoires on political broadcasts, 154–
    5, 161, 316–17, 318–19
Akenfield, 107, 264
Alcoholic drinks, advertising of, 200, 209
Ali, Mohammed, 122

336